Pitman Research Notes in Mathematics Series

Submission of proposals for consideration
Suggestions for publication, in the form of outlines and representative samples, are invited by the Editorial Board for assessment. Intending authors should approach one of the main editors or another member of the Editorial Board, citing the relevant AMS subject classifications. Alternatively, outlines may be sent directly to the publisher's offices. Refereeing is by members of the board and other mathematical authorities in the topic concerned, throughout the world.

Preparation of accepted manuscripts
On acceptance of a proposal, the publisher will supply full instructions for the preparation of manuscripts in a form suitable for direct photo-lithographic reproduction. Specially printed grid sheets can be provided and a contribution is offered by the publisher towards the cost of typing. Word processor output, subject to the publisher's approval, is also acceptable.

Illustrations should be prepared by the authors, ready for direct reproduction without further improvement. The use of hand-drawn symbols should be avoided wherever possible, in order to maintain maximum clarity of the text.

The publisher will be pleased to give any guidance necessary during the preparation of a typescript, and will be happy to answer any queries.

Important note
In order to avoid later retyping, intending authors are strongly urged not to begin final preparation of a typescript before receiving the publisher's guidelines. In this way it is hoped to preserve the uniform appearance of the series.

Longman Scientific & Technical
Longman House
Burnt Mill
Harlow, Essex, CM20 2JE
UK
(Telephone (0279) 426721)

Titles in this series. A full list is available from the publisher on request.

100 Optimal control of variational inequalities
V Barbu
101 Partial differential equations and dynamical systems
W E Fitzgibbon III
102 Approximation of Hilbert space operators Volume II
C Apostol, L A Fialkow, D A Herrero and D Voiculescu
103 Nondiscrete induction and iterative processes
V Ptak and F-A Potra
104 Analytic functions – growth aspects
O P Juneja and G P Kapoor
105 Theory of Tikhonov regularization for Fredholm equations of the first kind
C W Groetsch
106 Nonlinear partial differential equations and free boundaries. Volume I
J I Díaz
107 Tight and taut immersions of manifolds
T E Cecil and P J Ryan
108 A layering method for viscous, incompressible L_p flows occupying R^n
A Douglis and E B Fabes
109 Nonlinear partial differential equations and their applications: Collège de France Seminar. Volume VI
H Brezis and J L Lions
110 Finite generalized quadrangles
S E Payne and J A Thas
111 Advances in nonlinear waves. Volume II
L Debnath
112 Topics in several complex variables
E Ramírez de Arellano and D Sundararaman
113 Differential equations, flow invariance and applications
N H Pavel
114 Geometrical combinatorics
F C Holroyd and R J Wilson
115 Generators of strongly continuous semigroups
J A van Casteren
116 Growth of algebras and Gelfand–Kirillov dimension
G R Krause and T H Lenagan
117 Theory of bases and cones
P K Kamthan and M Gupta
118 Linear groups and permutations
A R Camina and E A Whelan
119 General Wiener–Hopf factorization methods
F-O Speck
120 Free boundary problems: applications and theory. Volume III
A Bossavit, A Damlamian and M Fremond
121 Free boundary problems: applications and theory. Volume IV
A Bossavit, A Damlamian and M Fremond
122 Nonlinear partial differential equations and their applications: Collège de France Seminar. Volume VII
H Brezis and J L Lions
123 Geometric methods in operator algebras
H Araki and E G Effros
124 Infinite dimensional analysis–stochastic processes
S Albeverio
125 Ennio de Giorgi Colloquium
P Krée
126 Almost-periodic functions in abstract spaces
S Zaidman
127 Nonlinear variational problems
A Marino, L Modica, S Spagnolo and M Degliovanni
128 Second-order systems of partial differential equations in the plane
L K Hua, W Lin and C-Q Wu
129 Asymptotics of high-order ordinary differential equations
R B Paris and A D Wood
130 Stochastic differential equations
R Wu
131 Differential geometry
L A Cordero
132 Nonlinear differential equations
J K Hale and P Martinez-Amores
133 Approximation theory and applications
S P Singh
134 Near-rings and their links with groups
J D P Meldrum
135 Estimating eigenvalues with *a posteriori/a priori* inequalities
J R Kuttler and V G Sigillito
136 Regular semigroups as extensions
F J Pastijn and M Petrich
137 Representations of rank one Lie groups
D H Collingwood
138 Fractional calculus
G F Roach and A C McBride
139 Hamilton's principle in continuum mechanics
A Bedford
140 Numerical analysis
D F Griffiths and G A Watson
141 Semigroups, theory and applications. Volume I
H Brezis, M G Crandall and F Kappel
142 Distribution theorems of L-functions
D Joyner
143 Recent developments in structured continua
D De Kee and P Kaloni
144 Functional analysis and two-point differential operators
J Locker
145 Numerical methods for partial differential equations
S I Hariharan and T H Moulden
146 Completely bounded maps and dilations
V I Paulsen
147 Harmonic analysis on the Heisenberg nilpotent Lie group
W Schempp
148 Contributions to modern calculus of variations
L Cesari
149 Nonlinear parabolic equations: qualitative properties of solutions
L Boccardo and A Tesei
150 From local times to global geometry, control and physics
K D Elworthy

151 A stochastic maximum principle for optimal control of diffusions
U G Haussmann
152 Semigroups, theory and applications. Volume II
H Brezis, M G Crandall and F Kappel
153 A general theory of integration in function spaces
P Muldowney
154 Oakland Conference on partial differential equations and applied mathematics
L R Bragg and J W Dettman
155 Contributions to nonlinear partial differential equations. Volume II
J I Díaz and P L Lions
156 Semigroups of linear operators: an introduction
A C McBride
157 Ordinary and partial differential equations
B D Sleeman and R J Jarvis
158 Hyperbolic equations
F Colombini and M K V Murthy
159 Linear topologies on a ring: an overview
J S Golan
160 Dynamical systems and bifurcation theory
M I Camacho, M J Pacifico and F Takens
161 Branched coverings and algebraic functions
M Namba
162 Perturbation bounds for matrix eigenvalues
R Bhatia
163 Defect minimization in operator equations: theory and applications
R Reemtsen
164 Multidimensional Brownian excursions and potential theory
K Burdzy
165 Viscosity solutions and optimal control
R J Elliott
166 Nonlinear partial differential equations and their applications: Collège de France Seminar. Volume VIII
H Brezis and J L Lions
167 Theory and applications of inverse problems
H Haario
168 Energy stability and convection
G P Galdi and B Straughan
169 Additive groups of rings. Volume II
S Feigelstock
170 Numerical analysis 1987
D F Griffiths and G A Watson
171 Surveys of some recent results in operator theory. Volume I
J B Conway and B B Morrel
172 Amenable Banach algebras
J-P Pier
173 Pseudo-orbits of contact forms
A Bahri
174 Poisson algebras and Poisson manifolds
K H Bhaskara and K Viswanath
175 Maximum principles and eigenvalue problems in partial differential equations
P W Schaefer
176 Mathematical analysis of nonlinear, dynamic processes
K U Grusa
177 Cordes' two-parameter spectral representation theory
D F McGhee and R H Picard
178 Equivariant K-theory for proper actions
N C Phillips
179 Elliptic operators, topology and asymptotic methods
J Roe
180 Nonlinear evolution equations
J K Engelbrecht, V E Fridman and E N Pelinovski
181 Nonlinear partial differential equations and their applications: Collège de France Seminar. Volume IX
H Brezis and J L Lions
182 Critical points at infinity in some variational problems
A Bahri
183 Recent developments in hyperbolic equations
L Cattabriga, F Colombini, M K V Murthy and S Spagnolo
184 Optimization and identification of systems governed by evolution equations on Banach space
N U Ahmed
185 Free boundary problems: theory and applications. Volume I
K H Hoffmann and J Sprekels
186 Free boundary problems: theory and applications. Volume II
K H Hoffmann and J Sprekels
187 An introduction to intersection homology theory
F Kirwan
188 Derivatives, nuclei and dimensions on the frame of torsion theories
J S Golan and H Simmons
189 Theory of reproducing kernels and its applications
S Saitoh
190 Volterra integrodifferential equations in Banach spaces and applications
G Da Prato and M Iannelli
191 Nest algebras
K R Davidson
192 Surveys of some recent results in operator theory. Volume II
J B Conway and B B Morrel
193 Nonlinear variational problems. Volume II
A Marino and M K V Murthy
194 Stochastic processes with multidimensional parameter
M E Dozzi
195 Prestressed bodies
D Iesan
196 Hilbert space approach to some classical transforms
R H Picard
197 Stochastic calculus in application
J R Norris
198 Radical theory
B J Gardner
199 The C^*-algebras of a class of solvable Lie groups
X Wang
200 Stochastic analysis, path integration and dynamics
K D Elworthy and J C Zambrini

201 Riemannian geometry and holonomy groups
S Salamon
202 Strong asymptotics for extremal errors and polynomials associated with Erdös type weights
D S Lubinsky
203 Optimal control of diffusion processes
V S Borkar
204 Rings, modules and radicals
B J Gardner
205 Two-parameter eigenvalue problems in ordinary differential equations
M Faierman
206 Distributions and analytic functions
R D Carmichael and D Mitrovic
207 Semicontinuity, relaxation and integral representation in the calculus of variations
G Buttazzo
208 Recent advances in nonlinear elliptic and parabolic problems
P Bénilan, M Chipot, L Evans and M Pierre
209 Model completions, ring representations and the topology of the Pierce sheaf
A Carson
210 Retarded dynamical systems
G Stepan
211 Function spaces, differential operators and nonlinear analysis
L Paivarinta
212 Analytic function theory of one complex variable
C C Yang, Y Komatu and K Niino
213 Elements of stability of visco-elastic fluids
J Dunwoody
214 Jordan decomposition of generalized vector measures
K D Schmidt
215 A mathematical analysis of bending of plates with transverse shear deformation
C Constanda
216 Ordinary and partial differential equations. Volume II
B D Sleeman and R J Jarvis
217 Hilbert modules over function algebras
R G Douglas and V I Paulsen
218 Graph colourings
R Wilson and R Nelson
219 Hardy-type inequalities
A Kufner and B Opic
220 Nonlinear partial differential equations and their applications: Collège de France Seminar. Volume X
H Brezis and J L Lions
221 Workshop on dynamical systems
E Shiels and Z Coelho
222 Geometry and analysis in nonlinear dynamics
H W Broer and F Takens
223 Fluid dynamical aspects of combustion theory
M Onofri and A Tesei
224 Approximation of Hilbert space operators. Volume I. 2nd edition
D Herrero
225 Operator theory: proceedings of the 1988 GPOTS–Wabash conference
J B Conway and B B Morrel
226 Local cohomology and localization
J L Bueso Montero, B Torrecillas Jover and A Verschoren
227 Nonlinear waves and dissipative effects
D Fusco and A Jeffrey
228 Numerical analysis 1989
D F Griffiths and G A Watson
229 Recent developments in structured continua. Volume II
D De Kee and P Kaloni
230 Boolean methods in interpolation and approximation
F J Delvos and W Schempp
231 Further advances in twistor theory. Volume I
L J Mason and L P Hughston
232 Further advances in twistor theory. Volume II
L J Mason, L P Hughston and P Z Kobak
233 Geometry in the neighborhood of invariant manifolds of maps and flows and linearization
U Kirchgraber and K Palmer
234 Quantales and their applications
K I Rosenthal
235 Integral equations and inverse problems
V Petkov and R Lazarov
236 Pseudo-differential operators
S R Simanca
237 A functional analytic approach to statistical experiments
I M Bomze
238 Quantum mechanics, algebras and distributions
D Dubin and M Hennings
239 Hamilton flows and evolution semigroups
J Gzyl
240 Topics in controlled Markov chains
V S Borkar
241 Invariant manifold theory for hydrodynamic transition
S Sritharan
242 Lectures on the spectrum of $L^2(\Gamma\backslash G)$
F L Williams
243 Progress in variational methods in Hamiltonian systems and elliptic equations
M Girardi, M Matzeu and F Pacella
244 Optimization and nonlinear analysis
A Ioffe, M Marcus and S Reich
245 Inverse problems and imaging
G F Roach
246 Semigroup theory with applications to systems and control
N U Ahmed
247 Periodic-parabolic boundary value problems and positivity
P Hess
248 Distributions and pseudo-differential operators
S Zaidman
249 Progress in partial differential equations: the Metz surveys
M Chipot and J Saint Jean Paulin
250 Differential equations and control theory
V Barbu

251 Stability of stochastic differential equations with respect to semimartingales
X Mao
252 Fixed point theory and applications
J Baillon and M Théra
253 Nonlinear hyperbolic equations and field theory
M K V Murthy and S Spagnolo
254 Ordinary and partial differential equations. Volume III
B D Sleeman and R J Jarvis
255 Harmonic maps into homogeneous spaces
M Black
256 Boundary value and initial value problems in complex analysis: studies in complex analysis and its applications to PDEs 1
R Kühnau and W Tutschke
257 Geometric function theory and applications of complex analysis in mechanics: studies in complex analysis and its applications to PDEs 2
R Kühnau and W Tutschke
258 The development of statistics: recent contributions from China
X R Chen, K T Fang and C C Yang
259 Multiplication of distributions and applications to partial differential equations
M Oberguggenberger
260 Numerical analysis 1991
D F Griffiths and G A Watson
261 Schur's algorithm and several applications
M Bakonyi and T Constantinescu
262 Partial differential equations with complex analysis
H Begehr and A Jeffrey
263 Partial differential equations with real analysis
H Begehr and A Jeffrey
264 Solvability and bifurcations of nonlinear equations
P Drábek
265 Orientational averaging in mechanics of solids
A Lagzdins, V Tamuzs, G Teters and A Kregers
266 Progress in partial differential equations: elliptic and parabolic problems
C Bandle, J Bemelmans, M Chipot, M Grüter and J Saint Jean Paulin
267 Progress in partial differential equations: calculus of variations, applications
C Bandle, J Bemelmans, M Chipot, M Grüter and J Saint Jean Paulin
268 Stochastic partial differential equations and applications
G Da Prato and L Tubaro
269 Partial differential equations and related subjects
M Miranda
270 Operator algebras and topology
W B Arveson, A S Mishchenko, M Putinar, M A Rieffel and S Stratila
271 Operator algebras and operator theory
W B Arveson, A S Mishchenko, M Putinar, M A Rieffel and S Stratila
272 Ordinary and delay differential equations
J Wiener and J K Hale
273 Partial differential equations
J Wiener and J K Hale
274 Mathematical topics in fluid mechanics
J F Rodrigues and A Sequeira
275 Green functions for second order parabolic integro-differential problems
M G Garroni and J F Menaldi
276 Riemann waves and their applications
M W Kalinowski
277 Banach C(K)-modules and operators preserving disjointness
Y A Abramovich, E L Arenson and A K Kitover
278 Limit algebras: an introduction to subalgebras of C*-algebras
S C Power
279 Abstract evolution equations, periodic problems and applications
D Daners and P Koch Medina
280 Emerging applications in free boundary problems
J Chadam and H Rasmussen
281 Free boundary problems involving solids
J Chadam and H Rasmussen
282 Free boundary problems in fluid flow with applications
J Chadam and H Rasmussen
283 Asymptotic problems in probability theory: stochastic models and diffusions on fractals
K D Elworthy and N Ikeda
284 Asymptotic problems in probability theory: Wiener functionals and asymptotics
K D Elworthy and N Ikeda
285 Dynamical systems
R Bamon, R Labarca, J Lewowicz and J Palis
286 Models of hysteresis
A Visintin
287 Moments in probability and approximation theory
G A Anastassiou
288 Mathematical aspects of penetrative convection
B Straughan
289 Ordinary and partial differential equations. Volume IV
B D Sleeman and R J Jarvis
290 *K*-theory for real *C**-algebras
H Schröder
291 Recent developments in theoretical fluid mechanics
G P Galdi and J Necas
292 Propagation of a curved shock and nonlinear ray theory
P Prasad
293 Non-classical elastic solids
M Ciarletta and D Ieşan
294 Multigrid methods
J Bramble
295 Entropy and partial differential equations
W A Day
296 Progress in partial differential equations: the Metz surveys 2
M Chipot
297 Nonstandard methods in the calculus of variations
C Tuckey
298 Barrelledness, Baire-like- and (LF)-spaces
M Kunzinger
299 Nonlinear partial differential equations and their applications. Collège de France Seminar. Volume XI
H Brezis and J L Lions
300 Introduction to operator theory
T Yoshino

301 Generalized fractional calculus and applications
V Kiryakova
302 Nonlinear partial differential equations and their applications. Collège de France Seminar Volume XII
H Brezis and J L Lions
303 Numerical analysis 1993
D F Griffiths and G A Watson
304 Topics in abstract differential equations
S Zaidman
305 Complex analysis and its applications
C C Yang, G C Wen, K Y Li and Y M Chiang
306 Computational methods for fluid-structure interaction
J M Crolet and R Ohayon
307 Random geometrically graph directed self-similar multifractals
L Olsen
308 Progress in theoretical and computational fluid mechanics
G P Galdi, J Málek and J Necas
309 Variational methods in Lorentzian geometry
A Masiello
310 Stochastic analysis on infinite dimensional spaces
H Kunita and H-H Kuo
311 Representations of Lie groups and quantum groups
V Baldoni and M Picardello
312 Common zeros of polynomials in several variables and higher dimensional quadrature
Y Xu
313 Extending modules
N V Dung, D van Huynh, P F Smith and R Wisbauer
314 Progress in partial differential equations: the Metz surveys 3
M Chipot, J Saint Jean Paulin and I Shafrir
315 Refined large deviation limit theorems
V Vinogradov
316 Topological vector spaces, algebras and related areas
A Lau and I Tweddle
317 Integral methods in science and engineering
C Constanda
318 A method for computing unsteady flows in porous media
R Raghavan and E Ozkan
319 Asymptotic theories for plates and shells
R P Gilbert and K Hackl
320 Nonlinear variational problems and partial differential equations
A Marino and M K V Murthy
321 Topics in abstract differential equations II
S Zaidman
322 Diffraction by wedges
B Budaev
323 Free boundary problems: theory and applications
J I Diaz, M A Herrero, A Liñan and J L Vazquez
324 Recent developments in evolution equations
A C McBride and G F Roach
325 Elliptic and parabolic problems: Pont-à-Mousson 1994
C Bandle, J Bemelmans, M Chipot, J Saint Jean Paulin and I Shafrir
326 Calculus of variations, applications and computations: Pont-à-Mousson 1994
C Bandle, J Bemelmans, M Chipot, J Saint Jean Paulin and I Shafrir
327 Conjugate gradient type methods for ill-posed problems
M Hanke

C Bandle

University of Basel, Switzerland

J Bemelmans

RWTH Aachen, Germany

M Chipot

J Saint Jean Paulin

I Shafrir

University of Metz, France

(Editors)

Calculus of variations, applications and computations

Pont-à-Mousson 1994

Longman Scientific & Technical

Copublished in the United States with
John Wiley & Sons, Inc., New York

Longman Scientific & Technical
Longman Group Limited
Longman House, Burnt Mill, Harlow
Essex CM20 2JE, England
and Associated companies throughout the world.

Copublished in the United States with
John Wiley & Sons Inc., 605 Third Avenue, New York, NY 10158

First published 1995

AMS Subject Classifications: (Main) 35Jxx, 35Kxx, 49xx
(Subsidiary) 65xx, 73xx, 65Mxx

ISSN 0269-3674

ISBN 0 582 23962 1

British Library Cataloguing in Publication Data

A catalogue record for this book is available from the British Library

Library of Congress Cataloging-in-Publication Data

A catalog record for this book is available

Printed and bound in Great Britain
by Biddles Ltd, Guildford and King's Lynn

Contents

Preface

This volume collects lectures given during the second European Conference on Elliptic and Parabolic Problems (Pont-à-Mousson, June 1994).

Various important topics in Partial Differential Equations are addressed : Calculus of Variations, Control Theory, Modelling, Numerical Analysis and various applications in Physics, Mechanics and Engineering.

We would like to thank all the participants of this meeting for their help in making it succesful. Special thanks go to the contributors for this volume.

The meeting has been made possible by grants from the "Caisse d'épargne de Lorraine Nord", the C.N.R.S., the "Region de Lorraine", the A.A.U.L., the RWTH Aachen, the University of Basel, the University of Metz and the "Gesellschaft von Freuden der Aachener Hochschule e.V.". We express our deep appreciation to them.

Finally, we wish to express our special thanks to Longman for helping us to publish these proceedings.

C. Bandle, J. Bemelmans, M. Chipot, J. Saint Jean Paulin and I. Shafrir

L BOUKRIM AND J MOSSINO

Isoperimetric inequalities for a generalized multidimensional Muskat problem

I - Introduction

In this introduction we first describe the mathematical problem and then we give its motivation as a generalization of the so-called Muskat problem in oil production.

Let $1 < p < \infty$; β is a locally integrable function defined in $\mathbb{R}^+$ which does not vanish and has constant sign.

Let $\Omega = \omega_3 \backslash \bar{\omega}_1$ be a given domain in $\mathbb{R}^N$ with regular boundary $\gamma_1 \bigcup \gamma_3$, $\gamma_1 = \partial\omega_1$, $\gamma_3 = \partial\omega_3$; m_1 and m_3 denote the respective measures of ω_1 and ω_3.

Let $\delta : (m_1, m_3) \to \mathbb{R}^+$ be a positive function such that $1/\delta$ belongs to $L^1(m_1, m_3)$.

Let $a_1, a_2 : \Omega \times [m_1, m_3] \to \mathbb{R}^+$ be positive functions such that, for every m in $[m_1, m_3]$, $a_i(.,m)$ and $1/a_i(.,m)$ are in $L^\infty(\Omega)$ for i = 1,2 and

$$\operatorname*{ess\,sup}_{x \in \Omega} a_1(x,m) \leq \operatorname*{ess\,inf}_{x \in \Omega} a_2(x,m). \tag{1.1}$$

We assume that $A_1, A_2 : \Omega \times [m_1, m_3] \to \mathbb{R}^{N \times N}$ are given and such that, for every m in $[m_1, m_3]$, $A_i(., m)$ belongs to $L^\infty(\Omega)^{N \times N}$ and

$$a.e.\ x \in \Omega, \quad \forall \xi \in \mathbb{R}^N, \ (A_i(x,m)\xi, \xi) \geq a_i(x,m) \|\xi\|^2 \ (i = 1,2) \tag{1.2}$$

where (.,.) denotes the scalar product in $\mathbb{R}^N$ and $\|.\|$ denotes the Euclidian norm.

Finally let $\omega_2(0)$ (at time 0) with regular boundary $\gamma_2(0)$ be such that $\omega_1 \subset \omega_2(0) \subset \omega_3$.

We assume that ω_2 depends on time in such a way that its boundary $\gamma_2(t)$ moves with normal velocity

$$v_\nu(x,t)\ (= (v(x,t),\ \nu(x,t))) = \beta(t)\,\delta(|\omega_2(t)|)\frac{\partial u_1}{\partial \nu_1}(x,t) = \beta(t)\,\delta(|\omega_2(t)|)\frac{\partial u_2}{\partial \nu_2}(x,t) \quad (1.3)$$

where $\nu(x,t)$ denotes the normal to $\gamma_2(t)$ pointing outside $\Omega_1(t) = \omega_2(t)\backslash\overline{\omega}_1$ and for i = 1, 2,

$$\frac{\partial u_i}{\partial \nu_i} = (\mathsf{A}_i u_i, \nu)$$

with

$$\mathsf{A}_i u_i = (B_i \nabla u_i,\ \nabla u_i)^{\frac{p}{2}-1} B_i \nabla u_i\ ,$$
$$B_i(x,t) = A_i(x,\ |\omega_2(t)|)\ ,$$

$\nabla = \nabla_x$, $u_1(.,t)$ and $u_2(.,t)$ are respectively defined in $\Omega_1(t)$ and $\Omega_2(t) = \omega_3\backslash\overline{\omega_2(t)}$ and satisfy (with $div = div_x$)

$$\begin{aligned} &div\ (\mathsf{A}_1 u_1) = 0 \quad in \quad \Omega_1(t),\\ &div\ (\mathsf{A}_2 u_2) = 0 \quad in \quad \Omega_2(t),\\ &u_1 = 1 \quad on \quad \gamma_1,\quad u_2 = 0 \quad on\ \gamma_3,\\ &u_1 = u_2 \quad and \quad \frac{\partial u_1}{\partial \nu_1} = \frac{\partial u_2}{\partial \nu_2} \quad on \quad \gamma_2(t). \end{aligned} \quad (1.4)$$

This problem (1.3), (1,4) is a natural generalization (in a quasilinear and anisotropic framework) of the Muskat problem, which is obtained for $p = 2$, $A_i\ (x,\ m) = a_i\ (x)\ Id$ (*Id* denotes the identity matrix). The Muskat problem (cf. [13], [14], [15], [4], [7], [10]) is a model (which is used for instance in oil production) for the flow of two immiscible fluids occupying the respective subdomains $\Omega_1(t)$ and $\Omega_2(t)$ in a porous medium Ω . As far as we know, the existence of solution, even for the "simple" Muskat problem is nonstandard (cf. [1], [7], [10]).

In this paper we give isoperimetric (optimal, explicit) inequalities by comparison with a problem with spherical symmetry. In particular we give optimal estimates for the critical time after which no "regular" solution may exist, for the respective volumes of fluids and, in some cases, for the diffusion speed $d|\omega_2(t)|/dt$. For similar problems with moving free boundaries, isoperimetric inequalities have been obtained by B. GUSTAFSSON and J. MOSSINO [8], [9]. The present situation is a different one since the front $\gamma_2(t)$ is not a level surface for $u_1(t)$ and $u_2(t)$. The research of a natural symmetrized problem passes through the rearrangement (with respect to x) of the function

$$b(x,t) = \begin{cases} (a_1(x,|\omega_2(t)|) \text{ for } x \in \Omega_1(t), \\ (a_2(x,|\omega_2(t)|) \text{ for } x \in \Omega_2(t), \end{cases} \tag{1.5}$$

a method introduced by A. ALVINO and G. TROMBETTI [2] in the context of degenerate elliptic problems (see also [5] for a different proof using the relative rearrangement defined by J. MOSSINO, R. TEMAM [11] and studied by J. M. RAKOTOSON and R. TEMAM ([16] [17])).

These results were announced in a short note [6] in a simpler framework (homogeneity, isotropy of the porous medium).

II - Explicit bounds for the maximal existence time, the respective volumes of fluids and the diffusion speed.

In this section we give explicit bounds for the maximal existence time of "regular" solutions, for the respective volumes of fluids, i.e. of $\Omega_1(t)$ and $\Omega_2(t)$, and for the diffusion speed $d|\omega_2(t)|/dt$. In section III we prove that these bounds are isoperimetric : all inequalities given in section II become equalities if ω_1, $\omega_2(0)$ and ω_3 are concentric balls, $A_i(x,m) = a_i(x,m)Id$ where $a_1(.,m)$ and $a_2(.,m)$ are radial for fixed m and satisfy suitable monotonicity conditions.

Our explicit bounds will be consequences of the following "abstract" inequality.

Theorem 1 : Let p' be such that $\frac{1}{p} + \frac{1}{p'} = 1$, α_N be the measure of the unit ball in $\mathbb{R}^N$, $s(\beta)$ $(= \pm 1)$ the sign of β. Let $m_2(t)$ be the measure of $\omega_2(t)$. Let

$$a_i^*(s,m) = Inf\{\ \theta \in \mathbb{R},\ |a_i(m) < \theta| \geq s\ \}\ (i = 1,2) \tag{2.1}$$

- with $|a_i(m) < \theta|$ = meas $\{x \in \Omega,\ a_i(x,m) < \theta\}$ - be the increasing (i.e. nondecreasing) rearrangement of a_i with respect to x. Let ψ be any antiderivative of the positive function

$$\psi'(m) = \frac{1}{\delta(m)}\left[\int_{m_1}^{m} s^{\frac{p'}{N}-p'} a_1^*(s-m_1,m)^{-p'/2}ds + \int_{m}^{m_3} s^{\frac{p'}{N}-p'} a_2^*(s-m,m)^{-p'/2}ds\right]^{p/p'}. \quad (2.2)$$

For any t such that $\gamma_2(t)$ is "regular" on $[0,t]$, we have

$$N^p \alpha_N^{p/N} \int_0^t |\beta(\tau)|\, d\tau \le s(\beta)\,[\psi(m_2(0)) - \psi(m_2(t))]. \quad (2.3)$$

Proof : Let us define

$$u(x,t) = u_i(x,t) \quad for\ x \in \Omega_i(t),$$
$$B(x,t) = B_i(x,t) = A_i(x,m_2(t)) \quad for\ x \in \Omega_i(t)$$

and (using (1.4))

$$C(t) = \int_\Omega (B\nabla u, \nabla u)^{p/2} dx = -\int_{\gamma_2(t)} \frac{\partial u_1}{\partial \nu_1} d\gamma = -\int_{\gamma_2(t)} \frac{\partial u_2}{\partial \nu_2} d\gamma. \quad (2.4)$$

Using (1.2), the technic of rearrangement developped in [5] (see also [2], [3]) leads to

for every θ, θ' with $0 \le \theta \le \theta' \le 1$, (2.5)

$$\theta' - \theta \le N^{-p'} \alpha_N^{-p'/N} C(t)^{p'/p} \int_{|u(t)>\theta'|}^{|u(t)>\theta|} (m_1+s)^{\frac{p'}{N}-p'} b^*(s-|u(t)>\theta'|,t)^{-p'/2} ds$$

where b^* is the increasing rearrangement (*w. r.* to x) of the function b defined in (1.5) :

$$b^*(s,t) = inf\{\theta \in \mathbb{R},\ |b(t) < \theta| \ge s\}$$

with $|b(t) < \theta| = meas\ \{x \in \Omega,\ b(x,t) < \theta\}$, and where (e.g) $|u(t) > \theta| = meas\ \{x \in \Omega,\ u(x,t) > \theta\}$. The reader is referred to [12] (for instance) for the classical properties of rearrangements. By (1.1)

$$b^*(s,t) = \begin{cases} (a_1(.,m_2(t))|_{\Omega_1(t)})^*(s) \ for\ 0 \le s < |\Omega_1(t)|, \\ (a_2(.,m_2(t))|_{\Omega_2(t)})^*(s - |\Omega_1(t)|) \ for\ |\Omega_1(t)| < s \le |\Omega|. \end{cases} \quad (2.6)$$

Hence, as for any measurable function f and measurable subset E, $(f|_E)^* \geq f^*|_{[0,|E|]}$,

$$b^*(s,t) \geq \begin{cases} a_1^*(s,m_2(t)) \text{ for } 0 \leq s < m_2(t)-m_1 , \\ a_2^*(s+m_1-m_2(t), m_2(t)) \text{ for } m_2(t)-m_1 < s \leq m_3-m_1 , \end{cases} \tag{2.7}$$

with $a_i^*(s,m)$ defined in (2.1). From (2.5) with $\theta' = 1$, $\theta = 0$, (2.7) and (2.2) it follows

$$N^p \alpha_N^{p/N} \leq C(t)\, \delta(m_2(t))\psi'(m_2(t)). \tag{2.8}$$

Now by (1.3), (2.4), as long as $\gamma_2(t)$ is "regular" in x and t

$$\frac{dm_2}{dt} = \int_{\gamma_2(t)} v_\nu d\gamma = -\beta(t)\, \delta(m_2(t))\; C(t). \tag{2.9}$$

It follows from (2.8), (2.9) that

$$N^p \alpha_N^{p/N} \leq -\frac{1}{\beta(t)}\psi'(m_2(t))\frac{dm_2}{dt} = \frac{-s(\beta)}{|\beta(t)|}\psi'(m_2(t))\frac{dm_2}{dt}$$

or

$$N^p \alpha_N^{p/N} |\beta(t)| \leq -s(\beta)\psi'(m_2(t))\frac{dm_2}{dt} \tag{2.10}$$

which gives (2.3) by integration.

As a corollary of Theorem 1, we get

<u>Theorem 2</u> : <u>With the notations defined in Theorem</u> 1, <u>let</u>

$$\bar{t}_C , \quad \bar{m}_2(t) , \quad \bar{C}(t)$$

<u>be explicitly given by</u>

$$N^p \alpha_N^{p/N} \int_0^{\bar{t}_C} |\beta(\tau)|\, d\tau = \begin{cases} \psi(m_2(0)) - \psi(m_1) \text{ if } \beta > 0 , \\ \psi(m_3) - \psi(m_2(0)) \text{ if } \beta < 0 , \end{cases} \tag{2.11}$$

$$N^p \alpha_N^{p/N} \int_0^{t} |\beta(\tau)|\, d\tau = s(\beta)\, [\psi(m_2(0)) - \psi(\bar{m}_2(t))] \text{ for } 0 \leq t \leq \bar{t}_C , \tag{2.12}$$

$$\bar{C}(t) = \frac{N^p \alpha_N^{p/N}}{\delta(\bar{m}_2(t))\psi'(\bar{m}_2(t))} \quad for \quad 0 \le t \le \bar{t}_C. \tag{2.13}$$

If the front $\gamma_2(t)$ remains "regular" on [0,t], the following inequalities hold

$$t \le \bar{t}_C \tag{2.14}$$

(that is no regular solution may exist after $\bar{t}_C$),

$$m_2(t) \le \bar{m}_2(t) \quad if \quad \beta > 0, \; m_2(t) \ge \bar{m}_2(t) \; if \; \beta < 0, \tag{2.15}$$

$$N^p \alpha_N^{p/N} = \bar{C}(t)\,\delta(\bar{m}_2(t))\psi'(\bar{m}_2(t)) \le C(t)\,\delta(m_2(t)\psi'(m_2(t)) \tag{2.16}$$

where $C(t)$ is defined in (2.4). If $\beta > 0$ and if for almost every x in Ω, the functions $a_1(x,.)$ and $a_2(x,.)$ are nonincreasing, it follows

$$\bar{C}(t) \le C(t) = -\frac{1}{\beta(t)\,\delta(m_2(t))}\frac{dm_2}{dt}(t) \tag{2.17}$$

and hence if, moreover, δ is nonincreasing

$$-\frac{dm_2}{dt}(t) \ge \bar{C}(t)\;\beta(t)\;\delta(\bar{m}_2(t)) \tag{2.18}$$

which is an estimate for the diffusion speed.

Proof : As $m_1 \le m_2(t) \le m_3$ and ψ is increasing

$$s(\beta)\;[\psi(m_2(0)) - \psi(m_2(t))] \le \begin{cases} \psi(m_2(0)) - \psi(m_1) & if\ \beta > 0, \\ \psi(m_3) - \psi(m_2(0)) & if\ \beta < 0 \end{cases}$$

$$= N^p\;\alpha_N^{p/N} \int_0^{\bar{t}_c} |\beta(\tau)|\,d\tau$$

by definition of $\bar{t}_C$ (see (2.11)). By (2.3) it follows

$$N^p \alpha_N^{p/N} \int_0^t |\beta(\tau)|\, d\tau \leq s(\beta)\, [\psi(m_2(0)) - \psi(m_2(t))] \leq N^p \alpha_N^{p/N} \int_0^{\bar{t}_C} |\beta(\tau)|\, d\tau$$

and we get (2.14); then (2.15) follows from the above and (2.12) (note that ψ is increasing). By (2.8), (2.13) one gets (2.16), and from (2.16)

$$\bar{C}(t) \leq C(t) \frac{\delta(m_2(t))}{\delta(\bar{m}_2(t))} \frac{\psi'(m_2(t))}{\psi'(\bar{m}_2(t))}.$$

Using (2.15) for $\beta > 0$, one gets (2.17) once checking that $\delta\psi'$ is nondecreasing, which we prove below. Actually, let us define

$$A_1(s,m) = s^{\frac{p'}{N}-p'} a_1^*(s-m_1, m)^{-p'/2},$$

$$A_2(s,m) := s^{\frac{p'}{N}-p'} a_2^*(s-m, m)^{-p'/2},$$

$$F(m) = [\psi'(m)\, \delta(m)]^{p'/p} = \int_{m_1}^{m} A_1(s,m)\, ds + \int_{m}^{m_3} A_2(s,m)\, ds$$

and let us prove that for $m \leq m'$, $F(m) \leq F(m')$. We have

$$F(m) - F(m') = \int_{m_1}^{m'} (A_1(s,m) - A_1(s,m'))\, ds + \int_{m'}^{m_3} (A_2(s,m) - A_2(s,m'))\, ds$$
$$+ \int_{m}^{m'} (A_2(s,m) - A_1(s,m))\, ds,$$

$A_1(s,m) \leq A_1(s,m')$ and $A_2(s,m) \leq A_2(s,m')$ since a_1 and a_2 are nonincreasing with respect to m and a_2^* is nondecreasing with respect to s; moreover $A_2(s,m) \leq A_1(s,m)$ by (1.1) since

$$a_1^*(s-m_1,m) \leq \operatorname*{ess\,sup}_x\, a_1(x,m) \leq \operatorname*{ess\,inf}_x\, a_2(x,m) \leq a_2^*(s-m,m).$$

It follows that $F(m) - F(m') \leq 0$: in other words F, and hence $\delta\psi'$, are non decreasing. Finally (2.18) follows from (2.17), (2.15) is δ is nonincreasing.

Now we interprete the bounds obtained in Theorem 2. Let ω_1^S, $\omega_2^S(0)$, ω_3^S be the balls of $\mathbb{R}^N$ with center 0 (for instance) having same respective measures as ω_1, $\omega_2(0)$, ω_3. We define below an evolution problem $P^{\mathbf{R}}$ (the rearrangement, in some sense, of the Muskat problem P enounced in (1.3), (1.4)) and denote by q and $q^{\mathbf{R}}$ two corresponding quantities in P and $P^{\mathbf{R}}$.

$P^{\mathbf{R}}$ is defined as follows : it has spherical symmetry and its geometrical and functional data are respectively

(*i*) $\omega_1^{\mathbf{R}} = \omega_1^S$, $\omega_2^{\mathbf{R}}(0) = \omega_2^S(0)$, $\omega_3^{\mathbf{R}} = \omega_3^S$,

(*ii*) $\beta^{\mathbf{R}} = \beta$, $\delta^{\mathbf{R}} = \delta$ *and for* $x \in \Omega^{\mathbf{R}}$, $m \in [m_1,m_3]$,

$$a_1^{\mathbf{R}}(x,m) = a_1^*(\alpha_N \|x\|^N - m_1, m) \;\; for \;\; m_1 \le \alpha_N \|x\|^N < m \tag{2.19}$$

and $a_1^{\mathbf{R}}(x,m)$ is unprescribed for $m < \alpha_N \|x\|^N \le m_3$,

$$a_2^{\mathbf{R}}(x,m) = a_2^*(\alpha_N \|x\|^N - m, m) \;\; for \;\; m < \alpha_N \|x\|^N \le m_3 \tag{2.20}$$

and $a_2^{\mathbf{R}}(x,m)$ is unprescribed for $m_1 \le \alpha^N \|x\|^N < m$,

$$A_i^{\mathbf{R}}(x,m) = a_i^{\mathbf{R}}(x,m) \; Id \;\; for \;\; i = 1, 2. \tag{2.21}$$

Moreover the evolution of the sphere $\gamma_2^{\mathbf{R}}(t)$ obeys the corresponding laws $(1.3)^{\mathbf{R}}$, $(1.4)^{\mathbf{R}}$ which are written :

$$(1.4)^{\mathbf{R}} \begin{cases} div\,(a_i^{\mathbf{R}}(x, m_2^{\mathbf{R}}(t))^{p/2} \, |\nabla u_i^{\mathbf{R}}|^{p-2} \, \nabla u_i^{\mathbf{R}}) = 0 \; in \; \Omega_i^{\mathbf{R}}(t) \; for \; i = 1,2, \\ u_1^{\mathbf{R}} = 1 \; on \; \gamma_1^{\mathbf{R}} \, , \; u_2^{\mathbf{R}} = 0 \;\; on \; \gamma_3^{\mathbf{R}} \, , \\ on \; \gamma_2^{\mathbf{R}}(t) \, : \, u_1^{\mathbf{R}} = u_2^{\mathbf{R}} \;\; and \\ a_1^{\mathbf{R}}(x, m_2^{\mathbf{R}}(t))^{p/2} \, |\nabla u_1^{\mathbf{R}}|^{p-2} \nabla u_1^{\mathbf{R}} = a_2^{\mathbf{R}}(x, m_2^{\mathbf{R}}(t))^{p/2} \, |\nabla u_2^{\mathbf{R}}|^{p-2} \nabla u_2^{\mathbf{R}} \end{cases}$$

and $(1.3)^{\mathbf{R}}$ simply tells that the last line is also $\dfrac{1}{\beta(t) \;\; \delta(m_2^{\mathbf{R}}(t))} \; v_\nu^{\mathbf{R}}(x,t)$.

Since $\Omega_i^{\mathbf{R}}$ have spherical symmetry, it is not difficult to check that $(1.4)^{\mathbf{R}}$ has one unique solution $u^{\mathbf{R}}$ ($=u_1^{\mathbf{R}}$ *in* $\Omega_1^{\mathbf{R}}(t)$, $u_2^{\mathbf{R}}$ *in* $\Omega_2^{\mathbf{R}}(t)$). It is independent of the definitions of $a_1^{\mathbf{R}}$ *on* $\{ m < \alpha_N \|x\|^N \le m_3\}$ *and* $a_2^{\mathbf{R}}$ *on* $\{ m_1 \le \alpha_N \|x\|^N < m\}$

and it is given by

$$u^{\mathbf{R}}(x,t) = 1 - \left[\int_{m_1}^{\alpha_N \|x\|^N} s^{\frac{p'}{p}-p'} d(s,t)^{-\frac{p'}{2}} ds\right]\left[\int_{m_1}^{m_3} s^{\frac{p'}{N}-p'} d(s,t)^{-p'/2} ds\right]^{-1}$$

where

$$d(s,t) = \begin{cases} a_1^*(s-m_1,\ m_2^{\mathbf{R}}(t)) & for\ \ m_1 \le s < m_2^{\mathbf{R}}(t), \\ a_2^*(s-m_2^{\mathbf{R}}(t),\ m_2^{\mathbf{R}}(t)) & for\ \ m_2^{\mathbf{R}}(t) < s \le m_3\ . \end{cases}$$

Moreover (see (2.4))

$$C^{\mathbf{R}}(t) \left(= \int_{\Omega^{\mathbf{R}}} (B^{\mathbf{R}}\nabla u^{\mathbf{R}},\ \nabla u^{\mathbf{R}})^{p/2} dx \right) = \frac{N^p \alpha_N^{p/N}}{\delta(m_2^{\mathbf{R}}(t))\psi'(m_2^{\mathbf{R}}(t))}.$$

Now using $(1.3)^{\mathbf{R}}$ it follows

$$C^{\mathbf{R}}(t) = \frac{N^p \alpha_N^{p/N}}{\delta(m_2^{\mathbf{R}}(t))\psi'(m_2^{\mathbf{R}}(t))} = -\frac{1}{\beta(t)\,\delta(m_2^{\mathbf{R}}(t))}\frac{dm_2^{\mathbf{R}}}{dt} \tag{2.22}$$

and hence

$$\psi(m_2^{\mathbf{R}}(t)) - \psi(m_2(0)) = \int_0^t \psi'(m_2^{\mathbf{R}}(\tau))\frac{dm_2^{\mathbf{R}}}{dt}(\tau)d\tau = -N^p \alpha^{p/N}\int_0^t \beta(\tau)d\tau$$

that is equality is achieved in (2.3) for $P^{\mathbf{R}}$ in place of P. Then from (2.12) one gets $m_2^{\mathbf{R}}(t) = \bar{m}_2(t)$ since ψ is strictly increasing. Finally from (2.13), (2.22) one gets $\bar{C}(t) = C^{\mathbf{R}}(t)$. In other words we have proved

<u>Theorem 3</u> : <u>The quantities</u> $\bar{t}_C$, $\bar{m}_2(t)$, $\bar{C}(t)\beta(t)\delta(\bar{m}_2(t))$ <u>defined in Theorem</u> 2 <u>are respectively</u> :

- <u>the critical (or stopping) time for</u> $P^{\mathbf{R}}$, <u>that the time for which</u> $\gamma_2^{\mathbf{R}}(t)$ reaches $\gamma_1^{\mathbf{R}}$ (<u>if</u> $\beta > 0$) <u>or</u> $\gamma_3^{\mathbf{R}}$ (<u>if</u> $\beta < 0$),

- <u>the measure of</u> $\omega_2^{\mathbf{R}}(t)$,

- the "speed of diffusion" $-\frac{dm_2^{\mathbf{R}}}{dt}(t) = C^{\mathbf{R}}(t)\beta(t)\delta(m_2^{\mathbf{R}}(t))$.

If $\beta > 0$, if, for every x in Ω, the functions $a_1(x,.)$ and $a_2(x,.)$ are nonincreasing and if δ is nonincreasing it follows (see (2.18)) that as long as the front $\gamma_2(t)$ remains regular

$$(0 \leq) - \frac{dm_2^{\mathbf{R}}(t)}{dt}(t) \leq -\frac{dm_2}{dt}(t),$$

that is the speed of diffusion is not smaller for P than for $P^{\mathbf{R}}$.

III - These bounds are isoperimetric

In this section we prove that the bounds given in Theorems 2 and 3 are isoperimetric.
Theorem 4 : The explicit bounds given in section II are achieved if ω_1, $\omega_2(0)$ and ω_3 are balls with center 0, if $A_i(x,m) = a_i(x,m)\ Id\ (i=1,2)$, if for every m in $[m_1,m_3]$, $a_1(.,m)$ is radial and nondecreasing with the radius on

$$\{m_1 \leq \alpha_N \|x\|^N < m\}$$

and $a_2(.,m)$ is radial and nondecreasing with the radius on $\{m < \alpha_N\|x\|^N \leq m_3\}$ and if a_1 and a_2 satisfy

$$\operatorname*{ess\,sup}_{m_1 \leq \alpha_N\|x\|^N < m} a_1(x,m) \leq \operatorname*{ess\,inf}_{m < \alpha_N\|x\|^N \leq m_3} a_1(x,m), \tag{3.1}$$

$$\operatorname*{ess\,sup}_{m < \alpha_N\|x\|^N \leq m_3} a_2(x,m) \leq \operatorname*{ess\,inf}_{m_1 \leq \alpha_N\|x\|^N < m} a_2(x,m). \tag{3.2}$$

Proof : It is enough to prove that $P^{\mathbf{R}}$ coincides with P under the conditions stated in the Theorem. Clearly (see the definition of $P^{\mathbf{R}}$) $P^{\mathbf{R}}$ coincides with P if and only if

$$\omega_1 = \omega_1^S,\ \omega_2(0) = \omega_2^S(0),\ \omega_3 = \omega_3^S,$$

$$a_1(x,m) = a_1^*(\alpha_N\|x\|^N - m_1, m) \quad for \quad m_1 \leq \alpha_N\|x\|^N < m, \tag{3.3}$$

$$a_2(x,m) = a_2^*(\alpha_N\|x\|^N - m,\ m) \quad for \quad m < \alpha_N\|x\|^N \leq m_3. \tag{3.4}$$

It is easy to check that (3.3) is fulfilled if $a_1(.,m)$ is radial and nondecreasing with the radius in $\{m_1 \leq \alpha_N \|x\|^N < m\}$ and if (3.1) holds true. The conditions on a_2 are derived similarly from (3.4).

Remark : If a_1 and a_2 are independent of m the conditions stated in Theorem 4 reduce to : a_1 is radial and nondecreasing with the radius and a_2 is constant.

IV Application to the Muskat problem in oil production

Let us consider the simple Muskat problem (see [13], [14], [15], [4], [7], [10]) modelling the displacement of oil by water in a porous medium occupying a given domain $\Omega = \omega_3 \setminus \bar{\omega}_1$ in $\mathbb{R}^2$. At time t, $\gamma_2(t)$ (surrounding ω_1) is a free boundary which represents the interface (abrupt front) between the subdomain $\Omega_1(t)$ of Ω (with boundary $\gamma_1 \cup \gamma_2(t)$) containing (viscositied) water only and the subdomain $\Omega_2(t) = \Omega \setminus \overline{\Omega_1(t)}$ containing oil only. The respective pressures of fluids in $\Omega_1(t)$ and $\Omega_2(t)$ are $p_1(x,t)$ and $p_2(x,t)$. They are given on γ_1 and γ_3 :

- either (case A)

$$p_1(x,t) = P_1(t) \text{ on } \gamma_1, \quad p_2(x,t) = P_2(t) < P_1(t) \text{ on } \gamma_3$$

- or (case B of the so-called "piezometric" conditions)

$$p_1(x,t) = \Pi_1(m_2(t)) \text{ on } \gamma_1, \quad p_2(x,t) = \Pi_2(m_2(t)) < \Pi_1(m_2(t)) \text{ on } \gamma_3.$$

Note that the inequalities above simply mean that water is used to displace oil.

From Darcy's law and conservation of mass it follows that

$$\begin{cases} -div \left(\dfrac{k_1}{\mu_1} \nabla p_1 \right) = 0 \text{ in } \Omega_1(t), \\ -div \left(\dfrac{k_2}{\mu_2} \nabla p_2 \right) = 0 \text{ in } \Omega_2(t), \\ p_1 = p_2 \text{ and } -\dfrac{k_1}{\mu_1}\dfrac{\partial p_1}{\partial \nu} = -\dfrac{k_2}{\mu_2}\dfrac{\partial p_2}{\partial \nu} = \phi v_\nu \text{ on } \gamma_2(t) \end{cases}$$

where div, ∇ and $\dfrac{\partial}{\partial \nu} = \nu.\nabla$ are taken with respect to the space variable x, k_1 and k_2

are functions ($k_1(x)$ and $k_2(x)$) representing the permeabilities of the porous medium with respect to water and oil respectively, μ_1 and μ_2 are viscosities, ν is the normal to $\gamma_2(t)$ pointing outside $\Omega_1(t)$, v_ν is the normal velocity of displacement of $\gamma_2(t)$ and ϕ is the porosity of the medium. Here we suppose that ϕ is constant.

We set

$$p(x,t) = \begin{cases} p_1(x,t) & \text{if } x \in \Omega_1(t) , \\ p_2(x,t) & \text{if } x \in \Omega_2(t) , \end{cases}$$

$$u(x,t) = \frac{p_2(t)|_{\gamma_3} - p(x,t)}{p_2(t)|_{\gamma_3} - p_1(t)|_{\gamma_1}} = \begin{cases} u_1(x,t) & \text{if } x \in \Omega_1(t) , \\ u_2(x,t) & \text{if } x \in \Omega_2(t) , \end{cases}$$

$$\alpha_1(x) = \frac{k_1(x)}{\mu_1} \quad , \quad \alpha_2(x) = \frac{k_2(x)}{\mu_2} \text{ and}$$

$$\text{in case A : } \beta(t) = \frac{P_2(t) - P_1(t)}{\phi} < 0 \quad , \quad \delta(m_2(t)) \equiv 1,$$

$$\text{in case B : } \beta(t) \equiv -1 \quad , \quad \delta(m_2(t)) = \frac{\Pi_1(m_2(t) - \Pi_2(m_2(t))}{\phi} > 0.$$

We get

$$\begin{cases} -div\ (\alpha_1 \nabla u_1) = 0 & \text{in } \Omega_1(t) , \\ -div\ (\alpha_2 \nabla u_2) = 0 & \text{in } \Omega_2(t) , \\ u_1 = 1 \text{ on } \gamma_1 , \quad u_2 = 0 \text{ on } \gamma_3 , \\ \left(\dfrac{\partial u_1}{\partial \nu_1} = \right) \alpha_1 \dfrac{\partial u_1}{\partial \nu} = \left(\dfrac{\partial u_2}{\partial \nu_2} = \right) \alpha_2 \dfrac{\partial u_2}{\partial \nu} = \dfrac{1}{\beta(t)\,\delta(m_2(t))} v_\nu & \text{on } \gamma_2(t) . \end{cases}$$

It is not reasonable to assume that the functions $\alpha_i(x)$ (or their rearrangements) are known but we may assume that for some constants $a_i > 0$, $\alpha_i(x) \geq a_i$ $(i = 1, 2)$. We assume $a_1 < a_2$. Let us comment this condition. In the case k_1 and k_2 constant, one can take $a_i = \dfrac{k_i}{\mu_i}$ and then $a_1 < a_2$ means $\dfrac{k_1}{k_2}\dfrac{\mu_2}{\mu_1} < 1$. This condition is satisfied for example if $k_1 \leq k_2$ (i. e. the permeability of the porous medium to oil is not smaller than its permeability to water) and $\mu_2 < \mu_1$ (i. e. the viscosity of viscositied water is greater than the viscosity of oil). Remember that if $\frac{k_1}{k_2}\frac{\mu_2}{\mu_1} > 1$ it is well known [4] that instabilities develop from the very beginning and fingers appear.

Below we apply the results of the section II to the simple Muskat problem, studying separately cases A and B

- case A :

We get from (2.2)

$$\psi'(m) = \frac{1}{a_1} Log\frac{m}{m_1} + \frac{1}{a_2} Log\frac{m_3}{m} ,$$
$$\psi(m) = C_1 m Log m + C_2 m$$

with

$$C_1 = \frac{1}{a_1} - \frac{1}{a_2} , \quad C_2 = \frac{1}{a_2}(Log m_3 + 1) - \frac{1}{a_1}(Log m_1 + 1).$$

From (2.11), (2.14), the maximal existence time of regular solution is $\bar{t}_C$ given by

$$\frac{4\pi}{\Phi}\int_0^{\bar{t}_C} (P_1(\tau) - P_2(\tau))d\tau = C_1(m_3 Log m_3 - m_2(0) Log m_2(0)) + C_2(m_3 - m_2(0)).$$

From (2.12), (2.15), as long as the front remains regular, the volume of water $m_2(t)$ (up to substracting m_1) is not smaller than the solution $\bar{m}_2(t)$ of

$$\frac{4\pi}{\Phi}\int_0^{t} (P_1(\tau) - P_2(\tau))d\tau = C_1(\bar{m}_2(t) Log \bar{m}_2(t) - m_2(0) Log m_2(0)) + C_2(\bar{m}_2(t) - m_2(0))$$

- case B :

We get from (2.2)

$$\psi'(m) = \frac{\Phi}{\Pi_1(m) - \Pi_2(m)}\left(\frac{1}{a_1} Log\frac{m}{m_1} + \frac{1}{a_2} Log\frac{m_3}{m}\right).$$

Let ψ be any antiderivative of ψ'. From (2.11), (2.14), the maximal existence time of regular solution is

$$\bar{t}_C = \frac{1}{4\pi} [\psi(m_3) - \psi(m_2(0))],$$

and from (2.12), (2.15), as long as the front remains regular, the volume of water is not smaller than the solution $\bar{m}_2(t)$ of

$$4\pi t = \psi(\bar{m}_2(t)) - \psi(m_2(0)).$$

- All these bounds (cases A and B) are achieved if ω_1, $\omega_2(0)$ and ω_3 are concentric balls and if $\frac{k_i}{\mu_i} = a_i$ $(i = 1, 2)$ are constant.

Acknowledgements : The authors thank B. GUSTAFSSON, L. JIANG, M. PRIMICERIO and H. VAN DAMME for stimulating discussions and interesting remarks.

References

[1] F. ABERGEL, J. MOSSINO, Caractérisation du problème de Muskat multidimensionnel et existence de solutions régulières, C.R. Acad. Sci. Paris, t.319, Série I, p.35-40, 1994.

[2] A. ALVINO, G. TROMBETTI, Sulle migliori costanti di maggiorazione per una classe di equazioni ellittiche degeneri, Ric. di Mat., 27, pp 413-428, 1978.

[3] A. ALVINO, G. TROMBETTI, Isoperimetric inequalities connected with torsion problem and capacity, Boll. U.M.I., (6), 4-B, pp. 773-787, 1985.

[4] J. BEAR, Dynamics of Fluids in Porous Media, American Elsevier Publishing company, inc., New york, London, Amsterdam, second printing, 1975.

[5] L. BOUKRIM, Inégalités isopérimétriques pour un problème d'électrostatique, C.R. Acad. Sci. Paris, t. 318, Série I, p. 435-438, 1994 ; see also Inégalités Isopérimétriques pour Certains Problèmes de Conductivité dans des Milieux Non Homogènes, Thésis to be presented at the University of Paris XI, Orsay, France, 1994.

[6] L. BOUKRIM, J. MOSSINO, Inégalités isopérimétriques pour un problème de Muskat, C.R. Acad. Sci. Paris, t. 317, Série I, p. 329-332, 1993.

[7] C. M. ELLIOT, J.R. OCKENDON, Weak and Variational Methods for Moving Boundary Problems, Pitman, London, 1982.

[8] B. GUSTAFSSON, J. MOSSINO, Some isoperimetric inequalities in electrochemistry and Hele Shaw flows, I. M. A. Journal of Appl. Math. , 39, pp. 33-49, 1987 ; see also J. MOSSINO, Inégalités isopérimétriques en électrolyse, C. R. Acad. Sci. Paris, I, 301, pp. 869-871, 1985.

[9] B. GUSTAFSSON, J. MOSSINO, Isoperimetric inequalities for the Stefan problem, SIAM J. Math. Anal., 20, n°5, pp. 1095-1108, 1989.

[10] L. JIANG, Z. CHEN, Weak formulation of a multidimensional Muskat problem, Proceedings of the Irsee Conference on Free Boundary Problems, in "Free Boundary Problems : Theory and Applications" , vol. II, K.H. HOFMANN and J. SPREKELS eds, Pitman Research Notes in Math. Series, n° 186, 1990.

[11] J. MOSSINO, R. TEMAM, Directional derivative of the increasing rearrangement mapping and application to a queer differential equation in plasma physics, Duke Math J. , Vol. 48, pp. 475-485, 1981.
[12] J. MOSSINO, Inégalités Isopérimétriques et Applications en Physique, Hermann, Travaux en cours, 1984.
[13] M. MUSKAT, The flow of Homogeneous Fluids through Porous Media, Mc Graw-Hill, New york, 1937; 2ne printing by Edwards, Arn. Arbor, Mich., 1946.
[14] M. MUSKAT, Physical Principles of Oil Production, Mc Graw-Hill, New york 1949.
[15] M. MUSKAT, R. D. WYCOFF, H. G. BOTSET and M.W. MERES, Flow of gas liquid mixtures through sands, Trans. A.I.M.E. Petrol, 123, pp. 69-96, 1937.
[16] J. M. RAKOTOSON, R. TEMAM, Relative rearrangement in quasilinear elliptic variational inequalities, Ind. Univ. Math. J., C, vol. 36, n°4, pp. 757-810 ,1987.
[17] J. M. RAKOTOSON, Some properties of the relative rearrangement, J. Math. An. Appl. , 135, pp. 488-500, 1987.

L.B. : Département de Mathématiques,
Bâtiment 425, Université Paris Sud
91405 ORSAY Cedex FRANCE

J.M. : C.N.R.S., Laboratoire de Mathématiques Appliquées et Physique Mathématique d'Orléans,
UFR de Sciences,
Université d'Orléans, BP 6759
45067 ORLEANS Cedex 2 FRANCE

M BOUKROUCHE

Solution of a free boundary problem of the Hele–Shaw type, in the Ovsjannikov scales

The classical Hele-Shaw flow problem have been studied extensively [3] [7] and more recently for exemple by [4] [5]. In this paper we consider the generalization of Hele-Shaw flow problem taking gap geometry and non-constant injection or suction rate into account. An incompressible fluid is contained between two surfaces, a first one containing the injection or the withdrawing orifice, and the second one, not necessary flat. Let the two surfaces have equations: $x_3 = 0$ and $x_3 = \varepsilon h(x) = \varepsilon h(x_1, x_2)$, where h is the distance between the two surfaces. If $\varepsilon h(x)$ is sufficiently small compared to the lateral dimensions of the surfaces, the slow flow equations can be averaged across the gap to give [1] the fluid velocity $v = (v_1, v_2)$ in terms of the pressure p(t, x) by

$$v = (v_1, v_2) = -\frac{h^2}{12\mu}\nabla p \tag{0.1}$$

where μ is the viscosity coefficient, and the equation governing this problem in the fluid region $\Omega(t)$ and away from prescribed singularities of p, is

$$\nabla(\frac{h^3}{12\mu}\nabla p) = 0 \tag{0.2}$$

We shall consider flows in which no rigid boundaries are present, and in which $\partial\Omega(t) := \Gamma(t)$ consists of a simple closed curve, so that either $\Omega(t)$ or $\mathbf{R}^2 \setminus \Omega(t)$ is finite and simply connected. We shall ignore surface tension effects, and assume that:

$$p = 0 \quad \text{on} \quad \Gamma(t) \tag{0.3}$$

and the second condition holding on $\Gamma(t)$ is

$$\frac{h^2}{12\mu}\frac{\partial p}{\partial n} = -vn \tag{0.4}$$

where n is the outward pointing unit normal vector to $\partial\Omega(t)$.
The motion of the fluid will be driven by a source or sink of strength Q(t) at the point source (x_1^0, x_2^0) in $\Omega(t)$.
At each fixed time $t > 0$, $Q(t) < 0$ gives extraction of fluid, and $Q(t) > 0$ gives injection of fluid. The problem consists of finding $\Gamma(t)$ for each time $t > 0$, given the

continuous flow Q(t) and assuming that

$$\partial\Omega(0) = \Gamma(0) \quad \text{is an analytic given curve} \tag{0.5}$$

The goal of the present paper is first to derive a differential equation arising in the generalized Hele-Shaw flow problem (0.2)-(0.5), using the quasiconformal mapping, and to study conditions of existence, uniqueness and smoothness of his solution.

1 What type is the singularity of p at $z^0 = (x_1^0, x_2^0)$

Let $D(z^0, \varepsilon)$ be a small neighbourhood of z^0 and $C(z^0, \varepsilon)$ is the boundary of $D(z^0, \varepsilon)$. Using the complex Green formula for $hu = h(v_1 - iv_2)$ in $C^1(\Omega(t))$, we have

$$\begin{aligned}
\Re\Big(\int_{\Gamma(t)} hu(-idz)\Big) &= \int_{\Gamma(t)} hv_1 dx_2 - hv_2 dx_1 \\
&= \int_{\Omega(t)\setminus D(z^0,\varepsilon)} \Big(\frac{\partial hv_1}{\partial x_1} + \frac{\partial hv_2}{\partial x_2}\Big) dx_1 dx_2 + \Re\Big(\int_{C(z^0,\varepsilon)} hu(-idz)\Big) \\
&= Q(t)
\end{aligned}$$

Hence

$$hu = \frac{Q(t)}{2\pi(z - z^0)} + \varphi$$

where φ is an analytic function.
As $p = 0$ on $\Gamma(t)$ and z^0 is a point source in $\Omega(t)$ then the complex potentiel

$$w(z) = \frac{Q(t)}{2\pi} Ln(z - z^0) + \varphi_1 = \frac{Q(t)}{2\pi} G(z, z^0, \Omega(t)) \tag{1.1}$$

is the multi-valued complex Green function of the equation (0.2).

Whence

$$\Re(w(z)) = p = \frac{Q(t)}{2\pi} Ln|z - z^0| + \varphi_2 = \frac{Q(t)}{2\pi} g(z, z^0, \Omega(t)) \tag{1.2}$$

where $g(z, z^0, \Omega(t)) = \Re(G(z, z^0, \Omega(t)))$ and $\varphi_2 = \Re(\varphi_1)$, which is a single-valued function.

2 The new moving boundary condition

Each point $z = (x_1, x_2)$ on $\Gamma(t)$ moves with the velocity:

$$\frac{\partial z}{\partial t} = \bar{u}.$$

Then from (0.1) and (1.2) we have

$$\frac{\partial z}{\partial t} = -\frac{h^2 Q(t)}{12\mu\pi}\frac{\partial g(z, z^0, \Omega(t))}{\partial \bar{z}}$$

and the velocity of $\Gamma(t)$ at the point z, in the direction of the outward normal to $\Omega(t)$ is given by

$$\frac{\partial n}{\partial t} = -\frac{h^2 Q(t)}{24\mu\pi}\frac{\partial g(z, z^0, \Omega(t))}{\partial n} \quad \text{on } \Gamma(t) \tag{2.1}$$

As $\forall (z,t)$, with $z \in \Gamma(t)$, we have $g(z, z^0, \Omega(t)) = 0$

thus

$$d_{(z,t)}g = \frac{\partial g(z, z^0, \Omega(t))}{\partial n}dn + \frac{\partial g(z, z^0, \Omega(t))}{\partial t}dt = 0$$

i.e

$$\frac{\partial n}{\partial t} = -\left(\frac{\partial g(z, z^0, \Omega(t))}{\partial t}\right) / \left(\frac{\partial g(z, z^0, \Omega(t))}{\partial n}\right) \tag{2.2}$$

which implies:

$$\frac{\partial g(z, z^0, \Omega(t))}{\partial t} = \frac{h^2 Q(t)}{24\mu\pi}\left(\frac{\partial g(z, z^0, \Omega(t))}{\partial n}\right)^2 \quad \forall z \in \Gamma(t). \tag{2.3}$$

We shall use this formula to obtain a differential equation for the quasiconformal mapping.

3 The quasiconformal mapping

We introduce a complex potential $W = p + i\Phi$ where Φ is a Stream function. The complex potential W may be seen to satisfy a differential equation of the form

$$\frac{\partial W}{\partial \bar{z}} - q(z)\frac{\partial \bar{W}}{\partial \bar{z}} = 0 \tag{3.1}$$

where

$$q(z) = \frac{1 - \frac{h^3}{12\mu}}{1 + \frac{h^3}{12\mu}}$$

and

$$h(x) = h\left(\frac{z+\bar{z}}{2}, \frac{z-\bar{z}}{2i}\right)$$

Theorm 1 [6] *There exists a unique generalized solution W of the equation (3.1) such that:* $W(z^0) = z^0$ *and* $W(z) = z + T(\phi)$ *is a homeomorphism of the entire plane* $\mathbb{C}$.

The operator T is given by:

$$T(\phi) = -\frac{1}{\pi}\int_{\Omega(t)}\left(\frac{\phi(\zeta)d\zeta d\eta}{\zeta - (z - z^0)}\right) + \frac{1}{\pi}\int_{\Omega(t)}\frac{\phi(\zeta)d\zeta d\eta}{\zeta}$$

Let $W(\Omega(t)) = \Omega^*(t)$ be the image of $\Omega(t)$ under this fixed homeomorphism, and let Ξ_t be the (unique up to the normalization conditions) conformal mapping of $\Omega^*(t)$ onto the unit disk D(0, 1) in the τ -plane.
The inverse function to $\Xi_t oW$, which we denote by f_t , is the quasiconformal mapping such that $f_t(S^1) \equiv \Gamma(t)$ where $S^1 \equiv \partial D(0.1)$

It is easy to see that

$$w_1(z) := \frac{Q(t)}{2\pi}Ln\left((\Xi_t oW)(z)\right) \tag{3.2}$$

is also a solution of the equation (3.1). We maintain that $w_1(z)$ is up a normalization

the complex Green's function for the equation (3.1).
Thus from (1.2) and (3.2), we have:

$$g(z, z^0, \Omega(t)) = Ln\left(|\Xi_t oW)(z)|\right) = Ln(|\tau|) \quad \forall \tau \in D(0.1)$$

or

$$g\left(f_t(\tau), f_t(0), \Omega(t)\right) = Ln(|\tau|) \quad \forall \tau \in D(0.1) \tag{3.3}$$

To obtain a differential equation for f_t , we differentiate this equation with respect to t , writing $f(t,\tau) := f_t(\tau)$
Whence

$$-\frac{\partial g}{\partial t} = 2\Re\left(\frac{\partial g}{\partial z}\frac{\partial f}{\partial t}\right) \tag{3.4}$$

and with respect to τ and $\bar{\tau}$

$$\frac{1}{2\tau} = \frac{\partial g}{\partial z}\frac{\partial f}{\partial \tau} + \frac{\partial g}{\partial \bar{z}}\frac{\partial \bar{f}}{\partial \tau} \tag{3.5}$$

$$\frac{1}{2\bar{\tau}} = \frac{\partial g}{\partial z}\frac{\partial f}{\partial \bar{\tau}} + \frac{\partial g}{\partial \bar{z}}\frac{\partial \bar{f}}{\partial \bar{\tau}} \tag{3.6}$$

From which we may deduce that

$$\frac{\partial g}{\partial z} = \frac{1}{J(f)}\left(\frac{1}{2\tau}\frac{\partial \bar{f}}{\partial \bar{\tau}} - \frac{1}{2\bar{\tau}}\frac{\partial \bar{f}}{\partial \tau}\right) \tag{3.7}$$

$$\frac{\partial g}{\partial \bar{z}} = \frac{1}{J(f)}\left(\frac{1}{2\bar{\tau}}\frac{\partial f}{\partial \tau} - \frac{1}{2\tau}\frac{\partial f}{\partial \bar{\tau}}\right) \tag{3.8}$$

where $J(f) = |\frac{\partial f}{\partial \tau}|^2 - |\frac{\partial f}{\partial \bar{\tau}}|^2$ is the Jacobien of f which does not vanishes on a neighbourhood of $\bar{D}(0,1)$.

4 The new formulation of the problem

The equation (2.3) can be written in the following form

$$\frac{\partial g}{\partial t} = \frac{h^2 Q(t)}{6\pi\mu}\frac{\partial g}{\partial z}\frac{\partial g}{\partial \bar{z}} \quad \text{on } \Gamma(t) \tag{3.1}$$

and from (3.4) we have

$$- \Re\left(\frac{\partial g}{\partial z}\frac{\partial f}{\partial t}\right) = \frac{h^2 Q(t)}{12\pi\mu}\frac{\partial g}{\partial z}\frac{\partial g}{\partial \bar{z}} \quad \text{on } \Gamma(t) \tag{3.2}$$

Hence

$$\Re\left(\frac{\partial g}{\partial \bar{z}})^{-1}\frac{\partial f}{\partial t}\right) = -\frac{h^2 Q(t)}{12\pi\mu} \tag{3.3}$$

The substitution of the expression for $(\frac{\partial g}{\partial \bar{z}})^{-1}$ yields the differential equation for f:

$$\Re\left(\frac{\frac{\partial f}{\partial t}}{\frac{\partial f}{2\bar{\tau}\partial\tau} - \frac{\partial f}{2\tau\partial\bar{\tau}}}\right) = -\frac{Q(t)h^2}{12\mu\pi J(f)} \quad \forall (t,\tau) \in \mathbb{R}_+^* \times S^1 \tag{3.4}$$

$$f(t,0) = z^0 \tag{3.5}$$

$$f(0,\tau) = f_0(\tau) \tag{3.6}$$

The initial problem (0.2)-(0.5) can be represented in the following form:

The new formulation of the problem:

H1) *Given the real-valued function* Q , *continuous* $\forall t \in]0,T]$, *and the real-valued function h holomorphic for* z *in* $\mathbb{C}$.

H2) *Considering the extension of* S^1 *to a circular ring* $K_c = b^{-1} < |\tau| < b$, $b > 1$, *for all t in* $]0,T]$, *and given* f_0 , *generalized analytic and univalent function defined in a neighbourhood of* $\bar{D}(0,1)$, *and two constants* $R > 0$ *and* $r_2 \in]1,b]$, *such that* $|f_0(\tau)| \leq R$ *in* $\bar{D}_{r_2} := \tau \in \mathbb{C} \ : \ |\tau| \leq r_2$.

Find f, generalized analytic and univalent function defined in a neighbourhood of $\bar{D}(0,1)$, *continuously differentiable with respect to t and satisfying (4.4)-(4.6)* .

5 Existence, uniqueness and smoothness results

We recall the hypothesis (H1) and (H2), we consider a fixed r_1 such that $r_1 \in]0.r_2]$, then let us choose the scale of Banach spaces:

$$(\mathcal{B}_s, ||.||_s)_{0<s\leq 1}$$

where

$$\mathcal{B}_s := \mathcal{H}(D_{r_1+s(r_2-r_1)}) \bigcap C(\bar{D}_{r_1+s(r_2-r_1)})$$

which is the space of all generalized analytic and univalent functions defined in $D_{r_1+s(r_2-r_1)}$, and which are in $C(\bar{D}_{r_1+s(r_2-r_1)})$

and

$$||.||_s := Sup_{D_{r_1+s(r_2-r_1)}} |.|$$

Proposition

Let us suppose that $f \in C^1\left([0, a_0[, \mathcal{H}(D_1) \bigcap C^1(\bar{D}_1)\right)$ *is for each* $t \in [0, a_0[$, *a univalent function in* $\bar{D}_1$, *and in* $[0, a_0[\times D_1$ *is a solution of the new formulation of the problem satisfying the condition:*

$$\Im\left(\frac{\frac{\partial f}{\partial t}}{\frac{\partial f}{2\bar{\tau}\partial\tau} - \frac{\partial f}{2\tau\partial\bar{\tau}}}\right)(t,0) = 0 \tag{5.1}$$

Then f is a solution of the following problem

$$\frac{\partial f}{\partial t}(t,\tau) = \mathcal{L}(t,f) \tag{5.2}$$

$$f(t,0) = z^0 \tag{5.3}$$

$$f(0,\tau) = f_0(\tau) \quad for\ |\tau| < 1 \tag{5.4}$$

where

$$\mathcal{L}(t,f) = -\frac{Q(t)}{6\mu J(f)}\left(\frac{\partial f}{2\bar{\tau}\partial\tau} - \frac{\partial f}{2\tau\partial\bar{\tau}}\right)\frac{1}{2i\pi}\int_{|\rho|=1}\left(h^2\frac{\rho+\tau}{\rho-\tau}\right)\frac{d\rho}{\rho}$$

Conversly, let us suppose that $f \in C^1\left([0, a_0[, \mathcal{H}(D_1)) \bigcap C^1(\bar{D}_1)\right)$, *is a solution of problem (5.2)-(5.4); Then f represents a univalent solution of the new formulation of the problem.*

The proof of this proposition follows from the application of the *Schwarz formula* see for exemple [8].

Introducing $Y(t,\tau) = f(t,\tau) - f_0(\tau)$ this implies a homogeneous initial condition.

Thus the problem (5.2)-(5.4) can be transformed to the following one:

$$\frac{\partial Y}{\partial t}(t,\tau) = \mathcal{L}(t, Y + f_0) \tag{5.5}$$

$$Y(0,\tau) = 0 \tag{5.6}$$

Theorm 2 *Under the assumptions (H1) and (H2), then for every* $r_1 \in]1, r_2[$ *there exists a positive constant* a_0 *and a uniquely determined univalent - in* $\bar{D}_{r_1}$ *- function* f *belonging to*

$$C^1\left([0, a_0[, \mathcal{H}(D_{r_1}) \bigcap C^1(\bar{D}_{r_1})\right)$$

and satisfying the new formulation of the problem (4.4)-(4.6).

Proof. We apply the nonlinear abstract Cauchy-Kovalevsky theorem (see [9] et [10]), to the problem (5.5)(5.6.), we have only to show in [2] that the operator $\mathcal{L}_0$ defined by $\mathcal{L}_0(t,Y) := \mathcal{L}(t, Y + f_0)$ satisfies in the above introduced scale of Banach spaces the following conditions:

$$\lim_{t_2 \to t_1} ||\mathcal{L}_0(t_2, Y(t_2) - \mathcal{L}_0(t_1, Y(t_1)||_{s'} = 0, \quad \forall t_2 \ \forall t_1 \text{ in }]0, T]$$

$$)||\mathcal{L}_0(t,0)||_{s'} \leq \frac{C_1.}{1-s'} \quad \forall s' \ \in]0,1[$$

$$||\mathcal{L}_0(t, Y_1) - \mathcal{L}_0(t, Y_2||_{s'} \leq \frac{C_2.}{s-s'}||Y_1 - Y_2||_s$$

$\forall\, s, s'$ such that $0 < s' < s \leq 1$, $t \in]0, T]$ *and* Y_1, Y_2 belonging to the set: $Y \in \mathcal{B}_s : ||Y||_s < R.$ ∎

References

[1] G.Bayada M.Boukrouche M.El-Alaoui Talibi "Generalized Hele-Shaw type problems and the transient lubrication problem" To appear in *Nonlinear Analysis TMA*.

[2] M.Boukrouche "The quasiconformal mappings methods to solve a free boundary problem for generalized Hele-Shaw flows" To appear in *Complex Analysis Theory and Applications.*

[3] B.Gustafsson "On a differentiel equation arising in Hele-Shaw flow moving boundary problem" *Trita-Mat.36, (1986); Math. Roy. Inst. Tech. Stockholm* (1981).

[4] Yu.E.Hohlov S.D.Howison "On the classification of solutions to the zero-surface -tension model for Hele-Shaw flows" *Quart.Appl.Math, 51* (1994), 777-789.

[5] S.D.Howison "Complex variable methods in Hele-Shaw moving boundary problems" *Euro. Jnl. of Applied Mathematics vol.3,* (1992), 209-224.

[6] V.N.Monakhov "Boundary-value problems with free boundaries for elliptic systems of equations" *Trans. of AMS vol.57, Providence, RI* (1983).

[7] S.Richardson "Hele-Shaw flows with a free boundary produced by the injection of fluid into a narrow channel" *J.Fluid Mech. 56*, (1972).

[8] I.N.Vekua "Generalized analytic functions" *Fizmatgiz, Moscow, 1959; English transl., Pergamon Press, Oxford, and Addison-Wesley, Reading, Mass.*, (1962).

[9] T.Nishida "A note on a theorem of Nirenberg" *J. Diff. Geom., 12*, (1977), 629-633.

[10] L.V.Ovsjannikov "A nonlinear Cauchy problem in a scale of Banach spaces" *Dokl. Akad. Nauk. SSSR, 200* (1971), 789-792.

Mahdi BOUKROUCHE
Centre de maths. URA 740 CNRS
INSA-de-LYON bât 401,
69621 Villeurbanne Cedex.

F BROCK

Continuous polarization and symmetry of solutions of variational problems with potentials

We investigate a moving reflection method to show symmetry properties of local minimizers of variational problems with potentials. To this to every function lying in the admissible class a continuous scale of its polarizations is defined. The original and the polarized functions satisfy well known inequalities for product integrals and convolutions.

Consider the minimum problem

$$J(v) \longrightarrow \min!, \quad v \in K, \tag{1}$$

where K is a closed subset of some function space X (e.g. $L^p(\mathbf{R}^n)$) and J is a real lower semicontinuous functional. If $v \in K$, we often also have $v^* \in K$, where v^* denotes the Steiner-symmetrization of v with respect to some hyperplane in $\mathbf{R}^n$ and:

$$J(v^*) \leq J(v). \tag{2}$$

The literature about inequalities of the form (2) is large. Many examples and applications can be found in the monography [7].

If (2) is valid with the equality sign only in the case $v = v^*$, then it could be proved for the *absolute* minimum U of (1), that $U = U^*$. However this argumentation fails for the *local* minima u of the functional J, since the symmetrization u^* generally lies not close to u in the norm of X.

Therefore the following question is natural: Is there to every $v \in K$ a closed family of functions $K_v \subseteq K$, which contains functions "close" to v (and v itself), such that for every $w \in K_v$:

$$J(w) \leq J(v)\,, \tag{3}$$

where the equality sign in (3) is valid only in the case $v = w$? In this case we could also prove symmetry properties of u.

An example of such a function family is the *continuous Steiner-symmetrization*. This is a homotopy, which connects v with its Steiner-symmetrization v^* on a (in the norm of $L^p(\mathbf{R}^n)$, $1 \leq p \leq +\infty$) continuous path (see [6,9,12] for different variants of this construction). We shall not go in detail here and mention only, that for those functionals, which will be considered in this article, the inequalities (3) were proved in [6]. On the other hand it seemed to be difficult to the author to discuss the equality sign in (3). This was the motivation to investigate a simplier function family.

In the present paper we use a moving reflection method. We define to each function $v \in K$ a continuous scale of its polarizations v_λ, $\lambda \in \mathbf{R}$. This elementary kind of rearrangement can often be used to show symmetrization lemmata (see [3,4,5,7,13]). In particular the integral inequalities (3) and their strict variants have easy proofs. We show that the local minimizers of some variational problems with potentials are symmetric. At the end we give an application on a model for self-gravitating axisymmetric rotating liquids.

First we introduce some notations. By $x, y, \ldots,$ we denote points in $\mathbf{R}^n$. If $n \geq 2$ we use the following partitions:

$$x = (x_1, x'), \quad y = (y_1, y'), \quad x', y' \in \mathbf{R}^{n-1},$$

and for convenience we write $x = x_1$, $y = y_1$ in the case $n = 1$. If λ is a real number we denote by T_λ and H_λ^+ the hyperplane $\{x_1 = \lambda\}$ and the halfspace $\{x_1 > \lambda\}$, respectively, and by x^λ and y^λ the reflection points of x and y with respect to T_λ. If M is a L-measurable - measurable in short - subset of $\mathbf{R}^n$ and f some measurable function on $\mathbf{R}^n$, then let $|M|$, M^* and f^* denote the n-dimensional L-measure of M and the Steiner-symmetrizations of M and f with respect to the variable x_1, respectively (for definitions see [9], pp. 8). The Steiner-symmetrization of a function can be defined for a wide class of functions, which we denote by $F(\mathbf{R}^n)$:

> $f \in F(\mathbf{R}^n) \Longleftrightarrow$
> f is measurable on $\mathbf{R}^n$ and there is a number $c_f \in [-\infty, +\infty)$, such that $f \geq c_f$ in $\mathbf{R}^n$ and the level sets $\{f > c\}$ have finite measure for all $c > c_f$.

Further let $F^+(\mathbf{R}^n)$ be the subclass of functions f from $F(\mathbf{R}^n)$, for which we have $c_f = 0$.

Definition 1: (*Polarization of functions*)
Let f be a measurable function on $\mathbf{R}^n$ and $\lambda \in \mathbf{R}$. Then the function f_λ defined by:

$$f_\lambda(x) := \begin{cases} \max\{f(x); f(x^\lambda)\} & \text{if } x_1 \geq \lambda \\ \min\{f(x); f(x^\lambda)\} & \text{if } x_1 < \lambda \end{cases}, \quad x \in \mathbf{R}^n, \tag{4}$$

is called the *polarization of f with respect to T_λ*.

It will also be useful for our proofs to consider the polarizations of measurable sets.

Definition 2: (*Polarization of sets*)
Let M be a measurable set and $f = \chi(M)$ its characteristic function. Then the set M_λ defined by the equation

$$f_\lambda = \chi(M_\lambda) \tag{5}$$

is called the *polarization of* M.

Lemma 1: Let f, g, f_k measurable functions on $\mathbf{R}^n$, M, N measurable sets in $\mathbf{R}^n$, $a, b, c, d, \lambda, \lambda_k \in \mathbf{R}$, $k = 1, 2, \ldots$, and φ a continuous nondecreasing function on $\mathbf{R}$. Then

$$(\varphi(f))_\lambda = \varphi(f_\lambda), \tag{6}$$
$$\{f > c\}_\lambda = \{f_\lambda > c\}, \tag{7}$$
$$|M_\lambda| = |M| \text{ and } |\{a \leq f_\lambda < b\}| = |\{a \leq f < b\}| \quad \text{(preservation of measure)}, \tag{8}$$
$$\text{if } M \subseteq N \text{ and } f \leq g \text{ a.e., then } M_\lambda \subseteq N_\lambda \text{ and } f_\lambda \leq g_\lambda \text{ a.e.} \quad \text{(monotonicity)}, \tag{9}$$
$$\text{if } f_k \longrightarrow f \text{ in measure or a.e., then } (f_k)_\lambda \longrightarrow f_\lambda \text{ in measure or a.e., respectively} \quad \text{(continuity of the mapping } f \longmapsto f_\lambda \text{)}, \tag{10}$$
$$\text{if } f \text{ is continuous, then so is } f_\lambda, \tag{11}$$
$$\text{if } f \text{ is Hölder-continuous with an exponent } \mu \in (0,1] \text{ and a constant } K > 0, \text{ then so is } f_\lambda, \tag{12}$$
$$\text{if } \lambda_k \longrightarrow \lambda, \text{ then } f_{\lambda_k} \longrightarrow f_\lambda \text{ in measure in any compact subset of } \mathbf{R}^n \quad \text{(continuity of the mapping } \lambda \longmapsto f_\lambda \text{)}. \tag{13}$$

Further, if in the last case (13) in addition $f \in L^p(\mathbf{R}^n)$, $1 \leq p < +\infty$, then:

$$f_{\lambda_k} \longrightarrow f \quad \text{in} \quad L^p(\mathbf{R}^n), \tag{14}$$

and if f is continuous, then:

$$f_{\lambda_k} \longrightarrow f_\lambda \text{ everywhere in } \mathbf{R}^n. \tag{15}$$

P r o o f : The proofs of the properties (6)-(12) are easy and we leave them to the reader. (13) and (14) follow from the continuity of the Lebesgue-integral with respect to translations. We refer also to the proof of the similar property (21) of Lemma 4. The property (15) is an immediate consequence of (13).

From recent papers a general convolution inequality is known, which contains some other inequalities as special cases (see [3,4,5]). Here we present a slightly generalized version which has no influence on the proof given in [4], p. 4818.

Lemma 2:
a) Let $f, g \in F^+(\mathbf{R}^n)$ and $h = h(z_1, x', y')$, $(z_1 \in \mathbf{R},\ x', y' \in \mathbf{R}^{n-1})$, a nonnegative measurable function which is even in z_1 and monotone nonincreasing in z_1 for $z_1 > 0$. Further let $j : \mathbf{R}_0^+ \longrightarrow \mathbf{R}_0^+$ be a monotone nondecreasing and convex function with

$j(0) = 0$ and $\lambda \in \mathbf{R}$. Then:

$$\iint\limits_{\mathbf{R}^{2n}} j\Big(|f_\lambda(x) - g_\lambda(y)|\Big) h(x_1 - y_1, x', y')dxdy \leq \iint\limits_{\mathbf{R}^{2n}} j\Big(|f(x) - g(y)|\Big) h(x_1 - y_1, x', y')dxdy, \tag{16}$$

if the right integral in (16) converges.
b) If we assume in addition, that h is *strictly* monotone decreasing with respect to z_1 for $z_1 > 0$ and that j is *strictly* convex, then (16) is valid with the equality sign only if one of the following conditions is satisfied :

(i) $f = f_\lambda$ and $g = g_\lambda$,

(ii) f_λ and g_λ are reflections of f and g, respectively, i.e. $-f = (-f)_\lambda$, $-g = (-g)_\lambda$,

(iii) one of the functions f or g is symmetric in x_1 with respect to T_λ.

Corollary: Let f, g, h and λ as above. Then:

$$\iint\limits_{\mathbf{R}^{2n}} f_\lambda(x)g_\lambda(y)h(x_1 - y_1, x', y')dxdy \geq \iint\limits_{\mathbf{R}^{2n}} f(x)g(y)h(x_1 - y_1, x', y')dxdy, \tag{17}$$

if one of the integrals in (17) converges.
Further if $f, g \in L^p(\mathbf{R}^n)$, $1 \leq p \leq +\infty$, then:

$$\|f_\lambda - g_\lambda\|_{L^p(\mathbf{R}^n)} \leq \|f - g\|_{L^p(\mathbf{R}^n)}. \tag{18}$$

P r o o f : (17) follows from (16) with $j(z) = z^2$, if we take into account, that $\int f_\lambda^2 = \int f^2$ (see (8)).
The inequalities (18) we derive in the cases $1 \leq p < +\infty$ by choosing $j(z) = z^p$, $h = \chi_{\{|x-y|<\varepsilon\}}$ and then passing to the limit $\varepsilon \to 0$. Finally in the case $p = +\infty$ (18) follows directly from Definition 1.

The following lemma is easy to prove, nevertheless it seems to be new.

Lemma 3:
a) Let be $f, g \in L^2(\mathbf{R}^n)$ and $\lambda \in \mathbf{R}$. Then:

$$\int\limits_{\mathbf{R}^n} f_\lambda(x)g_\lambda(x)\, dx \geq \int\limits_{\mathbf{R}^n} f(x)g(x)\, dx. \tag{19}$$

b) If we assume in addition, that g is even in x_1 and *strictly* monotone decreasing in x_1 for $x_1 > 0$, then (19) is valid with the equality sign only if one of the conditions (i)-(iii) of Lemma 2,**b)** is satisfied.

P r o o f :
a) From a simple partition into cases it follows, that for every $x \in H_\lambda^+$:

$$f_\lambda(x)g_\lambda(x) + f_\lambda(x^\lambda)g_\lambda(x^\lambda) \geq f(x)g(x) + f(x^\lambda)g(x^\lambda), \tag{20}$$

which yields (19) after an integration over all $x \in H_\lambda^+$.
b) Assume that (19) is valid with the equality sign. Then (20) holds with the equality sign for almost every $x \in H_\lambda^+$. Now if g is even in x_1 and strictly monotone decreasing in x_1 for $x_1 > 0$, then we can easily conclude the alternatives (i)-(iii) of Lemma 1.

The following continuity property closes the "link" in Lemma 1, (13)-(15), for the case $\lambda = -\infty$ and is basic for the applications:

Lemma 4: Let be λ_k, $k = 1, 2, \ldots$, any sequence converging to $-\infty$ and let $f \in F(\mathbf{R}^n)$ be a.e. finite and bounded below. Then:

$$f_{\lambda_k} \longrightarrow f \quad \text{in measure}\ . \tag{21}$$

Further if $f \in L^p(\mathbf{R}^n)$, $1 \leq p < +\infty$, then:

$$f_{\lambda_k} \longrightarrow f \quad \text{in } L^p(\mathbf{R}^n), \tag{22}$$

and if f is continuous, then:

$$f_{\lambda_k} \longrightarrow f \quad \text{everywhere in} \quad \mathbf{R}^n. \tag{23}$$

P r o o f : First we observe that, if M is any measurable set of finite measure, then it follows from Definition 2 and Lemma 1,(8) and (9):

$$|M_{\lambda_k} \setminus M| = |M \setminus M_{\lambda_k}| \leq |M \cap \{x | x_1 < \lambda_k\}| \longrightarrow 0 \text{ as } k \longrightarrow +\infty. \tag{24}$$

Since f is bounded below, we conclude that c_f is finite. We choose some number $\delta > 0$ and introduce the sets

$$\begin{aligned} M_{ik} &:= \Big\{c_f + i\frac{\delta}{2} < f_{\lambda_k} \leq c_f + (i+1)\frac{\delta}{2}\Big\} \cap \Big\{c_f + (i+2)\frac{\delta}{2} < f\Big\}, \\ N_i &:= \Big\{c_f + i\frac{\delta}{2} < f \leq c_f + (i+1)\frac{\delta}{2}\Big\}, \ i = 0, 1, \ldots, \ k = 1, 2, \ldots. \end{aligned}$$

From (9) it follows that $|M_{ik}| \leq |N_i|$, $i = 0, 1, \ldots$, $k = 1, 2, \ldots$, and:

$$|\Big\{f - f_{\lambda_k} \geq \delta\Big\}| \leq \sum_{i=0}^{+\infty} |M_{ik}| \leq |\Big\{c_f + \delta < f\Big\}| + \sum_{i=1}^{+\infty} |N_i| < +\infty.$$

Further we conclude from (24):

$$|M_{ik}| \leq |\Big\{c_f + (i+1)\frac{\delta}{2} < f\Big\} \setminus \Big\{c_f + (i+1)\frac{\delta}{2} < f_{\lambda_k}\Big\}| \longrightarrow 0 \text{ as } k \longrightarrow +\infty.$$

This shows, that also $|\{f - f_{\lambda_k} \geq \delta\}| \longrightarrow 0$ as $k \to +\infty$. Together with an analogue consideration for the sets $\{f_{\lambda_k} - f \geq \delta\}$ we derive (21).

The property (22) is an immediate consequence.
If we assume, that f is continuous, then:

$$\lim_{x_1 \to -\infty} f(x) = c_f,$$

and the assertion (23) follows from Definition 1.

Next we give a simple criterion to identify the "symmetrized" functions:

Lemma 5: Let $u \in F(\mathbf{R}^n)$ and assume that for every $\lambda \in \mathbf{R}$ we have $u_\lambda = u$ or $(-u)_\lambda = -u$. Then it follows $u^*(x) = u(\sigma x)$, where σ is some translation in the direction x_1.

P r o o f : We can assume that u is not constant, since otherwise there would be nothing to prove. Then we observe, that for λ small enough the possibility $(-u)_\lambda = -u$ is excluded and for very large λ we cannot have $u_\lambda = u$. From (13) it follows, that there are numbers $\mu \in \mathbf{R}$, such that simultaneously $u_\mu = u$ and $(-u)_\mu = -u$. Let be λ_0 the smallest number μ with this property. Then we conclude again, that λ_0 is finite and $u_{\lambda_0} = u$, $(-u)_{\lambda_0} = -u$. Since $u_\lambda = u$ for every $\lambda < \lambda_0$, it follows that for almost any $x \in H_\lambda^+$ with $\lambda < \lambda_0$:

$$u(x^\lambda) \leq u(x) = u(x^{\lambda_0}).$$

But this means, that u is monotone nondecreasing in x_1 for $x_1 < \lambda_0$ and monotone nonincreasing in x_1 for $x_1 > \lambda_0$, q.e.d.

Now we are able to prove the symmetry of local minimizers of some variational problems with potentials.

Theorem : Let K be a closed subset of $L^p(\mathbf{R}^n) \cap F^+(\mathbf{R}^n)$, $1 \leq p < +\infty$, and let K have the property, that if $v \in K$, then also $v^* \in K$ and $v_\lambda \in K$ for every $\lambda \in \mathbf{R}$. Further let φ, ψ real continuous functions on $\mathbf{R}_0^+$ with φ strictly increasing, j a real nonnegative and convex function on $\mathbf{R}_0^+$ with $j(0) = 0$, $g \in F^+(\mathbf{R}^n)$ with $g = g^*$, $h = h(z_1, x', y')$, $(z_1 \in \mathbf{R},\ x', y' \in \mathbf{R}^{n-1})$, a nonnegative measurable function which is even in z_1 and monotone nonincreasing in z_1 for $z_1 > 0$, and:

$$\begin{aligned} J(v) \ := \ & \iint_{\mathbf{R}^{2n}} j(|v(x) - v(y)|) h(x_1 - y_1, x', y')\, dx dy - \int_{\mathbf{R}^n} \varphi(v(x)) g(x)\, dx \\ & + \int_{\mathbf{R}^n} \psi(v(x))\, dx, \qquad v \in K. \end{aligned} \tag{25}$$

Finally let the functions g, h, j satisfy one of the following conditions:

(i) g is *strictly* decreasing in x_1 for $x_1 > 0$;

(ii) h is *strictly* decreasing in z_1 for $z_1 > 0$ and j is *strictly* convex.

Then if u is a local minimum of J in K, we have $u^*(x) = u(\sigma x)$, where σ is some translation in the direction x_1.

P r o o f : Assume that the assertion is not true. Then by Lemma 5 there are real numbers λ, such that neither $u_\lambda = u$ nor $(-u)_\lambda = -u$, i.e. both sets $\{x \in H_\lambda^+ : u(x) < u(x^\lambda)\}$ and $\{x \in H_\lambda^+ : u(x) > u(x^\lambda)\}$ have positive measure. From the continuity properties (14) and (22) we conclude then, that to every given $\varepsilon > 0$ there is a number λ, such that $0 < \|u_\lambda - u\|_{L^p(\mathbf{R}^n)} = \|(-u)_\lambda + u\|_{L^p(\mathbf{R}^n)} < \varepsilon$. Now if we replace u by the function u_λ in $J(u)$, then the sum of the first two integrals in (25) becomes smaller because of the conditions (i) and (ii) in the assumptions and the parts **b)** of the Lemmata 2 and 3, while the third integral in (25) remains unchanged in view of the properties (6),(7) of Lemma 1, a contradiction.

The minimum problems of the Theorem describe equilibrium states in continuum mechanics.

Example:

Self-gravitating axisymmetric rotating liquids (see [8], pp. 418, [1,2])
Consider an inviscid and irrotational fluid with density $v(x)$ rotating about the x_1 -axis, $(x \in \mathbf{R}^3,\ x = (x_1, x_2, x_3) = (x_1, x'))$. It is assumed to be axisymmetric, i.e. $v(x) = v(x_1, |x'|)$, ($|\cdot|$ - Euclidian distance in $\mathbf{R}^2$ or $\mathbf{R}^3$, respectively). The rotation law is

$$\vec{w} = |x'|^{-1} s(|x'|)(0, -x_3, x_2), \qquad \vec{w} \text{ - velocity vector,}$$

where $s(|x'|)$ is the angular velocity. If the fluid is compressible, then we assume the polytropic law

$$p = Cv^\gamma, \quad \gamma = 1 + \frac{1}{\beta}, \quad 0 < \beta < 3, \ C = \text{ const.}, \quad (v \text{ - pressure }).$$

In the incompressible case we assume that $v = 1$ in the interior of the fluid.
Further we set

$$\int_{\mathbf{R}^3} v \, dx \quad = \quad M, \quad (M > 0 \text{ given constant }),$$

$$m(|x'|) \quad = \quad \frac{1}{M} \int_{\{|y'|<|x'|\}} v(y) \, dy,$$

and introduce a function $A(m)$, $(m \in [0,1])$, the angular momentum per unit mass, such that

$$A(0) = 0, \quad A(m) \quad \text{monotone nondecreasing and } A^2(m) \in C^1[0,1].$$

Formally we have

$$A^2(m(|x'|)) = |x'| s(|x'|). \tag{26}$$

(26) is valid under the assumption, that the rotating fluid, during the evolution which led to an equilibrium, was moving in such a way that the angular momentum did not change for any fraction of the mass lying within a distance $|x'|$ from the axis of rotation.
The equilibrium figures are *local minima* of the following variational problems:

a) *Compressible case :*

$$
\begin{aligned}
J(v) &:= -\frac{1}{2}\iint\limits_{\mathbf{R}^6} \frac{v(x)v(y)}{|x-y|}\, dx dy + \frac{1}{2}\int\limits_{\mathbf{R}^3} \frac{A^2(m(x'))v(x)}{|x'|^2}\, dx \\
&\quad + C\beta \int\limits_{\mathbf{R}^3} v^\gamma(x)\, dx \longrightarrow \text{Min !}, \quad v \in K, \qquad (27) \\
K &:= \{v \in L^1(\mathbf{R}^3) \cap L^\gamma(\mathbf{R}^3) \ : \ \int\limits_{\mathbf{R}^3} v\, dx = M,\ v(x) = v(x_1, |x'|) \geq 0\}
\end{aligned}
$$

The three terms of the energy functional J represent, respectively, the gravitational potential energy, the rotational kinetic energy and the internal energy.

b) *Incompressible case :*

$$
\begin{aligned}
J_0(v) &:= -\frac{1}{2}\iint\limits_{\mathbf{R}^6} \frac{v(x)v(y)}{|x-y|}\, dx dy + \frac{1}{2}\int\limits_{\mathbf{R}^3} \frac{A^2(m(x'))v(x)}{|x'|^2}\, dx \longrightarrow \text{Min !}, \\
&\quad v \in K_0, \qquad (28) \\
K_0 &:= \{v \in L^1(\mathbf{R}^3) \cap L^\infty(\mathbf{R}^3) \ : \ \int\limits_{\mathbf{R}^3} v\, dx = M,\ v(x) = v(x_1, |x'|),\ 0 \leq v \leq 1\}
\end{aligned}
$$

Another model for the rotating fluid is obtained when one prescribes the angular velocity $s(|x'|)$ instead of the angular momentum $A(m)$. In this case the kinetic energy terms in (27) and (28) are replaced by

$$\int\limits_{\mathbf{R}^3} v(x)B(x')\, dx$$

where

$$B(x') := \int\limits_0^{|x'|} ts^2(t)\, dt\ .$$

Note that the problems (27),(28) can also be regarded as models of rotating stars.

We observe that the term $m(x')$ in the functionals J and J_0 does not change for any equimeasurable rearrangement of the admissible functions v in the direction x_1. Therefore it is easy to see, that with a suitable choice of the functions j, h, ϕ, g and ψ in (25) we can apply the Theorem.

Thus the local minima u of the above problems have the property, that there is a plane T orthogonal to the x_1 -axis, such that u is symmetric with respect to T and monotone nonincreasing in x_1 on the right side of T. This symmetry property was well known for the *absolute minima* of the variational problems. I could find only one analogue result for *other* solutions in the literature (see [11], pp.12):

Let u a *bounded* solution of the Euler equations to (28) (problem **b)**) in the case of prescribed *constant* angular velocity s. Using the equilibrium conditions and the smoothness of the set $\{u > 0\}$ one can show the symmetry of u.

It is interesting to find out, under which conditions the symmetry of weak solutions of the variational problems (25) of the Theorem could be proved.

REFERENCES

[1] Auchmuty, J.F.G., R. Beals : *Variational solutions of some nonlinear free boundary problems* . Arch. Ration. Mech. Anal. **43** (1971), 255-271.

[2] Auchmuty, J.F.G. : *Existence of axisymmetric equilibrium figures* . Arch. Ration. Mech. Anal. **65** (1977), 249-261.

[3] Baernstein, A., B.A. Taylor : *Spherical rearrangements, subharmonic functions and *-functions in n-space.* Duke Math, J. **43** (1976), 245-268.

[4] Beckner, W. : *Sobolev inequalities, the poisson semigroup and analysis on the sphere S^n.* Proc. Nat. Acad. Sci. U.S.A. **89** (1992), 4816-4819.

[5] Beckner, W : *Geometric inequalities in Fourier analysis.* to appear in the proceedings of a conference at Princeton in 1991 in honor of E.M. Stein.

[6] Brock, F. : *Continuous Steiner-symmetrization.* to appear in Math. Nachrichten.

[7] Dubinin, V.N. : *Transformations of condensers in space.* Soviet Math. Dokl. **36** (1988), 217-219.

[8] Friedman, A. : *Variational Principles and Free-boundary Problems.* Wiley-Interscience (1982).

[9] Kawohl, B. : *Rearrangements and Convexity of Level Sets in PDE.* Springer Lecture Notes **1150** (1985).

[10] Kawohl, B. : *On the simple shape of stable equilibria.* Symposia mathematica, Vol.XXX, (1989).

[11] Lichtenstein, L. : *Gleichgewichtsfiguren rotierender Flüssigkeiten.* Springer-Verlag, Berlin 1933.

[12] Solynin, A.Yu. : *Continuous symmetrization of sets.* Zapiski Nauchnykh Seminarov Leningradskogo Otdeleniya Matematicheskogo Instituta im. V.A. Steklova Akademii Nauk SSSR, Vol. 185 (1990), 125-139.

[13] Volontis, V. : *Properties of conformal invariants.* Amer. J. Math. **74** (1952), 587-606.

authors address:
Friedemann Brock
Universität Leipzig
Fakultät für Mathematik und Informatik
Augustusplatz 9-10
Leipzig 04109
e-mail: brock@mathematik.uni-leipzig.d400.de

A CAÑADA, J L GÁMEZ AND J A MONTERO

An optimal control problem for a nonlinear elliptic equation arising from population dynamics

In this work we study the profitability of biological growing species, modelled by the diffusive Volterra-Lotka equation:

$$\begin{aligned} -\Delta u(x) &= u(x)[a(x) - f(x) - b(x)u(x)], \quad && x \in \Omega, \\ u(x) &= 0, && x \in \partial\Omega. \end{aligned} \tag{0.1}$$

where function u represents the density of the biological species at different points of Ω, which is a bounded and regular domain in $\mathbb{R}^N$. Function a is the intrinsic growth rate of species u in Ω, b means the crowding effect, and f plays the role of control for the growth of species u. The Laplacian operator, Δ, denotes the diffusion of the species u in the domain Ω. The Dirichlet condition means that the species u can not survive on the boundary of the domain Ω.

Our goal is to give conditions on the control function f to obtain the maximal profitability of the harvest. The benefit is given by the expression

$$J(f) = \int_\Omega Kuf - Mf^2,$$

where u is a solution of problem (0.1) depending on f. K and M mean, respectively, the price of the species and the cost of the control. We define an optimal control (if it exists), in a suitable functional space A, to be a function $f \in A$, such that

$$J(f) = \sup_{g \in A} J(g).$$

Similar problems have been studied by Leung and Stojanovic [3,4,5]. In [4] these authors have studied the equation (0.1) with Neumann boundary conditions, and taking the space of the controls, A, as

$$C_\delta = \{g \in L^\infty(\Omega) : 0 \leq g(x) \leq \delta \text{ a.e. in } \Omega\},$$

where $0 < \delta < \inf_{x\in\Omega} a(x)$. In this particular situation they prove the existence of optimal control $f \in C_\delta$. Under suitable conditions, they describe the optimal control f in terms of the solution of an appropriated elliptic system (the optimality system). Here we study the case of Dirichlet boundary condition, obtaining the existence result and a priori bounds for the optimal control in the space $A = L^\infty_+(\Omega)$. We also obtain a necessary and sufficient condition for the positivity of the benefit.

This work has been partially supported by DGICYT, Ministry of Education and Science (Spain), under grant number PB92-0941.

In Section 1, we set the appropriate results and notation. In Section 2 we show the main results concerning to the existence of the optimal control, a priori bounds, and the characterization of positivity of the profit.

1. Notation and preliminary results

For a function $e \in L^\infty(\Omega)$, we mean $\underline{e} = \operatorname{ess}\inf_{x\in\Omega} e(x)$, $\overline{e} = \operatorname{ess}\sup_{x\in\Omega} e(x)$.

In the sequel, K and M are positive constants and

$$[\mathbf{H}] \qquad \begin{array}{l} a, b \in L^\infty(\Omega),\ \underline{b} > 0 \\ f \in L^\infty_+(\Omega) = \{g \in L^\infty(\Omega) : g(x) \geq 0 \text{ a.e. } in\ \Omega\}. \end{array}$$

For a function $q \in L^\infty(\Omega)$, we define $\lambda_1(q)$ to be the principal eigenvalue of the eigenvalue problem

$$\begin{array}{ll} -\Delta u(x) + q(x)u(x) = \lambda u(x), & x \in \Omega \\ u(x) = 0, & x \in \partial\Omega. \end{array}$$

Moreover, the principal eigenvalue can be expressed variationally as

$$\lambda_1(q) = \inf_{u\in H_0^1(\Omega)\setminus\{0\}} \frac{\int_\Omega |\nabla u|^2 + \int_\Omega q|u|^2}{\int_\Omega |u|^2}.$$

It is known that the multiplicity of $\lambda_1(q)$ is equal to one and it is possible to choose an associated eigenfunction $\phi_1(q)$ (where previous infimum is attained, becomming a minimum) such that $\phi_1(q) \in C^{1,\alpha}(\overline{\Omega})$, $\forall \alpha \in (0,1)$, $\phi_1(q) > 0$ in Ω, $\|\phi_1(q)\|_{L^\infty} = 1$.

Proposition 1.1. *The eigenvalue $\lambda_1(q)$ has the following properties:*

i) $\lambda_1(q)$ is increasing with respect to the weight function q.

ii) $\forall M \in \mathbb{R}$, $\lambda_1(q + M) = \lambda_1(q) + M$.

iii) $\lambda_1(q)$ is continuous with respect to $q \in L^\infty(\Omega)$.

Proof. It is a direct consequence of the variational characterization of $\lambda_1(q)$. ∎

Theorem 1.2 . [1,2] (Existence of nonnegative and nontrivial solution of (0.1)). *Equation* (0.1) *has a weak nontrivial and nonnegative solution, u_f, if and only if $\lambda_1(-a + f) < 0$. Moreover, in this case the solution u_f is the unique nontrivial and nonnegative solution of* (0.1), *and it verifies the estimates*

$$\frac{-\lambda_1(-a+f)}{\overline{b}}\, \phi_1(-a+f)(x) \leq u_f(x) \leq \frac{\overline{a} - \underline{f}}{\underline{b}}, \ \forall x \in \Omega$$

Definition 1.3. *For every $f \in L^\infty_+(\Omega)$ we define u_f, to be the maximal nonnegative solution of the equation* (0.1). *In fact, $u_f \equiv 0$ iff $\lambda_1(-a+f) \geq 0$ and u_f is strictly positive in Ω iff $\lambda_1(-a+f) < 0$.*

REMARKS.

1. By elliptic estimates and by $0 \leq u_f(x) \leq \frac{\overline{a}}{\underline{b}}$, $\forall x \in \Omega$, we have that, for fixed $p > N$, $\|u_f\|_{W^{2,p}} \leq c$ where c is independent of $f \in L^\infty_+(\Omega)$.
2. If we take $f \in L^\infty(\Omega)$, Theorem 1.2 remains true, and the definition of u_f is still possible (see e.g. [2]).
3. (Monotonicity of u_f with respect to f). By using technics of sub- and super-solutions it is easy to prove that, if $f, g \in L^\infty_+(\Omega)$, $f \leq g$, then $u_f \geq u_g$.

2. Existence of optimal control and positivity of the profit

Our goal now is to find an optimal control in $L^\infty_+(\Omega)$, i.e., $f \in L^\infty_+(\Omega)$ maximizing the functional $J : L^\infty_+(\Omega) \to \mathbb{R}$, given by the expression

$$J(g) = \int_\Omega K u_g g - M g^2$$

Theorem 2.1. (Existence of an optimal control). *Consider the problem* (0.1) *under hypotheses* [H]. *Then, the optimal control problem has solution in the space $L^\infty_+(\Omega)$, i.e. $\exists f \in L^\infty_+(\Omega)$ such that $J(f) = \sup_{g \in L^\infty_+(\Omega)} J(g)$.*

To prove the above theorem we need the following

Lemma 2.2. (A priori bounds on the control f). *For each $f \in L^\infty_+(\Omega)$, consider $g = \min\{f, \frac{\overline{a}K}{\underline{b}M}\}$. Then $J(g) \geq J(f)$. Moreover if the set $\Omega_1 = \{x \in \Omega : f(x) > \frac{\overline{a}K}{\underline{b}M}\}$ has positive measure, then $J(g) > J(f)$. (In particular if $f \in L^\infty_+(\Omega)$ is an optimal control then $0 \leq f \leq \frac{\overline{a}K}{\underline{b}M}$ a.e. in Ω.)*

Proof of Lemma 2.2. We will consider two separate cases:

– **Case** $u_f \equiv 0$:

$$\begin{aligned} J(f) &= -\int_\Omega M f^2 \leq -\int_\Omega M g^2 \\ &\leq \int_\Omega K u_g g - M g^2 \\ &= J(g). \end{aligned}$$

Observe that previous inequality is strict if Ω_1 has positive measure.

- **Case $u_f > 0$ in Ω:** In particular, in this case, by the monotonicity property of u_f we have $u_g \geq u_f > 0$.
  - Observe that $\forall x \in \Omega \setminus \Omega_1$,

$$Ku_g(x)g(x) - Mg^2(x) \geq Ku_f(x)f(x) - Mf^2(x).$$

  - In the same way, $\forall x \in \Omega_1$, $g(x) = \frac{\overline{a}K}{\underline{b}M} < f(x)$, and then

$$u_f(x) \leq \frac{\overline{a}}{\underline{b}} = g(x)\frac{M}{K} < \frac{M}{K}(g(x) + f(x))$$

  consequently

$$Ku_g(x)g(x) - Mg^2(x) - Ku_f(x)f(x) + Mf^2(x) \geq$$

$$\geq (f(x) - g(x))[-Ku_f(x) + M(f(x) + g(x))] > 0.$$

  Integrating previous inequalities, we obtain directly that $J(g) \geq J(f)$, with strict inequality if Ω_1 has positive measure. ∎

Proof of Theorem 2.1. Observe that, as $u_f(x) \leq \frac{\overline{a}}{\underline{b}}$, J is bounded from above. Let us take a maximizing sequence f_n in $L^\infty_+(\Omega)$ for J. By previous lemma we can suppose that

$$0 \leq f_n \leq \frac{\overline{a}K}{\underline{b}M}$$

and consequently, we can obtain a subsequence again denoted as f_n, verifying

$$f_n \rightharpoonup f \text{ weakly in } L^2(\Omega)$$

$$u_{f_n} \to u_f \text{ strongly in } W_0^{1,2}(\Omega)$$

Now by using that $\|f\|_2 \leq \liminf \|f_n\|_2$ and

$$\lim_{n\to\infty} \int_\Omega Ku_{f_n}f_n = \int_\Omega Ku_f f,$$

we conclude that

$$J(f) = \int_\Omega Ku_f f - \int_\Omega Mf^2 \geq \limsup J(f_n) = \sup_{g \in L^\infty_+(\Omega)} J(g)$$

and the proof is complete. ∎

Previous theorem assures the existence of optimal control, even with the posibility of zero benefit. Next theorem give conditions to warranty the positivity of the benefit associated to the optimal control.

Theorem 2.3. (Positive benefit). *Consider the problem* (0.1) *under hypotheses* [H]. *Then*

$$\sup_{g \in L^\infty_+(\Omega)} J(g) > 0 \Leftrightarrow \lambda_1(-a) < 0$$

Proof. If the profit is positive then there exists $g \in L^\infty_+(\Omega)$ such that $J(g) > 0$ and consecuently $u_g > 0$ and $\lambda_1(-a) \le \lambda_1(-a+g) < 0$.

Reciprocally, observe that

$$-Mf^2 + Ku_f f = -M\left[f - \frac{Ku_f}{2M}\right]^2 + \frac{K^2 u_f^2}{4M}. \tag{2.1}$$

We try to find $\hat{f} \in L^\infty_+(\Omega)$ such that $\hat{f} = \dfrac{Ku_{\hat{f}}}{2M}$. This is equivalent to find a solution $\hat{f}$ of the problem

$$\begin{aligned} -\Delta p(x) &= p(x)\left[a(x) - \tfrac{2Mb(x)+K}{K} p(x)\right], & x \in \Omega, \\ p(x) &= 0, & x \in \partial\Omega. \end{aligned}$$

By using Theorem 1.2 and our hypothesis $\lambda_1(-a) < 0$, there exists an unique positive solution $\hat{f}$ of the above equation. Moreover, $\hat{f}$ satisfies

$$-\frac{K\lambda_1(-a)}{2M\bar{b}+K}\phi_1(-a) \le \hat{f} \le \frac{\bar{a}K}{2M\underline{b}+K} \quad \text{in } \Omega. \tag{2.2}$$

Consecuently

$$J(\hat{f}) = M\int_\Omega \hat{f}^2 > 0,$$

and therefore

$$\sup_{g \in L^\infty_+(\Omega)} J(g) > 0.$$

This finishes the proof. ∎

REMARKS.

1. If $g \in L^\infty_+(\Omega)$ verifies $g \ge \hat{f}$ a.e. in Ω, $g \not\equiv \hat{f}$ then, taking into account the expression (2.1) and the monotonicity property of u_f, we obtain $J(g) < J(\hat{f})$, where $\hat{f}$ is defined as in Theorem 2.3. Hence, such a function g can not be an optimal control.

2. If the benefit is positive (or equivalently, if $\lambda_1(-a) < 0$), then, estimating the value $J(\hat{f})$ in previous theorem, and using the a priori bounds on u_f, we obtain

$$\frac{MK^2\lambda_1^2(-a)}{(2M\bar{b}+K)^2}\|\phi_1(-a)\|_2^2 \le J(\hat{f}) \le \sup_{g \in L^\infty_+(\Omega)} J(g) \le \frac{K^2\bar{a}^2}{4\underline{b}^2 M}|\Omega|.$$

   Previous formula gives estimates for the profit (whenever it is positive), in terms of the coefficients of the problem.

REFERENCES

[1] H. Berestycki and P. L. Lions, *Some applications of the method of super and subsolutions*, Lect. Not. Math., Vol. 782, Springer-Verlag, (1980), 16-42.

[2] P. Hess, *Periodic-parabolic boundary value problems and positivity*, Longman Group U.K. Limited, (1991).

[3] A. Leung and S. Stojanovic, *Direct methods for some distributed games*, Diff. and Int. Eqns. **3**, No. 6 (1990), 1113-1125.

[4] A. Leung and S. Stojanovic, *Optimal control for elliptic Volterra-Lotka equations*, J. Math. Anal. Appl., 173 (1993), 603-619.

[5] S. Stojanovic, *Optimal damping control and nonlinear elliptic systems*, SIAM J. Control Optim. **29**, No 3 (1991), 594-608.

Departamento de Análisis Matemático,
Universidad de Granada,
18071, Granada, SPAIN.

J A CARRILLO AND J SOLER

Global existence of functional solutions for the Vlasov–Poisson–Fokker–Planck system in 3-D with bounded measures as initial data

The Vlasov-Poisson-Fokker-Planck system describes the evolution of the statistical distribution of particles in a physical plasma in which collision effects between particles are considered. The particles are affected by Brownian motion's law. We refer to [22] for a more detailed physical introduction.

Let f be the statistical distribution of particles depending on the position x, on the velocity v and on the time t. The Vlasov-Fokker-Planck equation which determines f is given by

$$\frac{\partial f}{\partial t} + (v \cdot \nabla_x) f + div_v((E - \beta v) f) - \sigma \Delta_v f \; = \; 0 \qquad (1)$$

where $E(t,x)$ is the vector force field which is related to f by the Poisson's law

$$rot\ E \; = \; 0 \qquad div\ E \; = \; \theta \rho(f). \qquad (2)$$

$\rho(f)$ is the macroscopic density of particles, that is

$$\rho(f)(t,x) \; = \; \int_{\mathbb{R}^3} f(t,x,v)\, dv.$$

θ values 1 or -1 depending on the considered case: electrostatic ($\theta = 1$) or gravitational ($\theta = -1$). The constants β and σ are related to the collision between the particles.

From the mathematical point of view the equation (1) is an hyperbolic equation with respect to x and parabolic with respect v. The nonlinearity comes from the coupling with the Poisson's law.

In order to solve the elliptic problem (2) is neccesary to introduce a decreasing condition at infinity. In the classical setting this condition becomes

$$\lim_{\|x\| \to \infty} E(t,x) \; = \; 0. \qquad (3)$$

We will consider the Cauchy problem for the problem (1)-(3), being f_o the initial distribution of particles.

Let us summarize the previous literature about this problem. Classical solutions have been studied under the hypothesis

$$f_o \in C_b^2(\mathbb{R}^6) \;\; , \;\; (1 + |v|^2)^{\frac{7}{2}} f_o \in L^1(\mathbb{R}^6) \cap L^\infty(\mathbb{R}^6)$$

and

$$(1 + |v|^2)^{\frac{7}{2}} D f_o \in L^1(\mathbb{R}^6) \cap L^\infty(\mathbb{R}^6)$$

where $\gamma > 3$. H. D. Victory and B. P. O'Dwyer in [22] proved the local in time existence of a classical solution in this framework. Later, G. Rein and J. Weckler in [20] gave sufficient conditions in order to obtain global in time classical solutions.

The first proof of a unique classical global solution for the three dimensional Vlasov-Poisson-Fokker-Planck equations was done by F. Bouchut in [4] under the hypothesis $f_o \in L^1(\mathbb{R}^6) \cap L^\infty(\mathbb{R}^6)$ and the moments in v bounded for $k < m$ with $m > 6$,

$$\int_{\mathbb{R}^6} |v|^k f_o(x,v)\, d(x,v) \;<\; \infty.$$

Note that these hypotheses imply more regularity on f_o and its macroscopic density by means of classical interpolation lemmas (see [18]). Also, F. Bouchut has proved a smoothing effect under the previous hypothesis in [5]. In fact, there exists a regularizing effect on the macroscopic density and on the force field for any positive time.

Let us state the concept of weak solution that we are going to deal with.

Setting $Q_T = \mathbb{R}^6 \times [0,T[$, $Q'_T = \mathbb{R}^3 \times [0,T[$, $\mathcal{M}^+(\mathbb{R}^N)$ the set of positive measures of bounded total variation in $\mathbb{R}^N$ and $L^1_{loc}(Q_T; f)$ the space of locally integrable functions with respect to the measure f. When we look for weak solutions, the Vlasov-Poisson-Fokker-Planck equations will be considered in the following sense: We will say that a pair (E, f) is a weak solution of the problem (1)-(3) if $f \in C_w([0,T], \mathcal{M}^+(\mathbb{R}^6))$, $E \in L^1_{loc}(Q_T; f)$ and verifies

$$\int_{Q_T} \left[f\frac{\partial \Phi}{\partial t} + f(v\cdot\nabla_x)\Phi + f((E-\beta v)\cdot\nabla_v)\Phi + f\sigma\Delta_v\Phi \right] d(t,x,v) \, +$$

$$\int_{\mathbb{R}^6} f_o\Phi_o\, d(x,v) \;=\; 0,$$

$$\int_{Q'_T} [E\cdot\nabla_x\varphi \;+ \theta\rho(f)\varphi]\; d(t,x) \;=\; 0$$

and

$$\int_{Q'_T} E\cdot rot_x\eta\; d(t,x) \;=\; 0$$

for any $\Phi \in C_o^\infty(Q_T)$, $\varphi \in C_o^\infty(Q'_T)$ and $\eta \in C_o^\infty(Q'_T)^3$ where $\Phi_o(x,v) = \Phi(0,x,v)$. These equations are considered together with the weak decreasing condition at infinity

$$\lim_{n\to\infty} \frac{1}{n^3} \int_{n\le|x|\le 2n} |E(t,x)|\; dx \;=\; 0. \tag{4}$$

We introduce this weaker decreasing condition in order to have a result of uniqueness for the elliptic problem (2) which is verified by (E, f).

Lemma 1 *Let $\rho \in \mathcal{M}^+(\mathbb{R}^3)$. Let us consider $E \in L^1_{loc}(\mathbb{R}^3)^3$ such that*

$$div\ E \;=\; \rho \;\;,\;\; rot\ E \;=\; 0 \;\; in\ \mathcal{D}'(\mathbb{R}^3),$$

where $\mathcal{D}'(\mathbb{R}^3)$ is the space of distributions on $\mathbb{R}^3$. If E satisfies the decreasing condition (4), then $E = K * \rho$ where K is the gradient of the fundamental solution of Δ.

The proof of this lemma can be seen in [8].

In this framework global existence results in three dimensions were obtained by H. D. Victory in [21] under the hypothesis $f_o \in L^1(\mathbb{R}^6) \cap L^\infty(\mathbb{R}^6)$ with the initial energy and initial inertial momentum bounded. Also, the global existence of weak solutions have been proved by the authors in [6] under the hypothesis $f_o \in L^1(\mathbb{R}^6) \cap L^p(\mathbb{R}^6)$ for some suitable p with the initial energy and initial inertial momentum bounded.

Finally, renormalized solutions were defined by R. DiPerna and P. L. Lions in [10]. A Renormalized Solution is a pair (E, f) with $f \in L^\infty([0,\infty), L^1(\mathbb{R}^6))$ and $E \in L^\infty([0,\infty), L^2(\mathbb{R}^3)^3)$ such that

$$\frac{\partial \alpha(f)}{\partial t} + (v \cdot \nabla_x)\alpha(f) + div_v(E\alpha(f)) - \beta div_v(v\alpha'(f)f) - \sigma\Delta_v\alpha(f) =$$

$$= -\alpha''(f)\left[\sigma|\nabla_v f|^2 + \beta v f \nabla_v f\right]$$

$$rot\ E = 0 \quad , \qquad div\ E = \rho(f)$$

where $\alpha(f)$ is a regular function, usually an approximation of $\log(1+f)$.

The existence of renormalized solutions was proved under the hypothesis

$$\int_{\mathbb{R}^6} (1 + |v|^2 + |\log f_o|) f_o \, d(x,v) < \infty$$

and the initial force field E_o belonging to $L^2(\mathbb{R}^3)^3$. Besides, if we consider the stellar dynamic system $(\theta = -1)$ it is assumed that $f_o \in L^{9/7}(\mathbb{R}^6)$.

DiPerna and Lions also proved that if $f \in L^\infty([0,\infty[, L^{p_o}(\mathbb{R}^6))$, for some p_o, then the renormalized solution is a weak solution.

Motivated by the energy concentrations (see [2]) and the singular behaviour of the density of particles (see [19]), in this communication we introduce a new concept of solution, called functional solution, to study the Cauchy problem for the equations (1)-(2) with a bounded measure as initial data. Let us remark that the initial value problem with bounded measures as initial data has been studied in the one dimensional case by Y. Zheng and A. Majda in [23] given a new concept of weak solution based on Volpert's average. Also, A. Majda, G. Majda and Y. Zheng in [19] have given examples of this type of solutions showing the singular behaviour of the distribution of particles and the force field. Concentration effects are produced even in the one dimensional case. Nonuniqueness of solution also happens in this case.

The interest of this kind of initial data is motivated by the large and relevant phenomena of the nature which they represent. For instance, when we consider a particle sheet as initial data, that is the concentration of the distribution of particles on a curve or on a surface

$$f_0 = \alpha\mu_S, \quad f_0 = \alpha\delta_S,$$

where μ_S is a measure defined on a curve or surface S, δ is the Dirac function and α is the strength on S.

Our first research direction was the analysis of this kind of initial data in the framework of weak solutions. We proved local and global existence, uniqueness and stability of solutions for some special measures as initial data. This work is contained in [7]. Hovewer, generally we obtain only local existence of weak solutions. In order to avoid this problem and to take into account the concentration effects, we introduce a new concept of solution which assure the global in time existence of solutions for bounded measures as initial data without other restrictions. This study was done for kinetic equations of Vlasov-type in [8].

The communication is structured as follows. In the first section we introduce the concept of functional solution for the Vlasov-Poisson-Fokker-Planck system and we study the relation between this concept and the weak and renormalized notions. Finally, in section two we will prove the main existence results and we will study the properties of the functional solutions when we assume more hypothesis on the initial data. Also, we analyze as an example of functional solutions the weak solutions obtained in [7].

1. Functional Solutions

Let us assume that $E \in L^1_{loc}(Q_T; f)$ and $f \in \mathcal{M}^+(Q_T)$ then, if the pair (E, f) is a weak solution of (1)-(3) we deduce

$$\int_{Q_T} \left[f\{\frac{\partial \Phi}{\partial t} + (v \cdot \nabla_x)\Phi + ((E - \beta v) \cdot \nabla_v)\Phi + \sigma \Delta_v \Phi + \theta \varphi\} \right] d(t, x, v)$$

$$+ \int_{Q'_T} [E \cdot \nabla_x \varphi \; + E \cdot rot_x \eta] \; d(t, x) \; + \; \int_{\mathbb{R}^6} f_o \Phi_o \; d(x, v) \; = \; 0$$

as we wrote above.

Let us define $\mathcal{E}$ to be the set of symbols of the form

$$F = F(f, E, t, x, v) = (F^1_{\psi,\Phi,\varphi}, F^2_{\varphi,\eta}) =$$

$$(f(\psi \; + \; (v \cdot \nabla_x)\Phi \; + \; ((E - \beta v) \cdot \nabla_v)\Phi \; + \; \sigma \Delta_v \Phi \; + \; \theta \varphi) \; , \; E(\nabla_x \varphi \; + \; rot_x \eta)),$$

where $f \in \mathbb{R}$, $E \in \mathbb{R}^3$, $\psi \in C_o(Q_T)$, $\Phi \in C^\infty_o(Q_T)$, $\varphi \in C^\infty_o(Q'_T)$ and $\eta \in C^\infty_o(Q'_T)^3$.

Let us consider $\mathcal{E}^\star$ the algebraic dual of $\mathcal{E}$. The dual pair $(\mathcal{E}^\star, \mathcal{E})$ is endowed with the weak-$\star$ topology $\sigma(\mathcal{E}^\star, \mathcal{E})$.

Let us define the following applications

$$\Pi : \mathcal{E} \longrightarrow \mathcal{E}$$

$$\Pi(F) = (\Pi_1(F), \Pi_2(F)) = (F^1_{\frac{\partial \Phi}{\partial t},\Phi,\varphi}, F^2_{\varphi,\eta})$$

and

$$\Pi_o : \mathcal{E} \longrightarrow \mathcal{E}_o$$

$$\Pi_o(F) = F^1_{\Phi_o,0,0},$$

where we have denoted by $\mathcal{E}_o$ the set $\Pi_o(\mathcal{E})$.

Thus, if the pair (E, f) is a weak solution of (1)-(3) we define

$$l_{(E,f)}(F) = \int_{Q_T} \Pi_1(F)(f(t,x,v), E(t,x), t, x, v)\ d(t,x,v) +$$

$$+ \int_{Q'_T} \Pi_2(F)(E(t,x), t, x)\ d(t,x)$$

and (E, f) is a weak solution of (1)-(3) if and only if

$$l_{(E,f)}(F) + \int_{\mathbb{R}^6} f_o \Phi_o\ d(x,v) \ = \ 0$$

for any $F \in \mathcal{E}$.

We consider by M the set of all the operators $l_{(E,f)}$ defined by pairs (E, f) with $E \in L^1_{loc}(Q_T; f)$, $f \in \mathcal{M}^+(Q_T)$ and E verifying the decreasing condition

$$\lim_{n \to \infty} \frac{1}{n^3} \int_{n \leq |x| \leq 2n} |E(t,x)|\ dx \ = \ 0.$$

The set of initial data is $M_o = \mathcal{M}^+(\mathbb{R}^6)$.

Associated to every $(E, f) \in M$ we can define the operator $l_{(E,f)} \in \mathcal{E}^\star$ by

$$l_{(E,f)}(F) = \int_{Q_T} \Pi_1(F)(f(t,x,v), E(t,x), t, x, v)\ d(t,x,v) +$$

$$+ \int_{Q'_T} \Pi_2(F)(E(t,x), t, x)\ d(t,x). \qquad (5)$$

Therefore, we can consider that M is imbedding in $\mathcal{E}^\star$. In fact, this correspondence is one-to-one (see [8]).

Let $\bar{M}$ and $\bar{M}_o$ be the closure of M and M_o in $\mathcal{E}^\star$ and $\mathcal{E}_o^\star$ respectively with the weak-$\star$ topologies.

Definition 1 *We will say that $l \in \bar{M}$ is a functional solution of the problem* (1)-(2) *with initial condition $l^o \in \bar{M}_o$ if*

$$l(\Pi(F)) + l^o(\Pi_o(F)) = 0$$

for any $F \in \mathcal{E}$.

This concept was studied previously in the framework of conservation laws by V. A. Galkin in [13] and [14] and for the Euler equation by V. A. Galkin and V. V. Russkikh in [15]. Also, in the context of Euler equation D. Chae and P. Dubovskii studied in [9] the relation between the functional solutions and the measure-valued solutions given by R. DiPerna and A. Majda in [11].

From the development shown above it is clear that any classical and weak solution define a functional solution.

Also, any renormalized solution (E,f) define a functional solution in the following sense. Let us consider any sequence $\alpha_\delta(f)$, for $\delta > 0$, such that $\alpha_\delta(f) \in C_b^1([0,\infty))$ and $\alpha_\delta(f) \to f$ in $L^\infty([0,\infty); L^1(\mathbb{R}^6))$ when $\delta \to 0$.

We define the operator $l_{(E,f)} \in \mathcal{E}^\star$ by

$$l_{(E,f)}(F) = \lim_{\delta \to 0} l_{(E,\alpha_\delta(f))}(F).$$

and this operator is a functional solution for the problem (1)-(2).

In the one dimensional case A. Majda and Y. Zheng have introduced a new concept of solution based on the Volpert symmetric average for which they proved global existence with bounded measures as initial data. This concept of solution coincides with the functional one when we apply this for the one dimensional Vlasov-Poisson-Fokker-Planck system.

2. Existence Results

Consider a sequence of mollified problems of the system whose regularized solution are (E^ϵ, f^ϵ) for every ϵ. Let $l_\epsilon = l_{(E^\epsilon,f^\epsilon)}$ the functional operator defined by (E^ϵ, f^ϵ).

- Let us say that the method is *approximate* for $l^o \in \bar{M}_o$ if and only if there exists a sequence $\epsilon_n \to 0$ such that

  $$|l_{\epsilon_n}(\Pi(F)) + l^o_{\epsilon_n}(\Pi_o(F))| + |l^o(\Pi_o(F)) - l^o_{\epsilon_n}(\Pi_o(F))| \to 0$$

  for any $F \in \mathcal{E}$ when $\epsilon_n \to 0$.

- Let us say that a method is *stable* if

  $$\sup_{\epsilon>0} \left| \int_{Q_T} f^\epsilon(t,x,v)\psi(t,x,v)\ d(t,x,v) \right| < \infty$$

  for any $\psi \in C_o^\infty(Q_T)$.

- The method is said to be *convergent* to a functional solution $l \in \bar{M}$ if there exists a sequence ϵ_n such that $l_{\epsilon_n} \to l$ in the weak-$\star$ topology.

In the following result we obtain the relation between these concepts.

Theorem 1 *A method approximate for $l^o \in \bar{M}_o$ and stable is convergent to a functional solution with initial condition $l^o \in \bar{M}_o$.*

The proof of the previous theorem relies on the fact that every weakly bounded set is relatively compact with the weak-$\star$ topology and can be seen in [8]. We will use it to obtain the existence results.

In order to obtain existence of functional solutions we must prove that a method of regularization of the system (1)-(2) is approximate and stable.

We are going to use the method of mollification due to E. Horst [16] in which the kernel K of the integral solution of the Poisson's law, which is the gradient of the fundamental solution of the laplacian, is regularized by means of a mollifier sequence ω_ϵ in position space

$$K^\epsilon = \omega_\epsilon * K.$$

Also, we regularize the initial condition. The mollified system becomes

$$\frac{\partial f^\epsilon}{\partial t} + (v \cdot \nabla_x) f^\epsilon + div_v((E^\epsilon - \beta v) f^\epsilon) - \sigma \Delta_v f^\epsilon \;=\; 0 \quad \text{on } (0,T) \times \mathbb{R}^6 \tag{6}$$

$$E^\epsilon(t,x) \;= \theta \; (K^\epsilon * \rho(f^\epsilon)(t,x)) \tag{7}$$

$$\rho(f^\epsilon)(t,x) \;=\; \int_{\mathbb{R}^3} f^\epsilon(t,x,v) \; dv \tag{8}$$

$$f(0,\cdot) \;=\; f_o^\epsilon. \tag{9}$$

The problem (6)-(9) has a unique solution (E^ϵ, f^ϵ) which satisfies

i) $f^\epsilon \in C^\infty([0,\infty[\times\mathbb{R}^6)$, $E^\epsilon \in C^\infty([0,\infty[\times\mathbb{R}^3)^3$.

ii) $f^\epsilon(t,\cdot)$ goes to f_o^ϵ when t goes to zero in $L^1(\mathbb{R}^6) \cap L^\infty(\mathbb{R}^6)$.

iii) f^ϵ is positive for all $t \geq 0$.

iv) The conservation of mass property holds, i.e.,

$$\|f^\epsilon(t,\cdot)\|_{L^1(\mathbb{R}^6)} = \|f_o\|_{\mathcal{M}^+(\mathbb{R}^6)}.$$

Let $l_\epsilon \;=\; l_{(E^\epsilon, f^\epsilon)}$ the operator defined by the solutions (E^ϵ, f^ϵ) of the problems (6)-(9). It is easy to deduce

$$l_\epsilon(\Pi(F)) + l^o_{f_o^\epsilon}(\Pi_o(F)) = \theta \int_{Q'_T} [\rho(f^\epsilon) - \omega_\epsilon * \rho(f^\epsilon)] \, \varphi \; d(t,x).$$

By means of properties of convolution we obtain that

$$\left| \int_{Q'_T} \rho(f^\epsilon) \, [\varphi - \check{\omega}_\epsilon * \varphi] \;\; d(t,x) \right| \leq \; \|\rho(f^\epsilon)\|_{L^\infty([0,T], L^1(\mathbb{R}^3))} \; \|\varphi - \check{\omega}_\epsilon * \varphi\|_{L^1([0,T], L^\infty(\mathbb{R}^3))}.$$

where $\check{\omega}_\epsilon(x) = \omega_\epsilon(-x)$. Then,

Therefore, using the conservation of the L^1 norm of the mass density and that $\check{\omega}_\epsilon * \varphi \to \varphi$ in $L^1(0,T,L^\infty(\mathbb{R}^3))$, we have

$$|l_\epsilon(\Pi(F)) + l^o_{f_o^\epsilon}(\Pi_o(F))| \to 0.$$

Combining this fact with the convergence of f_o^ϵ to f_o weakly as measures we obtain that the method defined by the solutions (E^ϵ, f^ϵ) is approximate for the Vlasov-Poisson-Fokker-Planck system with initial condition $f_o \in \mathcal{M}^+(\mathbb{R}^6)$.

The stability condition is clear from the conservation of mass in the regularized system. In fact,

$$\int_{Q_T} f^\epsilon(t,x,v) \psi(t,x,v) \; d(t,x,v) \;\leq\; T \; \|f_o\|_{\mathcal{M}^+(\mathbb{R}^6)} \; \|\psi\|_{L^\infty(Q_T)}.$$

Theorem 2 *Let $f_o \in \mathcal{M}^+(\mathbb{R}^6)$. Then, there exists a functional solution $l \in \bar{M}$ for the initial condition $l^o_{f_o}$ associated to f_o.*

In order to obtain more properties of the functional solution we will assume that

- $f_o \in \mathcal{M}^+(\mathbb{R}^6)$ a positive bounded measure.
- The initial force field $E_o \in L^2(\mathbb{R}^3)^3$.
- The initial kinetic energy is bounded, $|v|^2 f_o \in \mathcal{M}^+(\mathbb{R}^6)$.
- The initial inertial momentum is bounded, $|x|^2 f_o \in \mathcal{M}^+(\mathbb{R}^6)$.

We need a lemma due to E. Horst in [17] about the boundedness of the kinetic energy.

Lemma 2 *Let (E^ϵ, f^ϵ) the solution of the regularized system, then the kinetic energy and inertial momentum for the system remains bounded independently of ϵ for any time interval $[0, T]$. In the case of gravitational forces ($\theta = -1$) is necesary to assume that $f_o \in L^{\frac{9}{7}}(\mathbb{R}^6)$.*

Now, we will prove that f^{ϵ_n} converges weakly as measures to f in $\mathcal{M}^+(Q_T)$ and E^{ϵ_n} converges weakly in $L^2(Q'_T)$ to E for some subsequence ϵ_n, when $n \to \infty$. We are going to use a compactness result due to Prohorov in $\mathcal{M}^+(Q_T)$, endowed with the weak-$\star$ topology, that can be seen in [3].

Proposition 1 *Let $B > 0$ a constant and $\mathcal{S} \subset \mathcal{M}(Q_T)$ such that*

$$\|\mu\|_{\mathcal{M}(Q_T)} = B$$

for any $\mu \in \mathcal{S}$. Then, $\mathcal{S}$ is weak-$\star$ relatively compact in $\mathcal{M}(Q_T)$ if and only if for any $\delta > 0$ there exists a compact set $X \subset Q_T$ such that

$$\mu(X) > B - \delta$$

for any $\mu \in \mathcal{S}$. In addition, the weak-$\star$ limits verifies that the $\mathcal{M}(Q_T)$ norm is B.

Using this compactness result let us prove that the set of solutions of the problems (6)-(9) is relatively compact in suitable spaces.

Lemma 3 *The following assertions hold.*

i) The set of approximate solutions $\mathcal{S} = \{f^\epsilon, \epsilon > 0\}$ is relatively compact with respect to the weak-$\star$ topology in $\mathcal{M}(Q_T)$. Also, $\{\rho(f^\epsilon), \epsilon > 0\}$ is relatively compact with respect to the weak-$\star$ topology in $\mathcal{M}(Q'_T)$.

ii) The set of approximate force fields $\mathcal{S}' = \{E^\epsilon, \epsilon > 0\}$ is relatively compact with respect to the weak topology in $L^2(Q'_T)$.

iii) If $f^{\epsilon_n} \to f$ and $\rho(f^{\epsilon_n}) \to \rho$ weakly as measures in the respective spaces, then $\rho = \rho(f)$.

Proof. It is clear that

$$\int_{Q_T} f^\epsilon(t,x,v)\; d(t,x,v) \;=\; T\;\|f_o\|_{\mathcal{M}^+(\mathbb{R}^6)}.$$

Applying lemma 2 we can consider a constant C depending only on the initial data and on T such that

$$\int_{\mathbb{R}^6} (|x|^2+|v|^2)\; f^\epsilon(t,x,v)\; d(x,v) \;\le\; C.$$

Let us consider any $\delta > 0$ and $R > 0$ such that $CTR^{-2} < \delta$, then

$$R^2 \int_{\mathbb{R}^6/B_R} f^\epsilon(t,x,v)\; d(x,v) \;\le$$

$$\le \int_{\mathbb{R}^6/B_R} (|x|^2+|v|^2) f^\epsilon(t,x,v)\; d(x,v) \;\le\; C \;<\; T\delta R^2.$$

Therefore, we obtain

$$\int_0^T \int_{\mathbb{R}^6/B_R} f^\epsilon(t,x,v)(x,v)\; dt \;<\; \delta$$

for any $\delta > 0$. Thus, using theorem 1 the set $\mathcal{S}$ is weak relatively compact in $\mathcal{M}(Q_T)$. The last part of *i)* can be proved with the same arguments.

The second statement *ii)* is an easy consequence of Banach-Alaoglu's theorem.

The third statement is obtained by cutting off the density f, applying the same ideas of part *i)* and using the Lebesgue's theorem. ∎

Applying the same arguments as in the proof of theorem 2 we can obtain that for some subsequence, that we will denote with the same index, the operator l_{ϵ_n} converges to a functional solution l of the problem (1)-(3), i.e.,

$$l_{\epsilon_n}(F) \to l(F)$$

for any $F \in \mathcal{E}$.

Since $f^{\epsilon_n} \to f$ weakly as measures we obtain

$$\int_{Q_T} f^{\epsilon_n}\psi\; d(t,x,v) \to \int_{Q_T} \psi\; df(t,x,v)$$

for any $\psi \in C_o(Q_T)$.

Also, using that $E^{\epsilon_n} \to E$ weakly in $L^2(Q'_T)$ and that $\rho(f^{\epsilon_n}) \to \rho(f)$ weakly as measures, we deduce that

$$\int_{Q'_T} [E^{\epsilon_n}\cdot\nabla_x\varphi \;+ E^{\epsilon_n}\cdot rot_x\eta]\; d(t,x) \to \int_{Q'_T} [E\cdot\nabla_x\varphi \;+ E\cdot rot_x\eta]\; d(t,x)$$

and

$$\int_{Q_T} f^{\epsilon_n}\varphi \; d(t,x,v) \rightarrow \int_{Q_T} \varphi \; df(t,x,v)$$

for any $\varphi \in C_o^\infty(Q'_T)$ and $\eta \in C_o^\infty(Q'_T)^3$.

As a consequence, we obtain the following main result.

Theorem 3 *Let $f_o \in \mathcal{M}^+(\mathbb{R}^6)$ such that $(|x|^2+|v|^2)f_o \in \mathcal{M}^+(\mathbb{R}^6)$ and $E_o \in L^2(\mathbb{R}^3)^3$. Assume that the kinetic energy remains bounded independtly of ϵ in $[0,T]$. Then, there exists a functional solution $l \in \bar{M}$ for the initial condition $l^o_{f_o}$ associated to f_o that satisfies*

$$l(F^1_{\psi,0,0}, F^2_{0,0}) = \int_{Q_T} \psi \; df(t,x,v)$$

and

$$l(F^1_{0,0,\varphi}, F^2_{\varphi,\eta}) = \int_{Q'_T} [E\cdot\nabla_x\varphi \; + E\cdot rot_x\eta] \; d(t,x) + \int_{Q_T} \varphi \; df(t,x,v).$$

with $f \in \mathcal{M}^+(Q_T)$ and $E \in L^2(Q'_T)$. Moreover, the kinetic energy of the functional solution and the inertial momentum are finite, i.e., $(|x|^2+|v|^2) \in L^1(Q_T;f)$.

Using lemma 2 we obtain the following corollary.

Corollary 1 *The following assertions hold.*

i) In the electrostatic case ($\theta = 1$) if $f_o \in \mathcal{M}^+(\mathbb{R}^6)$ is such that $(|x|^2+|v|^2)f_o \in \mathcal{M}^+(\mathbb{R}^6)$ and $E_o \in L^2(\mathbb{R}^3)^3$, then the above theorem holds.

ii) In the gravitational case ($\theta = -1$) if $f_o \in L^1(\mathbb{R}^6)\cap L^{\frac{9}{7}}(\mathbb{R}^6)$ is such that $(|x|^2+|v|^2)f_o \in L^1(\mathbb{R}^6)$ and $E_o \in L^2(\mathbb{R}^3)^3$, then the above theorem holds.

Moreover, the functional solution (E,f) verifies the following equations in the sense of distributions

$$\frac{\partial f}{\partial t} + (v\cdot\nabla_x)f - div_v(\beta v f) - \sigma\Delta_v f + \Lambda \;=\; 0 \quad \text{on } (0,T)\times\mathbb{R}^6$$

$$rot\; E \;=\; 0 \quad , \quad div\; E \;=\; \theta\;\rho(f)$$

where Λ is a distribution of order one on Q_T defined by

$$\Lambda(\Phi) \;=\; \lim_{n\rightarrow\infty} \int_{Q_T} f^{\epsilon_n}(E^{\epsilon_n}\cdot\nabla_v)\Phi \; d(t,x,v)$$

for any $\Phi \in C_o^\infty(Q_T)$.

Let us remark that the concept of functional solution is useful in order to get a framework in which all the concepts of solutions known nowadays are included. Also it provides us a method to obtain solutions with only some L^1 apriori bounds. One of the open problems is to determine conditions for which the limit is exactly a solution in the sense of distributions or to study which is the structure of the limit. In this direction in the setting of corollary 1 it will be interesting to study the properties and the structure of the distribution Λ. In fact, the distribution Λ could split in something related to the solution (E, f) and a defect measure in the same sense as in [11] where R. DiPerna and A. Majda proved this result for the Euler equations in the framework of measure-valued solutions.

In this direction, let us note that if we assume that the initial condition belongs to a special space of measures (see [7]) we obtain local or global existence of weak solutions depending on the size of the initial data in the corresponding norms and the exponent of the space.

Let us introduce the classical Morrey spaces. A Radon measure μ belongs to the Morrey space of exponent p, $\tilde{L}_p(\mathbb{R}^N)$, if μ satisfies

$$TV(\mu)(B(a,r)) \leq Cr^{\frac{N}{p'}}$$

independtly of a and r. $\tilde{L}_p(\mathbb{R}^N)$ is a Banach space endowed with the norm

$$\|\mu\|_{\tilde{L}_p(\mathbb{R}^N)} = sup\{ r^{-\frac{N}{p'}} TV(\mu)(B(a,r)) \ : \ a \in \mathbb{R}^N, r > 0\}$$

with $1 \leq p \leq \infty$ and $TV(\mu) = |\mu|$ is the total variation of the measure μ. Let us note that $\tilde{L}_1(\mathbb{R}^N) = \mathcal{M}(\mathbb{R}^N)$ and $\tilde{L}_\infty(\mathbb{R}^N) = L^\infty(\mathbb{R}^N)$.

It is not enough to assume that the initial density is in a Morrey space because the operator of the linear part of (1) is not symmetric. Let us deduce which is the main idea to define the modified Morrey spaces. Assume that f is a smooth solution of the system, then

$$f(t,x,v) = \int_{\mathbb{R}^6} G(t,x,v,\xi,\nu)\ f_o(\xi,\nu)\ d\xi\ d\nu +$$
$$+ \int_0^t \int_{\mathbb{R}^6} G(t-s,x,v,\xi,\nu)\ F(s,\xi,\nu)\ d\xi\ d\nu\ ds$$

where $F = (E \cdot \nabla_v) f$ is the nonlinear part of the equation (1) and G is the fundamental solution of the linear part of the equation (1) (see [7]).

If we integrate in v variables the first part of the right hand side of the above inequality we obtain

$$\rho(f_1)(t,x) = \int_{\mathbb{R}^3} \frac{1}{(2\sigma d(t))^{\frac{3}{2}}} N(\frac{x-\xi}{\sqrt{2\sigma d(t)}}) \int_{\mathbb{R}^3} f_o(\xi - \mu(t)\nu, \nu) d\nu d\xi$$

where $N(x)$ is the heat kernel. Hence, in order to bound $\rho(f_1)$ in a Morrey space we need a uniform control of the norms of

$$\rho_h(f_o)(x) = \int_{\mathbb{R}^3} f_o(x - hv, v)\ dv, \qquad con\ h \in \mathbb{R}$$

in a Morrey space. Let us point out that x and v are not independent.

Let us denote by $B(a,r,h)$ the set of pairs $(x,v) \in \mathbb{R}^6$ that satisfies $\|x+hv-a\| \leq r$. Making a change of variables we can define a modified Morrey space in the following way. Let $M\tilde{L}_p(\mathbb{R}^6)$ be the space of all the Radon measures μ such that

$$\|\mu\|_{M\tilde{L}_p(\mathbb{R}^6)} = sup\{ r^{-\frac{3}{p'}} TV(\mu)(B(a,r,h)) \ : \ h \in \mathbb{R}, a \in \mathbb{R}^3, r > 0\}$$

is finite, with $1 \leq p \leq \infty$.

With this type of space of measures we are able to prove the local existence of weak solution for the problem (1)-(2). More exactly, we summarize the results in the following theorem.

Theorem 4 *If f_o belongs to $\tilde{L}_1(\mathbb{R}^6) \cap M\tilde{L}_{p_o}(\mathbb{R}^6)$, with $p_o \geq \frac{7+\sqrt{85}}{6}$, then the problem* (1)-(2) *has a unique weak solution (E,f) in a time interval $[0,T]$ with $T < T^*$.*

T^* depends on f_o and on the constants σ, β. We obtain global existence, $T^* = \infty$, when $p_o = \frac{7+\sqrt{85}}{6}$ for small enough initial data.

Besides, it appears some smoothing effect because $f \in BC((0,T], ML_r(\mathbb{R}^6))$, for some r close enough to 3 and $E(t,\cdot) \in L^\infty(\mathbb{R}^3)^3$ for any positive time.

As we have said above these weak solutions can be considered as functional solutions. Thus, these are examples in which the distribution Λ is exactly $(E \cdot \nabla_v)f$ with no concentration effects until unless some finite time T^*.

Also, we will remark that the boundedness of kinetic energy for the functional solution can be included in the definition of functional solution in order to consider only functional solutions meaningful from this physical point of view.

The concept of functional solution and the results can be introduced and proved for the Vlasov-Poisson system. The same concept can be defined for other kinetic equations of Vlasov type such as Vlasov-Maxwell system, Vlasov-Poisson-Boltzmann, etc... This will be included in [8]

It can be proved easily that the functional solutions of the Vlasov-Poisson-Fokker-Planck system converges to the functional solutions of the Vlasov-Poisson system when the constants σ and β go to zero.

Examples of functional solutions for the Vlasov-Poisson system that are neither renormalized nor weak solutions can be obtained. These examples comes from the fact that the Vlasov-Poisson system can be considered as a limit of the problem of N bodies (see [8]). For instance, if we consider a single particle at position x_o and velocity v_o the corresponding initial data is a delta function on the point (x_o, v_o), $\delta_{(x_o,v_o)}$. Therefore, we test $f(t,x,v) = \delta_{(x_o,x_o+tv_o)}$ to be a functional solution of Vlasov-Poisson system with initial condition $f_o(x,v) = \delta_{(x_o,v_o)}$. Assuming that the definitions of section one are carried on for the Vlasov-Poisson system, we define the following operator

$$l(F) = \int_0^T \left[\Psi(t, x_o + tv_o, v_o) + (v_o \cdot \nabla_x)\Phi(t, x_o + tv_o, v_o)\right] dt$$

for any $F \in \mathcal{E}$ with $\mathcal{E}$ the corresponding space of symbols for the Vlasov-Poisson system. It can be check easily that l is a functional solution of the Vlasov-Poisson system with initial condition defined by $\delta_{(x_o,v_o)}$.

REFERENCES

[1] P. Benilan, H. Brezis and M. Crandall, *A semilinear equation in* $L^1(\mathbb{R}^N)$, Analli della Scuola Norm. Sup. di Pisa, **4** (1975), pp. 523-555.

[2] Y. Brenier and E. Grenier, *Limite singulière du système de Vlasov-Poisson dans le régime de quasi neutralité: le cas indepéndant du temps*, C. R. Acad. Sci. Paris, Serie I, **318** (1994), pp. 121-124.

[3] P. Billingsley, *Convergence of Probability Measures*, Wiley, New York, 1968.

[4] F. Bouchut, *Existence and uniqueness of a global smooth solution for the Vlasov-Poisson-Fokker-Planck system in three dimensions*, J. Func. Anal., **111** (1993), pp. 239-258.

[5] F. Bouchut, *Smoothing effect for the non-linear Vlasov-Poisson-Fokker-Planck system*, to appear in J. Diff. Eq.

[6] J. A. Carrillo and J. Soler, *On the initial value problem for the Vlasov-Poisson-Fokker-Planck system with initial data in* L^p *spaces*, to appear in Math. Meth. in the Appl. Sci.

[7] J. A. Carrillo and J. Soler, *On the Vlasov-Poisson-Fokker-Planck equations with measures in Morrey spaces as initial data*, preprint.

[8] J. A. Carrillo and J. Soler, *On functional solutions for the three dimensional kinetic equations of Vlasov type with bounded measures as initial data*, preprint.

[9] D. Chae and P. Dubovskii, *Functional and Measure-Valued Solutions of the Euler Equations for Incompressible Fluid Flows*, preprint.

[10] R. DiPerna and P. L. Lions, *Global weak solutions of kinetic equations*, Sem. Matematico Torino, **46** (1988), pp. 259-288.

[11] R. DiPerna and A. Majda, *Oscillations and concentrations in weak solutions of the incompressible fluid equations*, Comm. Math. Phys., **108** (1987), pp. 667-689.

[12] V. A. Galkin, *Functional Solutions of Conservation Laws*, Sov. Phys. Dokl., **35(2)** (1990), pp. 133-135.

[13] V. A. Galkin, *The global solutions of conservation laws*, The Researches of nonlinear and random models of mathematical physics, (1990), Obninsk, pp. 4-22.

[14] V. A. Galkin and V. V. Russkikh, *Approximation solutions for Euler and KdV equations*, The Researches of nonlinear and random models of mathematical physics, (1992), Obninsk, pp. 14-35.

[15] E. Horst, *On the classical solutions of the initial value problem for the unmodified monlinear Vlasov Equations. I: General theory.*, Math. Meth. in the Appl. Sci., **3** (1981), pp. 229-248.

[16] E. Horst, *On the classical solutions of the initial value problem for the unmodified monlinear Vlasov Equations. II: Special cases.*, Math. Meth. in the Appl. Sci., **4** (1982), pp. 19-32.

[17] E. Horst and R. Hunze, *Weak solutions of the initial value problem for the unmodified nonlinear Vlasov equation*, Math . Meth. in the Appl. Sci., **6** (1984), pp. 262-279.

[18] A. Majda, G. Majda and Y. Zheng, *Concentrations in the one-dimensional Vlasov-Poisson equations, I:temporal development and non-unique weak solutions in the single component case*, preprint.

[19] G. Rein and J. Weckler, *Generic global classical solutions of the Vlasov-Fokker-Planck-Poisson system in three dimensions*, J. Differential Equations, **99** (1992), pp. 59-77.

[20] H. D. Victory, *On the existence of global weak solutions for Vlasov-Poisson-Fokker-Planck Systems*, J. Math. Anal. and Appl., **160** (1991), pp. 515-553.

[21] H. D. Victory and B. P. O'Dwyer, *On classical solutions of Vlasov-Poisson-Fokker-Planck systems*, Indiana Univ. Math. J., **39** (1990), pp. 105-157.

[22] Y. Zheng and A. Majda, *Existence of global weak solutions to one-component Vlasov-Poisson and Fokker-Planck-Poisson systems in one space dimension with measures as initial data*, to appear in Comm. Pure. Appl. Math.

Departamento de Matemática Aplicada, Facultad de Ciencias, Universidad de Granada, 18071 Granada (Spain).

Partially sponsored by Dirección General de Investigación Científica y Técnica. M.E.C. (Spain), Proyecto PB92-0953. Also, the second author thanks the sponsor of E.C. human capital and mobility programme contract n. ERBCHRXCT930413.

R ČIEGIS

Numerical methods for a forward-backward heat equation

1. Equations

We study the numerical solution of the following boundary value problem [1, 3]

$$v(x)\frac{\partial u}{\partial t} = \frac{\partial}{\partial x}(k(x)\frac{\partial u}{\partial x}) - d(x)u + f(x,t), \quad (x,t) \in Q_T, \tag{1a}$$
$$Q_T = \{(x,t)| \quad -1 < x < 1, \quad 0 < t < T\},$$
$$u(x,T) = g_1(x) \quad \text{for} \quad x \leqslant 0, \quad u(x,0) = g_2(x) \quad \text{for} \quad x \geqslant 0, \tag{1b}$$
$$u(-1,t) = u_1(t), \quad u(1,t) = u_2(t), \quad t > 0, \tag{1c}$$

where v, k, d, and f are given sufficiently smooth functions on $[0, 1]$ for which

$$0 < K_0 \leqslant k(x) \leqslant K_1, \ 0 \leqslant d(x) \leqslant D_1, \ |v(x)| \leqslant V_1, \tag{1d}$$

and $v(x) < 0$ for $x < 0, v(x) > 0$ for $x > 0$. As a special case we also consider the Fokker - Planck equation [6]

$$x\frac{\partial u}{\partial t} = \frac{\partial}{\partial x}((1-x^2)\frac{\partial u}{\partial x}), \quad (x,t) \in Q_T \tag{2a}$$

with $u(x,t)$ satisfying initial conditions

$$u(x,T) = g_1(x) \text{ for } x < 0, \quad u(x,0) = g_2(x) \text{ for } x > 0. \tag{2b}$$

The degeneracy of the differential operator in the x variable as $|x| \to 1$ accounts for the lack of boundary conditions at $|x| = 1$.

Physical situations giving rise to this problem are described in detail in [1, 3]. Initial conditions (1b) are specified for different time moments, therefore problem (1) must be treated as a boundary value problem with nonlocal boundary conditions. Numerical methods for (1) are investigated in [4 – 6]. We note that the analysis of convergence of such methods is based on the maximum principle,hence only strictly implicit difference schemes are investigated by these authors. The aim of this paper is to study the convergence of more general difference schemes.

First we subdivide the region Q_T into two subregions $Q^+ = \{(x,t)|(x,t) \in Q_T,\ x > 0\}$, $\quad Q^- = \{(x,t)|(x,t) \in Q_T,\ x < 0\}$ and rewrite problem (1) as

$$v(x)\frac{\partial u^\pm}{\partial t} = \frac{\partial}{\partial x}(k(x)\frac{\partial u^\pm}{\partial x}) - d(x)u^\pm + f(x,t), \qquad (x,t) \in Q^\pm, \tag{3a}$$

$$u^-(x,T) = g_1(x), \quad u^+(x,0) = g_2(x), \tag{3b}$$

$$u^-(-1,t) = u_1(t), \quad u^+(1,t) = u_2(t), \quad t > 0, \tag{3c}$$

where $u^\pm$ is the solution of (1) restricted to $Q^\pm$. For $x = 0$ we have ordinary conditions of continuity of the solution

$$u^+(0,t) = u^-(0,t), \quad k(0)\frac{\partial u^+(0,t)}{\partial x} = k(0)\frac{\partial u^-(0,t)}{\partial x}, \quad 0 \leqslant t \leqslant T. \tag{3d}$$

2. Difference scheme

A difference grid $\Omega_{h\tau} = \omega_\tau \times \omega_h$ is defined by

$$\omega_\tau = \{t_j = j\tau,\ j = 1,2,\cdots,K,\ t_K = T\},$$
$$\omega_h = \{x_i = -1 + ih,\ i = 1,2,\cdots,N-1,\ x_N = 1,\ N = 2M+1\}.$$

We use a slightly changed space grid for the Fokker - Planck equation

$$\omega_h = \{x_i = -1 + (i-0.5)h,\ i = 1,2,\cdots,N-1,\ x_{N-1} = 1 - 0.5h,\ N = 2M+1\}.$$

We also define subgrids $\Omega^-_{h\tau} = \{(x,t) \in \Omega_{h\tau},\ x < 0\}$, $\Omega^+_{h\tau} = \{(x,t) \in \Omega_{h\tau},\ x > 0\}$. Introduce some notation for functions defined on $\omega_\tau, \omega_h, \Omega^\pm_{h\tau}$:

$$y_i = y_i(t_j) = y(x_i,t_j), \quad \hat{y} = y_i(t_{j+1}), \quad y_t = (\hat{y} - y)/\tau,$$
$$y_i = y_t^-, \quad (x,t) \in \Omega^-_{h\tau},\ y_i = y_i^+, \quad (x,t) \in \Omega^+_{h\tau},$$
$$y_{\bar{x}} = (y_i - y_{i-1})/h, \quad y_x = (y_{i+1} - y_i)/h,$$
$$y^\sigma_\pm = (0.5 \pm \sigma)\hat{y}^\pm + (0.5 \mp \sigma)y^\pm, \quad 0 \leqslant \sigma \leqslant 0.5.$$

For vectors $y,\ w$ we define discrete products and norms

$$(y,w)_0 = \sum_{i=1}^{N-1} |v(x_i)| y_i w_i h, \quad (y,v)_1 = -\sum_{i=1}^{M} v(x_i) y_i w_i h,$$
$$(y,w)_2 = \sum_{i=M+1}^{N-1} v(x_i) y_i w_i h, \quad ||y||_j^2 = (y,y)_j, \quad j = 0,1,2,$$

$$(ky_{\bar{x}}, w_{\bar{x}}] = \sum_{i=1}^{N} k_{i-0.5} y_{\bar{x},i} w_{\bar{x},i} h, \quad ||ky_{\bar{x}}]|^2 = (ky_{\bar{x}}, y_{\bar{x}}],$$

$$||y||_{C(Q)} = \max_{(x,t)\in Q} |y(x,t)|, \quad (y,w) = \sum_{i=1}^{N-1} y_i w_i h.$$

We approximate problem (3) by the following family of difference schemes

$$v(x) y_t^{\pm} = (k_{i-0.5}(y_{\pm}^{\sigma})_{\bar{x}})_x - d y_{\pm}^{\sigma} + f(x_i, t_{j+0.5+\sigma}), \quad (x,t) \in \Omega_{h,\tau}^{\pm}, \tag{4a}$$

$$y^+(x,0) = g_2(x), \quad y^-(x,T) = g_1(x), \quad x \in \omega_h^{\pm}, \tag{4b}$$

$$y_0^- = u_1(t_j), \ y_N^+ = u_2(t_j), \ t \in \omega_\tau. \tag{4c}$$

For $\sigma = 0$ or $\sigma = 0.5$ we replace the continuity conditions (3d) by equalities

$$y^+(x_m) = y^-(x_m), \quad m = M, M+1, \tag{5a}$$

and for $0 < \sigma < 0.5$ we use the following continuity conditions

$$y_+^{\sigma}(x_m) = y_-^{\sigma}(x_m), \quad m = M, \ M+1. \tag{5b}$$

In the error analysis we assume the solution of (3) is smooth enough for our arguments to hold. Our main goal is to investigate the convergence of iterative methods for solving the difference problem. But first we prove some results on the accuracy of the solution of finite difference scheme (4). We define the error function by $z_i = u(x_i, t_j) - y_i$, where $u(t)$ and $y(t)$ denote the solutions of (3) and (4), respectively. It satisfies the difference problem

$$v(x) z_t^{\pm} = (k_{i-0.5}(z_{\pm}^{\sigma})_{\bar{x}})_x - d z_{\pm}^{\sigma} + \psi, \quad (x,t) \in \Omega_{h\tau}^{\pm}, \tag{6a}$$

$$z^+(x,0) = 0, \quad x \in \omega_h^+, \quad z^-(x,T) = 0, \quad x \in \omega_h^-, \tag{6b}$$

$$z_0^-(t_j) = 0, \quad z_N^+(t_j) = 0, \quad t \in \omega_\tau, \tag{6c}$$

where ψ_i is the local truncation error. Boundary conditions (6c) must be excluded from the system for the error function if we solve the Fokker - Planck equation. Using Taylor's series we obtain the following estimate

$$|\psi_i| \leqslant C(\tau^\alpha + h^2), \quad \alpha = \begin{cases} 1, & \text{if} \quad 0 < \sigma \leqslant 0.5, \\ 2, & \text{if} \quad \sigma = 0. \end{cases}$$

For $\sigma = 0.5$ the difference scheme (4) is strictly implicit and the accuracy estimate follows directly from the maximum principle (see, also [5])

$$||z||_{C(\Omega_{h,\tau})} \leqslant C(\tau + h^2).$$

Let $0 \leqslant \sigma < 0.5$, then the maximum principle is valid for the solution of scheme (4) if time and space steps τ, h satisfy the inequality

$$\tau \leqslant h^2 \min_{x \in \omega_h} |v(x)|/((1 - 2\sigma) \max_{x \in \omega_h} k(x)). \tag{7}$$

This requirement is very restrictive in many cases. For example, we cannot use this method of investigation for the Fokker - Planck equation, because one may easily verify that $\min |v(x)| = 0$.

We shall use the energy method to prove the convergence of difference scheme (4) for $0 \leqslant \sigma < 0.5$.

Theorem 1. *Let $0 \leqslant \sigma \leqslant 0.5$. Then there exists a unique solution of the difference scheme(4), (5b) such that*

$$||z(0)||_1{}^2 + ||z(T)||_2{}^2 + \sum_{j=0}^{K-1} \tau(||kz^{\sigma}_{\bar{x}}]|^2 + 2(dz^{\sigma}, z^{\sigma}) +$$

$$+2\sigma\tau||z_{\tau}||_0{}^2 \leqslant \sum_{j=0}^{K-1} \tau||\eta(t_j)||^2,$$

where

$$\eta_i(t_j) = -\sum_{l=i}^{N-1} k^{-0.5}_{l-0.5}\psi(x_i, t_l)h.$$

In case $\sigma = 0$ or $\sigma = 0.5$ the discrete continuity condition (5b) can be replaced by condition (5a).

P r o o f. We multiply equation (4a) by a function z^{σ}

$$(z_t, z^{\sigma})_0 + (dz^{\sigma}, z^{\sigma}) = ((kz^{\sigma}_{\bar{x}})_x, z^{\sigma}) + (\psi, z^{\sigma}). \tag{8}$$

Next we use the equality

$$z^{\sigma}_{\pm} = 0.5(\hat{z} + z) \pm \tau\sigma z_t^{\pm},$$

to obtain

$$(z_t, z^\sigma)_0 = \frac{1}{2\tau}((1, \hat{z}^2 - z^2)_2 + (1, z^2 - \hat{z}^2)_1 + 2\sigma\tau^2(1, {z_t}^2)_0). \tag{9a}$$

Applying Green's formula and using boundary conditions (6c) gives

$$((kz^\sigma_{\bar{x}})_x, z^\sigma_+)_2 = -(k, ((z^\sigma_+)_{\bar{x}})^2]_2 - k_{M+0.5} z^\sigma_+(x_M)(z^\sigma_+)_{x,M},$$

and

$$((kz^\sigma_{\bar{x}})_x, z^\sigma_-) = -(k, ((z^\sigma_-)_{\bar{x}})^2]_1 + k_{M+0.5} z^\sigma_-(x_M)(z^\sigma_-)_{\bar{x},M+1}.$$

These two equalities also hold in the case of the Fokker - Planck equation, since $k(x) = 0$ for $|x| = 1$. Hence, using continuity conditions (5b) (or (5a), if $\sigma = 0$ or $\sigma = 0.5$), we obtain that

$$((kz^\sigma_{\bar{x}})_x, z^\sigma) = -(kz^\sigma_{\bar{x}}, z^\sigma_{\bar{x}}]. \tag{9b}$$

Next we introduce a function $\eta = \eta(x_i, t_j)$ such that

$$(k_{i-0.5}\eta)_{x,i} = \psi, \quad i = 1, 2, \cdots, N-1, \quad \eta_N = 0.$$

Now it follows that

$$(\psi, z^\sigma) = ((k\eta)_x, z^\sigma) = -(\eta, k{z_{\bar{x}}}^\sigma].$$

Using the Schwarz inequality gives

$$|(\psi, z^\sigma)| \leqslant \frac{1}{2}(||\eta||^2 + ||kz^\sigma_{\bar{x}}]|). \tag{9c}$$

Substituting (9) into (8) and summing over j yields the needed estimate. The uniqueness of the solution follows from the obtained stability inequality. □

3.Iterative method

We use the Schwarz alternating method to solve the system of discrete equations (see, also [4 – 6])

$$v(x)\overset{s}{y}{}^{\pm}_t = (k(\overset{s}{y}{}^\sigma_\pm)_{\bar{x}})_x - d\overset{s}{y}{}^\sigma_\pm + f, \qquad (x,t) \in \Omega_{h,\tau}, \tag{10a}$$

$$\overset{s}{y}{}^+(x_i, 0) = g_2(x_i), \quad \overset{s}{y}{}^-(x_i, T) = g_1(x_i), \quad x \in \omega_h^\pm \tag{10b}$$

$$\overset{s}{y}{}^+_M(t_j) = \overset{s-1}{y}{}^-_M(t_j), \quad \overset{s}{y}{}^+_N(t_j) = u_2(t_j), \quad t \in \omega_\tau, \tag{10c}$$

$$\overset{s}{y}{}^-_0(t_j) = u_1(t_j), \quad \overset{s}{y}{}^-_{M+1}(t_j) = \overset{s}{y}{}^+_{M+1}(t_j). \tag{10d}$$

The method of investigation, which is proposed in [4 – 6], is based on the maximum principle and it leads to the restriction on the parameters τ and h in the case $\sigma \neq 0.5$ (see the condition (7)). Our goal is to prove the convergence of the iterative method for $0 \leqslant \sigma < 0.5$ with less restrictive requirements on τ and h.

First we consider the difference scheme obtained by the method of lines

$$v(x)\frac{\partial \overset{s}{Y}{}^{\pm}}{\partial t} = (k\overset{s}{Y}{}^{\pm}_{\bar{x}})_x - d(x)\overset{s}{Y}{}^{\pm} + f(x,t), \quad x \in \omega_h^{\pm} \tag{11a}$$

$$\overset{s}{Y}{}^{+}(x_i,0) = g_2(x_i), \quad \overset{s}{Y}{}^{-}(x_i,T) = g_1(x_i), \quad x \in \omega_h^{\pm}, \tag{11b}$$

$$\overset{s}{Y}{}^{+}_M(t) = \overset{s-1}{Y}{}^{-}_M(t), \quad \overset{s}{Y}{}^{+}_N(t) = u_2(t), \quad t > 0, \tag{11c}$$

$$\overset{s}{Y}{}^{-}_0(t) = u_1(t), \qquad \overset{s}{Y}{}^{-}_{M+1}(t) = \overset{s}{Y}{}^{+}_{M+1}(t). \tag{11d}$$

We investigate the convergence of semidiscrete iterative method (11) by using a well–known technique based on upper solutions and the maximum principle.

Theorem 2. *Iterative method (11) converges linearly and the following error estimate is valid for its solution*

$$\| \overset{s}{Y}(t) - Y(t)\|_{C(\omega_h)} \leqslant q^s\|\overset{o}{Y} - Y\|_{C(\omega_h)}, \qquad q < 1. \tag{12}$$

Proof. We denote the error by $\overset{s}{z}(t) = \overset{s}{Y}(t) - Y(t)$. It satisfies the following boundary value problem

$$v(x)\frac{\partial \overset{s}{z}{}^{\pm}}{\partial t} = (k\overset{s}{z}{}^{\pm}_{\bar{x}})_x - d(x)\overset{s}{z}{}^{\pm}, \quad (x,t) \in \omega_h^{\pm} \times (0,T),$$
$$\overset{s}{z}{}^{+}(x_i,0) = 0, \quad \overset{s}{z}{}^{-}(x_i,T) = 0, \quad x \in \omega_h^{\pm},$$
$$\overset{s}{z}{}^{+}_M(t) = \overset{s-1}{z}{}^{-}_M(t), \quad \overset{s}{z}{}^{+}_N(t_j) = u_2(t), \quad t > 0,$$
$$\overset{s}{z}{}^{-}_0(t) = u_1(t), \qquad \overset{s}{z}{}^{-}_{M+1}(t) = \overset{s}{z}{}^{+}_{M+1}(t), \quad t > 0.$$

We choose a barrier function $\overset{s}{Z} = Y\|\overset{s-1}{z}{}^{-}_M\|_{C[0,T]}$, where Y is defined by

$$v(x)\frac{\partial Y^{\pm}}{\partial t} = (kY^{\pm}_{\bar{x}})_x - d(x)Y^{\pm}, \quad (x,t) \in \omega_h^{\pm} \times (0,T),$$
$$Y^{-}(x,T) = 0, \quad Y^{+}(x,0) = 0, \quad x \in \omega_h^{\pm},$$
$$Y^{+}_M(t) = 1, \quad Y^{+}_N(t) = 0, \quad t > 0,$$
$$Y^{-}_0(t) = 0, \qquad Y^{-}_{M+1}(t) = Y^{+}_{M+1}(t).$$

It follows from the maximum principle that $|\overset{s}{z}{}_M^-| \leqslant \overset{s}{Z}$. Hence we have error estimate (12) with $q = ||Y_M^-||_{C[0,T]}$. □

It is easily verified that the maximum principle is still valid for the solution of strictly implicit difference scheme (10), $\sigma = 0.5$. The convergence of the proposed iterative method can be proved analogously to the case of the semidiscrete scheme.

Theorem 3. *Iterative method (10), $\sigma = 0.5$ converges linearly and the following error estimate is valid for its solution*

$$|| \overset{s}{y} (t_j) - y(t_j)||_{C(\omega_h)} \leqslant q^s || \overset{o}{y} (t_j) - y(t_j)||_{C(\omega_h)}, \qquad q = ||Y_M^-||_{C(\omega_\tau)} < 1,$$

where a barrier function $Y^{\pm}(x_i, t_j)$ is defined by

$$\begin{aligned} &v(x)Y_t^+ = (k\hat{Y}_{\bar{x}}^+)_x - d(x)\hat{Y}^+, \quad (x,t) \in \Omega_{h,\tau}^+, \\ &v(x)Y_t^+ = (kY_{\bar{x}}^-)_x - d(x)Y^-, \quad (x,t) \in \Omega_{h,\tau}^-, \\ &Y^-(x,T) = 0, \quad Y^+(x,0) = 0 \quad x \in \omega_{h,\tau}^{\pm}, \\ &Y_M^+(t) = 1, \quad Y_N^+(t) = 0, \quad t \in \omega_\tau, \\ &Y_0^-(t) = 0, \qquad Y_{M+1}^-(t) = Y_{M+1}^+(t). \end{aligned} \tag{13}$$

The estimate of parameter q is obtained by solving two stationary problems

$$\begin{aligned} &-(kW_{\bar{x}})_x + dW = 0, \quad x \in \omega_h^{\pm}, \\ &W_M^+ = 1, \quad W_N^+ = 0, \\ &W_0^-(t) = 0, \qquad W_{M+1}^- = 1. \end{aligned} \tag{14}$$

Then it follows from the maximum principle that

$$q \leqslant q_1 = W_{M+1}^+ W_M^-.$$

We see that the convergence rate of the iterative method depends explicitly only on the space step h, and is independent of the time step τ.

It is easily verified that for model problem (3) with $k(x) = 1$, $d(x) = 0$ we have

$$q_1 = \left(\frac{1 - 0.5h}{1 + 0.5h}\right)^2 \approx 1 - 2h.$$

Hence the convergence rate of the iterative method decreases linearly for small values of h and the estimate of q is independent of the time interval [0,T]. The same result holds for general problem (1) with the coefficients satisfying (1d).

The situation is quite different in the case of the Fokker - Planck equation. Solving (14) gives $W^{\pm}(x_i) \equiv 1, \quad x \in \omega_h^{\pm}$; hence the barrier function Y defined by (13) estimates the error of the iterative method. We see that the convergence rate of the iterative method depends not only on the space step h, but also on the time interval [0,T].

Next we present some computational results. We only present the results for one fixed value of the time step $\tau = 0.1$, though computations were performed for different values of τ. In full accordance with Theorem 3 no significant dependence was established between the number of iterations and the time step τ.

Example 1. Consider model problem (3) with the coefficients

$$k(x) = 1, \qquad d(x) = 0, \qquad f(x,t) = 0, \qquad x \in [-1,1],$$
$$v(x) = 1, \qquad x > 0, \qquad v(x) = -1, \qquad x < 0, \qquad t > 0,$$
$$u_1(t) = 0, \qquad u_2(t) = 0, \qquad g_1(x) = 0, \qquad g_2(x) = 0,$$

so the exact solution is $u(x,t) = 0$. The initial condition for $\bar{y_M}(t)$ is taken to be

$$\overset{0}{\bar{y_M}}(t_j) = 1 - \delta_{j0} - \delta_{jk} - \delta_{jK}, \; k = (K+1)/2, \; t \in \omega_\tau,$$

where δ_{ij} is the Kronecker function. Table 1 gives the number of iterations carried out in each case for different values of space grid point numbers N and two fixed values of T, $T = 2$ and $T = 4$, respectively. In all computations the convergence criterion is taken as follows

$$||\overset{s}{\bar{y_M}}(t)||_{C(\omega_\tau)} \leqslant \varepsilon, \quad \varepsilon = 0.01.$$

Table 1. *The number of iterations for Example 1.*

N	21	41	81
T=2	22	44	90
T=4	23	45	89

Numerical results confirm that the convergence rate is linear with respect to the space step h and independent of T.

Example 2. Consider the Fokker - Planck problem with homogeneous initial and boundary conditions

$$u_1(t) = 0, \qquad u_2(t) = 0, \qquad g_1(x) = 0, \qquad g_2(x) = 0.$$

Just like in Example 1 the time step was chosen $\tau = 0.1$

Table 2. *The number of iterations for Example 2.*

N	21	41	81
T=2	109	220	439
T=4	300	599	1206

From the results in Table 2 it is clear that the number of iterations carried out for the convergence of the iterations depends not only on h, but also on T. These numerical results clearly illustrate that the estimates of Theorem 3 are good measures of the convergence rate.

4. Convergence of the iterative method for $\sigma \neq 0.5$

In this section we consider the iterative method (10) with $\sigma = 0$. Similar results can be proved for other values of the parameter $\sigma \neq 0.5$, but the symmetrical finite difference scheme is most interesting in applications. As mentioned above, the maximum principle yields the convergence only when the grid parameters τ and h satisfy inequality (7). In order to relax this condition we use the well - known convergence estimates for boundary - value parabolic difference schemes. We first define the following semidiscrete problem

$$
\begin{aligned}
& v(x)\frac{\partial \overset{s}{Y}}{\partial t} = (k\overset{s}{Y}_{\bar{x}})_x - dY, \quad (x,t) \in \omega_h^{\pm} \times (0,T), \qquad (15)\\
& \overset{s}{Y}(x_i,0) = 0, \quad x \in \omega_h^{+}, \qquad \overset{s}{Y}(x_i,T) = 0, \quad x \in \omega_h^{-},\\
& \overset{s}{Y}{}_M^{+}(t) = \overset{s-1}{Z}_M, \quad \overset{s}{Y}{}_N^{+}(t) = 0, \quad t > 0,\\
& \overset{s}{Y}{}_0^{-}(t) = 0, \qquad \overset{s}{Y}{}_{M+1}^{-}(t) = \overset{s}{Z}_{M+1}(t),
\end{aligned}
$$

where $\overset{s-1}{Z}_M(t)$ interpolates the error $\overset{s-1}{z}{}_M^{-}(t_j)$ of the iterative method (10), $\overset{s}{Z}_{M+1}(t)$ interpolates the error $\overset{s}{z}{}_{M+1}^{-}(t_j)$. We assume that these functions satisfy

$$||\overset{s-1}{Z}_M(t)||_{C[0,T]} \leqslant (1 + C_0\tau^p)||\overset{s-1}{z}{}_M^{-}(t_j)||_{C(\omega_\tau)},$$

$$||\overset{s}{Z}_{M+1}(t)||_{C[0,T]} \leqslant (1 + C_0\tau^p)||\overset{s-1}{z}{}_M^{+}(t_j)||_{C(\omega_\tau)},$$

where $p_0 \geqslant 1$. One can easily prove that

$$||\overset{s}{Y}{}_{M+1}^{+}(t)||_{C[0,T]} \leqslant W_{M+1}^{+}||\overset{s-1}{Z}_M(t)||_{C[0,T]}, \qquad (16)$$

$$||\overset{s}{Y}{}^{-}_{M}(t)||_{C[0,T]} \leqslant W^{-}_{M} ||\overset{s}{Z}_{M+1}(t)||_{C[0,T]},$$

where $W^{\pm}$ is the solution of stationary problem (14). Now we can prove

Theorem 4. *Let* $\tau \leqslant Ch^{1/\mu}$, $\mu = \min(p_0, 2)$. *Then iterative method (10) with* $\sigma = 0$ *converges linearly and the following error estimate is valid*

$$||\overset{s}{z}{}^{-}_{M}(t_j)||_{C(\omega_\tau)} \leqslant q_2 ||\overset{s-1}{z}{}^{-}_{M}(t_j)||_{C(\omega_\tau)}, \quad 0 < q_2 < 1.$$

P r o o f. Following the standard convergence analysis the accuracy estimate can easily be proved for the solution of symmetrical difference scheme (10) (see [2])

$$||\overset{s}{z} - \overset{s}{Y}||_{C(\Omega_{h,\tau})} \leqslant C_1 \tau^2 ||\overset{s}{Y}||_{C(\omega_h \times [0,T])}, \tag{17}$$

where $\overset{s}{z}$ is the error function of iterative method (10) and $\overset{s}{Y}$ is the solution of (15). Using the maximum principle and inequalities (16) in (17) gives

$$||\overset{s}{z}{}^{-}_{M}(t_j)||_{C(\omega_\tau)} \leqslant q_1 (1 + C_1\tau^2)^2 (1 + C_0 \tau^p)^2 ||\overset{s-1}{z}{}^{-}_{M}(t_j)||_{C(\omega_\tau)}. \tag{18}$$

We have proved in Theorem 3 that $q_1 = 1 - C_2 h$. Then for a sufficiently small $\tau \leqslant \tau_0$, $h \leqslant h_0$ it follows from (18) that

$$q_2 = q_1 (1 + C_1\tau^2)^2 (1 + C_0\tau^p)^2 \leqslant 1 - C_2 h + C_3(\tau^p + \tau^2).$$

For $2C_3\tau^\mu \leqslant 0.5C_2 h$ we have that $q_2 = 1 - 0.5C_2 h < 1$. The proof is completed. □

5. Relaxation method

In this section we consider a modification of iterative method (10). Now we denote the solution of (10) by $\overset{s-0.5}{y}{}^{-}_{M}(t_j)$. A new iteration $\overset{s}{y}{}^{-}_{M}$ is obtained from the formula

$$\overset{s}{y}{}^{-}_{M}(t_j) = (1 - \omega) \overset{s-1}{y}{}^{-}_{M}(t_j) + \omega \overset{s-0.5}{y}{}^{-}_{M}(t_j), \quad t \in \omega_\tau,$$

where ω is a relaxation parameter.

The problem of finding the optimal value of ω is very difficult since the operator of iterative method (10) is nonsymmetrical. We propose a method for determining a quasioptimal value of the relaxation parameter ω. The max norm of the error $\overset{s}{z}_M(t)$ is minimized in our considerations. It is easily verified that such a selection of the norm is not optimal, and the L_2 norm would be more accurate.

In order to find the value of ω for which the max norm of the error $\overset{s}{z}_M(t)$ is minimal we must investigate two different cases. We first consider the case when the error $\overset{s-0.5}{z}_M(t)$ changes minimally after implementing one iteration of method (10), then consider the opposite case when it changes maximally. In fact, the case of minimal change is investigated in the proof of Theorem 3 and we have that

$$q_{min} = ||Y_M^-||_{C(\omega_\tau)},$$

where $Y_M^-(t)$ is the solution of (13). Now we estimate the maximal possible change of the error after one iteration. The stability of finite difference scheme (4) with respect to boundary conditions is defined by the following barrier function

$$\begin{aligned} &|v(x)|\frac{Z}{\tau} - (kZ_{\bar{x}})_x + dZ = 0, \qquad x \in \omega_h^{\pm}, \\ &Z_0^- = 0, \quad Z_{M+1}^- = 1, \\ &Z_M^+ = 1, \qquad Z_N^+ = 0, \end{aligned}$$

and the stability with respect to initial conditions is defined by the function Y

$$\begin{aligned} &|v(x)|\frac{Y - W}{\tau} - (kY_{\bar{x}})_x + dY = 0, \qquad x \in \omega_h^{\pm}, \\ &Y_0^- = 0, \quad Y_N^- = 0, \\ &Y_M^+ = 0, \qquad Y_N^+ = 1, \end{aligned}$$

where $W(x)$ is the solution of (14). Then the maximal change of the error after one iteration can be estimated as

$$1 - q_{max} = c(1 - Z_{M+1}^+ Z_M^+) + Y_M^- + Y_{M+1}^+ Z_M^-,$$

where we have denoted $c = |\overset{s-0.5}{Z_M}(t_j)|$. The parameter $\omega = \omega(c)$ is defined from the equality

$$1 - (1 - q_{min})\omega = \omega(1 - q_{max}) - c.$$

After simple calculations we have that

$$\omega_{opt} = \min_{0 \leqslant c \leqslant 1} \omega(c) = \omega(0) = 1/(1 + Y_M^- + Y_{M+1}^+ Z_M^- - q_{min}).$$

Now we present some numerical results. We have solved the model problem stated in Example 1. In Table 3 the number of iterations is given for different values of ω, and for different values of N. The first row contains the number of iterations for $\omega = 1$, the second row contains the number of iterations for $\omega = \omega_{opt}$, the third row gives the number of iterations for the optimal value of ω^*. Approximately optimal values ω^* have been chosen by running a number of problems with different values of ω. In all computations we have fixed $\tau = h$.

Table 3. *The number of iterations for different values of* ω.

N	21	41	81
ω | IT	1.0 | 21	1.0 | 44	1.0 | 90
ω_0 | IT	1.98 | 10	2.67 | 15	3.61 | 23
ω^* | IT	2.5 | 8	3.2 | 12	5.0 | 18

We see that the quasioptimal value ω_{opt} is a sufficiently accurate approximation of ω^*.

The proposed method for determining ω becomes even more effective if the grid parameters τ and h satisfy the relation $\tau = h^p$, $p < 1$. Table 4 presents the values of ω_{opt} for two different sets of τ and h, $p = 1$ and $p = 2$, respectively.

Table 4. *The value of* ω_{opt} *for different* N *and* p.

N	21	51	201
p=1	1.98	2.93	5.47
p=2	2.95	5.80	16.13

References

1. J.L. Lions, *Quelques methodes de resolution des problemes aux limites non lineaires,* Dunold Gauthier – Villars, Paris, 1969.
2. A.A. Samarskij, *The theory of difference schemes,* Nauka, Moscow, 1983 (in Russian).
3. D. Stein, I. Bernstein, *Boundary value problem involving a simple Fokker – Planck equation*, Phys. Fluids, 19 (1976), pp. 811 – 814.
4. P.N. Vabishchevich, *Numerical solution of a boundary value problem for a parabolic equation with a varying time direction*, Zh. Vychisl. Mat. i Mat. Fiz., 32 (1992), pp. 434 – 442 (in Russian).
5. V. Vanaja, R. Kellogg, *Iterative methods for a forward – backward heat equation*, SIAM J. Numer. Anal., 27 (1990), pp. 622 – 635.
6. V. Vanaja. *Numerical solution of a simple Fokker – Planck equation*, Appl. Numer. Math., 9 (1992), pp. 533 – 540.

Institute of Mathematics and Informatics
Akademijos Str. 4,
2600 Vilnius, Lithuania
e-mail: R. Ciegis @ mii.lt

Ph CLÉMENT, R HAGMEIJER AND G SWEERS*

On a Dirichlet problem related to the invertibility of mappings arising in 2D grid generation

1 Introduction

The present paper deals with the invertibility of mappings that transform simply connected two-dimensional domains into a convex domain. The mapping is defined by a system of second order elliptic equations. These mappings are used to generate so called structured grids in the physical domain to solve Computational Fluid Dynamics (CFD) problems. These grids are generated by mapping a uniform rectangular mesh from a rectangle onto the physical domain. To enable a consistent discretization of the flow equations, it is necessary that the mesh in the physical domain be non-overlapping. Hence it is necessary that the mapping be invertible.

A typical example of 2D grid generation is illustrated by the diagram in Figure 1. The boundary conforming mesh around a 2D airfoil is obtained as the image of a uniform mesh in rectangle R under a mapping T.

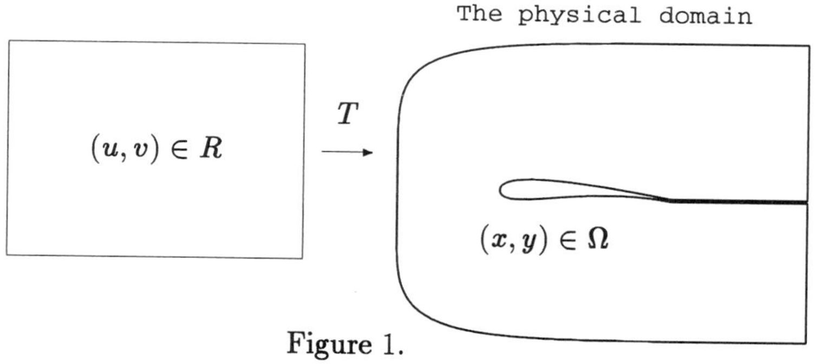

Figure 1.

*email: clement@twi.tudelft.nl hagm@nlr.nl sweers@twi.tudelft.nl

An elementary way to construct the basic mapping T is to define the parametric coordinates u and v as solutions of the Laplace equation in Ω:

$$\Delta u = 0 \qquad \Delta v = 0, \tag{1.1}$$

with $\Delta = \frac{\partial^2}{\partial x^2} + \frac{\partial^2}{\partial y^2}$ together with appropriate boundary conditions. The mapping T is then defined by

$$T^{inv} = (u, v)\,.$$

Mastin and Thompson considered such a problem in [14]. Winslow [21] replaced the Laplace equation (1.1) by isotropic diffusion equations

$$\nabla \cdot \frac{1}{w} \nabla u = 0 \qquad \nabla \cdot \frac{1}{w} \nabla v = 0, \tag{1.2}$$

with $\nabla = \left(\frac{\partial}{\partial x}, \frac{\partial}{\partial y}\right)$. The weight function $w(x, y)$ enables more direct control over the mesh spacing. .

The present paper deals with the mapping T that is defined by a system of two elliptic partial differential equations with Dirichlet boundary conditions. The physical domain, a simply connected two-dimensional domain, is the image under T of a convex domain. Existence, regularity and invertibility of the mapping (u, v) is established in Corollary 3.

An alternative way to enable mesh spacing control is to apply an additional mapping A, see [20] and [8]. The regularity of such an additional mapping is studied in earlier work of the present authors ([2]). That paper is concerned with a mapping T from the unit square onto itself that is defined by a similar elliptic system but with mixed Dirichlet and Neumann boundary conditions. Both problems are relevant for grid adaptation and generation problems.

The result in this paper depends strongly on a theorem of Carleman-Hartman-Wintner. This theorem is only true in two dimensional domains. In fact a straightforward generalization to more than two dimensional domains cannot be true. A counterexample to the proof of [15] for the three dimensional case can be found by using a special harmonic function due to Kellogg [12]. This function is shown in [2]. A direct counterexample can be found in [13].

2 Main result on smooth domains

Let the operator L be given by

$$L = a_{11}(x)\left(\frac{\partial}{\partial x_1}\right)^2 + a_{12}(x)\frac{\partial}{\partial x_2}\frac{\partial}{\partial x_1} + a_{22}(x)\left(\frac{\partial}{\partial x_2}\right)^2 + b_1(x)\frac{\partial}{\partial x_1} + b_2(x)\frac{\partial}{\partial x_2}, \tag{2.1}$$

where the coefficients satisfy for some $c > 0$ and $\gamma \in (0,1)$

$$\begin{aligned} &a_{ij} \in C^{0,1}(\bar{\Omega}) \quad 1 \leq i \leq j \leq 2, \\ &\textstyle\sum_{1\leq i\leq j\leq 2} a_{ij}\xi_i\xi_j \geq c\,|\xi|^2 \quad \text{on } \bar{\Omega}, \end{aligned} \tag{2.2}$$

$$b_i \in C^{\gamma}(\bar{\Omega}), \quad 1 \leq i \leq 2. \tag{2.3}$$

The problem is as follows. For appropriate boundary values find $(u, v) \in C(\bar{\Omega}) \cap C^2(\Omega)$, satisfying

$$\text{(a)} \left[\begin{array}{ll} Lu = 0 & \text{in } \Omega, \\ u = \varphi & \text{on } \partial\Omega, \end{array} \right. \quad \text{and} \quad \text{(b)} \left[\begin{array}{ll} Lv = 0 & \text{in } \Omega, \\ v = \psi & \text{on } \partial\Omega, \end{array} \right. \tag{2.4}$$

such that $(u, v) : \bar{\Omega} \to I\!R^2$ is injective.

The physical problem in general involves non smooth domains and in most cases the mesh is defined by mapping a rectangle to the physical domain. The singularities for (u, v) that occur because of corners come in two ways. For the physical domain having corners the elliptic p.d.e. has to be solved on a non smooth domain. Corners of the mesh domain, often a rectangle, that do not coincide with corners in the physical domain give rise to singularities of $\det(\nabla u, \nabla v)$. We will start with the case that both the physical and the mesh domain are smooth.

First we show the existence of an appropriate algebraic mapping between bounded, simply connected domains in $I\!R^2$ which have a Hölder smooth boundary.

Proposition 1 *Let Ω and Σ be Jordan domains in $I\!R^2$ with $\partial\Omega, \partial\Sigma \in C^{1,\gamma}$. Suppose h is a $C^{1,\gamma}$ diffeomorphism from $\partial\Omega$ onto $\partial\Sigma$ that preserves the orientation. Then there is an extension Φ of h such that*

$$\Phi \in C^{1,\gamma}\left(\bar{\Omega}; \bar{\Sigma}\right), \tag{2.5}$$

$$\Phi : \bar{\Omega} \to \bar{\Sigma} \textit{ is a bijection} \tag{2.6}$$

and

$$\det \begin{pmatrix} \Phi_{1,x_1} & \Phi_{1,x_2} \\ \Phi_{2,x_1} & \Phi_{2,x_2} \end{pmatrix} > 0 \quad \textit{on } \bar{\Omega}. \tag{2.7}$$

Remark 1.1 Let us recall some definitions. We denote

$$\mathbf{T} = \left\{ x \in I\!R^2; \|x\| = 1 \right\}, \tag{2.8}$$

$$\mathbf{D} = \left\{ x \in I\!R^2; \|x\| < 1 \right\}. \tag{2.9}$$

i. For the definition of a Jordan domain see the appendix. For open bounded set $\Omega \subset I\!R^2$ to be a Jordan domain, it is sufficient that there exists a injective function $\omega \in C(\mathbf{T})$ with $\partial\Omega = \omega(\mathbf{T})$.

ii. The boundary $\partial\Omega$ satisfies $\partial\Omega \in C^{1,\gamma}$, if there is a parameterization $\omega \in C^{1,\gamma}(\mathbf{T})$ of $\partial\Omega$ with $|\omega'| \neq 0$.

iii. The function h is a $C^{1,\gamma}$ diffeomorphism from $\partial\Omega$ onto $\partial\Sigma$ with $\partial\Omega, \partial\Sigma \in C^{1,\gamma}$, if $\tilde{h} = \sigma^{inv} \circ h \circ \omega \in C^{1,\gamma}(\mathbf{T};\mathbf{T})$ and $\left|\tilde{h}'\right| \neq 0$. Here ω (resp. σ) is a $C^{1,\gamma}$ parameterization of $\partial\Omega$ (resp. $\partial\Sigma$) as in ii.

Proof. By Carathéodory's extension of the Riemann Mapping Theorem (see page 18 of [16]) there exists a mapping $f_\Omega \in C(\bar{\mathbf{D}};\bar{\Omega})$ that is conformal from $\mathbf{D}$ onto Ω, with $f_\Omega : \mathbf{T} \to \partial\Omega$ injective. Conformal includes $f_\Omega : \mathbf{D} \to \Omega$ being injective. By Theorems of Kellogg-Warschawski (see page 48 and 49 of [16]) we find $f_\Omega \in C^{1,\gamma}(\bar{\mathbf{D}};\bar{\Omega})$ and $|f'_\Omega| \neq 0$ on $\bar{\mathbf{D}}$. Also a function f_Σ exists with similar properties. Hence we may restrict ourselves to the case that $\Omega = \Sigma = \mathbf{D}$. Let $\tilde{h} \in C^{1,\gamma}(\mathbf{T};\mathbf{T})$ be as in Remark 1.1.iii. If we have an appropriate $\hat{\Phi} : \bar{\mathbf{D}} \to \bar{\mathbf{D}}$, with $\hat{\Phi} = \tilde{h}$ on $\mathbf{T}$, then $\Phi := f_\Sigma \circ \hat{\Phi} \circ f_\Omega^{inv}$ will be an extension of h. The claims (2.5-2.6) will be immediate and, indeed, (2.7) follows from

$$\det(\nabla\Phi) = |f'_\Sigma|^2 \det\left(\nabla\hat{\Phi}\right) |f'_\Omega|^{-2}.$$

It remains to show that there exists such a function $\hat{\Phi}$.

For $\tilde{h} \in C^{1,\gamma}(\mathbf{T};\mathbf{T})$ as above, there exists a $C^{1,\gamma}$ function α, with the orientation preserving property implying $\alpha' > 0$, such that $\tilde{h}(\cos\varphi, \sin\varphi) = (\cos\alpha(\varphi), \sin\alpha(\varphi))$. Setting $\vartheta(r) = \exp(1 - r^{-1})$, we define an extension $\hat{\Phi} \in C^{1,\gamma}\left(\bar{\mathbf{D}};\bar{\mathbf{D}}\right)$ by

$$\hat{\Phi}\begin{pmatrix} r\cos\varphi \\ r\sin\varphi \end{pmatrix}^T = \begin{pmatrix} r\cos((1-\vartheta(r))\ \varphi + \vartheta(r)\ \alpha(\varphi)) \\ r\sin((1-\vartheta(r))\ \varphi + \vartheta(r)\ \alpha(\varphi)) \end{pmatrix}^T.$$

The function $\hat{\Phi} : r\mathbf{T} \to r\mathbf{T}$ is a bijection for every $r \geq 0$ and a direct computation shows that

$$\det\left(\nabla\hat{\Phi}\right) = (1-\vartheta(r)) + \vartheta(r)\ \alpha'(\varphi) > 0 \quad \text{on } \bar{\mathbf{D}}.$$

Notice that even if $\alpha' = 0$ somewhere we find $\det\left(\nabla\hat{\Phi}\right) > 0$ on $\mathbf{D}$. □

The function Φ that we obtain above can be used for the assumptions in the next theorem.

Theorem 2 *Let Ω be a simply connected domain in $I\!R^2$ with $\partial\Omega \in C^{1,\gamma}$ and let Σ be a bounded, open and convex set in $I\!R^2$. Let $\Phi \in C^{1,\gamma}(\bar{\Omega}; I\!R^2)$ be such that*

$$\Phi : \bar{\Omega} \to \bar{\Sigma} \text{ is a bijection,} \tag{2.10}$$

and

$$\det\begin{pmatrix} \Phi_{1,x_1} & \Phi_{1,x_2} \\ \Phi_{2,x_1} & \Phi_{2,x_2} \end{pmatrix} > 0 \quad on\ \bar{\Omega}. \tag{2.11}$$

Set $(\varphi, \psi) = \Phi_{|\partial\Omega}$. *Then problem* (2.4) *possesses exactly one solution* $u, v \in C^{1,\gamma}(\bar{\Omega}) \cap C^2(\Omega)$ *with*

$$(u, v) : \bar{\Omega} \to \bar{\Sigma} \text{ is a bijection} \tag{2.12}$$

and

$$\det\begin{pmatrix} u_{x_1} & u_{x_2} \\ v_{x_1} & v_{x_2} \end{pmatrix} > 0 \quad on\ \bar{\Omega}. \tag{2.13}$$

Corollary 3 *Let* Ω *and* Σ *be Jordan domains in* $\mathbb{R}^2$ *with* $\partial\Omega, \partial\Sigma \in C^{1,\gamma}$. *Suppose* h *is a* $C^{1,\gamma}$ *diffeomorphism from* $\partial\Omega$ *onto* $\partial\Sigma$ *that preserves the orientation. Set* $(\varphi, \psi) = h$.
Then problem (2.4) *possesses exactly one solution* $u, v \in C^{1,\gamma}(\bar{\Omega}) \cap C^2(\Omega)$, *and* (u, v) *satisfies* (2.12) *and* (2.13).

Remark 3.1 From (2.11) it follows that the $C^{1,\gamma}$ smoothness of $\partial\Omega$ is transferred to $\partial\Sigma$. That is, also Σ will have a $C^{1,\gamma}$ boundary. In fact (2.11) would also transfer corners of Ω to corners of Σ.

Remark 3.2 If Hölder smoothness is replaced by Dini smoothness the results in this section remain true. One may also replace $C^{1,\gamma}$ with $C^{k,\gamma}$ for $k > 1$.

Remark 3.3 A necessary condition on (φ, ψ) to find a function Φ that satisfies (2.10) and (2.11) is the following. The boundary $\partial\Omega$ is the union of four counterclockwise ordered closed curves $\Gamma_1, \Gamma_2, \Gamma_3, \Gamma_4$ such that

$$\begin{array}{lll} \varphi_\tau \geq 0, & \psi_\tau \geq 0 & \text{on } \Gamma_1 \\ \varphi_\tau \leq 0, & \psi_\tau \geq 0 & \text{on } \Gamma_2 \\ \varphi_\tau \leq 0, & \psi_\tau \leq 0 & \text{on } \Gamma_3 \\ \varphi_\tau \geq 0, & \psi_\tau \leq 0 & \text{on } \Gamma_4 \end{array} \tag{2.14}$$

and

$$\varphi_\tau^2 + \psi_\tau^2 > 0 \quad \text{on } \partial\Omega. \tag{2.15}$$

Here τ denotes the counterclockwise tangential direction. These conditions however do not imply that Σ is convex. See Remark 5.3. Assuming convexity and regularity the conditions are sufficient for Proposition 1 and hence for Theorem 2.

Proof of Theorem 2: The proof will be done in several steps.

i. *Existence:* By Theorem 6.13 of [6] there exist unique solutions u and v in $C^0(\bar{\Omega}) \cap C^{2,\gamma}(\Omega)$. A theorem in [5] (see also page 111 of [6]) yields $u, v \in C^{1,\gamma}(\bar{\Omega})$. Let us denote

$$F(x) = (u(x), v(x)) \quad \text{for } x \in \bar{\Omega}. \tag{2.16}$$

ii. *Σ contains $F(\Omega)$:* By assumption we have $F(\partial\Omega) = \Phi(\partial\Omega) = \partial\Sigma$. We will use two Theorems from the appendix for convex domains that use closed half spaces. Every closed half space in $I\!R^2$ can be written as

$$S = \left\{ y \in I\!R^2; w \cdot y \geq a \right\} \tag{2.17}$$

with some $w \in I\!R^2 \backslash \{0\}$ and $a \in I\!R$.

Let S, as in (2.17), be a closed half space containing $\bar{\Sigma}$. For $x \in \partial\Omega$ we have $F(x) = \Phi(x) \in \bar{\Sigma}$ and hence

$$w \cdot F(x) \geq a. \tag{2.18}$$

We also have

$$L(w \cdot F(x)) = w \cdot (Lu(x), Lv(x)) = 0 \quad \text{for } x \in \Omega.$$

Since $w \cdot F(\cdot) \not\equiv a$ on $\partial\Omega$ the strong maximum principle implies that

$$w \cdot F(x) > a \quad \text{for all } x \in \Omega. \tag{2.19}$$

Hence $F(\Omega) \subset S$. Since it holds for all appropriate S, Theorem B yields $F(\Omega) \subset \text{co}\left(\bar{\Sigma}\right) = \bar{\Sigma}$. Now suppose there is $x^* \in \Omega$ such that $F(x^*) \in \partial\Sigma$. We use Theorem A with $A = \bar{\Sigma}$ and $y = F(x^*)$ to get to a contradiction. By this theorem there is a closed half space S such that $F(x^*) \in \partial S$ and $\bar{\Sigma} \subset S$. By (2.19) one finds $F(x^*) \in S \backslash \partial S$, a contradiction. Hence $F(\Omega) \subset \Sigma$ holds.

iii. *Σ equals $F(\Omega)$:* Since Σ is convex it is a Jordan domain. Then Theorem D.*i* (see the appendix) can be applied to show that $F(\Omega) = \Sigma$.

iv. *The Jacobian is positive on the boundary of Ω:* We show that

$$\det(\mathcal{J}_F) > 0 \quad \text{on } \partial\Omega, \tag{2.20}$$

where we denote

$$\mathcal{J}_F(x) = \begin{pmatrix} u_{x_1}(x) & u_{x_2}(x) \\ v_{x_1}(x) & v_{x_2}(x) \end{pmatrix}.$$

By assumption we have $\det(\mathcal{J}_\Phi) > 0$ on Ω and hence on $\partial\Omega$. Although $F = \Phi$ on $\partial\Omega$ it does not straightforwardly imply that $\det(\mathcal{J}_F) > 0$ on $\partial\Omega$. Nevertheless, the result in (2.20) is true. Indeed, fix $y \in \partial\Omega$. From $F = \Phi$ on $\partial\Omega$ one deduces that

$$\frac{\partial}{\partial\tau} u = \frac{\partial}{\partial\tau} \Phi_1 \text{ and } \frac{\partial}{\partial\tau} v = \frac{\partial}{\partial\tau} \Phi_2,$$

where

- τ denotes the (counterclockwise) tangential direction of $\partial\Omega$ at y.

We will also use:

- n for the interior normal direction of $\partial\Omega$ at y,
- ζ for the (counterclockwise) tangential direction of $\partial\Sigma$ at $\Phi(y)$,
- η for the interior normal direction of $\partial\Sigma$ at $\Phi(y)$.

Notice that $\tau_1 = n_2$ and $\tau_2 = -n_1$. Since Σ is convex, we have

$$\langle \eta, \Phi(x) - \Phi(y) \rangle > 0 \quad \text{for all } x \in \Omega, \tag{2.21}$$

and hence

$$\langle \eta, \mathcal{J}_\Phi(y)\, n \rangle = \frac{\partial}{\partial n} \langle \eta, \Phi(y) \rangle \geq 0. \tag{2.22}$$

Let $y(t)$ be a parameterization of $\partial\Omega$ near $y = y(0)$ with $y'(t) \neq 0$. We may assume that such a parameterization exists since $\partial\Omega$ is $C^{1,\gamma}$ near y. Then there is $c_1 \neq 0$ such that we have

$$\tau = c_1\, y'(0).$$

Since $\Phi(y(t))$ parameterizes $\partial\Sigma$ near $\Phi(y)$ and $\frac{d}{dt}(\Phi(y(t))) = \mathcal{J}_\Phi(y(t))\; y'(t)$ with $\det(\mathcal{J}_\Phi) \neq 0$ we find for some $c_2 \neq 0$ that

$$\zeta = c_2 \mathcal{J}_\Phi(y)\, \tau.$$

Hence it follows that

$$\langle \eta, \mathcal{J}_\Phi(y)\, \tau \rangle = c_2^{-1} \langle \eta, \zeta \rangle = 0, \tag{2.23}$$

and since $\det(\mathcal{J}_\Phi) \neq 0$ it follows then that

$$\langle \eta, \mathcal{J}_\Phi(y)\, n \rangle \neq 0,$$

and hence together with (2.22) that

$$\langle \eta, \mathcal{J}_\Phi(y)\, n \rangle > 0. \tag{2.24}$$

Now we will derive similar results for F. Since Σ is convex we have that

$$\langle \eta, z - F(y) \rangle > 0 \quad \text{for all } z \in \Sigma.$$

Since $F(\Omega) \subset \Sigma$ we find

$$\langle \eta, F(x) - F(y) \rangle > 0 \quad \text{for all } x \in \Omega. \tag{2.25}$$

From (2.25) and the fact that $L\langle\eta, F(\cdot) - F(y)\rangle = 0$ in Ω, it follows by Hopf's boundary point lemma that

$$\langle\eta, \mathcal{J}_F(y)\ n\rangle = \frac{\partial}{\partial n}\langle\eta, F(y)\rangle > 0. \tag{2.26}$$

From the boundary conditions and the assumption that Ω is smooth near y it follows that

$$\langle\eta, \mathcal{J}_F(y)\ \tau\rangle = \frac{\partial}{\partial\tau}\langle\eta, F(y)\rangle = 0. \tag{2.27}$$

Rewrite (2.26-2.27) as

$$\langle\eta, (u_n(y), v_n(y))\rangle > 0 \tag{2.28}$$

and

$$\langle\eta, (u_\tau(y), v_\tau(y))\rangle = 0 \tag{2.29}$$

Set $\xi_1 = \frac{\partial}{\partial n}\Phi_2(y)$ and $\xi_2 = -\frac{\partial}{\partial n}\Phi_1(y)$. We obtain by (2.24) that $\xi_1\eta_2 - \xi_2\eta_1 > 0$ and with (2.29) that

$$\det(\mathcal{J}_F) = (\xi_1\eta_2 - \xi_2\eta_1)^{-1}\det\left(\begin{pmatrix}\xi_1 & \xi_2\\ \eta_1 & \eta_2\end{pmatrix}\begin{pmatrix}u_\tau & u_n\\ v_\tau & v_n\end{pmatrix}\right) =$$

$$= (\xi_1\eta_2 - \xi_2\eta_1)^{-1}\det\begin{pmatrix}\xi_1 u_\tau + \xi_2 v_\tau & \xi_1 u_n + \xi_2 v_n\\ \eta_1 u_\tau + \eta_2 v_\tau & \eta_1 u_n + \eta_2 v_n\end{pmatrix} =$$

$$= (\xi_1\eta_2 - \xi_2\eta_1)^{-1}\det\begin{pmatrix}\xi_1 u_\tau + \xi_2 v_\tau & \xi_1 u_n + \xi_2 v_n\\ 0 & \eta_1 u_n + \eta_2 v_n\end{pmatrix} =$$

$$= (\xi_1\eta_2 - \xi_2\eta_1)^{-1}(\xi_1 u_\tau + \xi_2 v_\tau)(\eta_1 u_n + \eta_2 v_n)$$

Since

$$\xi_1 u_\tau + \xi_2 v_\tau = \xi_1\Phi_{1,\tau} + \xi_2\Phi_{2,\tau} = \det(\mathcal{J}_\Phi) > 0$$

we find, with (2.28), that

$$\det(\mathcal{J}_F) > 0 \quad \text{on } \partial\Omega. \tag{2.30}$$

v. *The global degrees related with the gradients are zero:* Let $\binom{\alpha}{\beta} \in \mathbb{R}^2\setminus\{0\}$ and define

$$\phi(x) = \alpha u(x) + \beta v(x) \quad \text{for } x \in \partial\Omega.$$

We will show that

$$\deg(\nabla\phi, \Omega) = 0.$$

Because of (2.30) one finds $\nabla\phi \neq 0$ on $\partial\Omega$, and hence that the degree is well defined. Since this holds for all $\binom{\alpha}{\beta} \in \mathbb{R}^2\setminus\{0\}$ we may define a homotopy

$h_1(t,x)$ with $h_1(0,x) = \nabla\phi(x)$ and $h_1(1,x) = \nabla u(x)$ and such that $h(t,x) \neq 0$ on $\partial\Omega$. Hence

$$\deg(\nabla\phi,\Omega) = \deg(\nabla u,\Omega).$$

As before notice that $\frac{\partial}{\partial\tau}\Phi_1 = \frac{\partial}{\partial\tau}u$ and moreover, if $\frac{\partial}{\partial\tau}\Phi_1 = 0$, then $\frac{\partial}{\partial n}\Phi_1$ and $\frac{\partial}{\partial n}u$ are both nonzero. They even have the same sign since both $F(\Omega) \subset \Sigma$ and $\Phi(\Omega) \subset \Sigma$ hold. It shows that

$$t\begin{pmatrix}\Phi_{1,\tau}\\ \Phi_{1,n}\end{pmatrix} + (1-t)\begin{pmatrix}u_\tau\\ u_n\end{pmatrix} \neq 0 \quad \text{for all } t \in [0,1] \text{ and } x \in \partial\Omega.$$

Hence we may use the homotopy $h_2(t,x) = t\nabla\Phi_1 + (1-t)\nabla u$ and we find that

$$\deg(\nabla u,\Omega) = \deg(\nabla\Phi_1,\Omega).$$

Since $\nabla\Phi_1 \neq 0$ on $\bar{\Omega}$ we have $\deg(\nabla\Phi_1,\Omega) = 0$.

vi. *The Jacobian is positive inside Ω:* It remains to prove that

$$\det(\mathcal{J}_F) > 0 \quad \text{on } \bar{\Omega}. \tag{2.31}$$

Indeed, if (2.31) holds then F on $\bar{\Omega}$ is locally injective and Theorem D.*ii* yields that $F : \bar{\Omega} \to \bar{\Sigma}$ is a bijection.
We will show (2.31) by a contradiction argument. Suppose that $\det(\mathcal{J}_F(\bar{x})) = 0$ for some $\bar{x} \in \Omega$. Then there are $(\alpha,\beta) \neq (0,0)$ such that $\alpha\nabla u(\bar{x}) + \beta\nabla v(\bar{x}) = 0$. Set

$$\phi(x) = \alpha u(x) + \beta v(x) \quad \text{for } x \in \Omega.$$

As a consequence of Theorem C, we find that the zeros of $\nabla\phi$ are isolated, and that the local degree at a zero of $\nabla\phi$ is negative. That is, if $\nabla\phi(a) = 0$ there is $\varepsilon > 0$ such that $\nabla\phi \neq 0$ on $\partial B_\varepsilon(a) \subset \Omega$ and $\deg(\nabla\phi, B_\varepsilon(a)) < 0$. The additivity property of the degree shows that $\deg(\nabla\phi,\Omega) < 0$. Since we already showed that this degree is zero we have a contradiction. □

3 Domains with corners

We restrict ourselves to domains Ω and Σ with Lipschitz boundary consisting of finitely many sufficiently differentiable curves and finitely many corners. Problems involving corners of these domains can be roughly distinguished into four different types. For the sake of simplicity we will leave out the case that on the boundary of Ω or Σ two curves meet in a C^1 way (angle equals π). Notice that the convexity of Σ implies that its corners are convex.

I. The case that the corners (all being convex) of Ω are mapped to the corners of Σ. Then it is still possible to have $\Phi \in C^{1,\gamma}(\bar{\Omega}; I\!R^2)$ mapping Ω onto Σ with $\det(\nabla\Phi) > 0$ on $\bar{\Omega}$.

II. The boundary near a convex corner x^* of Ω is mapped to a smooth part of $\partial\Sigma$. Then there will be two possibilities. Either one has $\Phi \in C^1(\bar{\Omega}; I\!R^2)$ and $\det(\nabla\Phi(x)) \to 0$ when $x \to x^*$, or $\det(\nabla\Phi(x))$ is bounded on Ω near x^* but $\Phi \notin C^1(\bar{\Omega}; I\!R^2)$.

III. The boundary near a concave corner x^* of Ω is mapped to a smooth part of $\partial\Sigma$. Then there will not be a $\Phi \in C^1(\bar{\Omega}; I\!R^2)$ that maps Ω to Σ. However it is possible to have $\Phi \in C^0(\bar{\Omega}; I\!R^2) \cap C^1(\Omega; I\!R^2)$ with $\det(\nabla\Phi)$ bounded on Ω near x^*.

IV. A smooth part of $\partial\Omega$ is mapped on a neighborhood of a corner on $\partial\Sigma$. Similar features appear as in III.

In the next proposition and theorem we consider the first case. The proposition shows an algebraically defined regular mapping from Ω onto a rectangle.

Proposition 4 *Let Ω be a Jordan domain such that $\partial\Omega$ consists of four $C^{k,\gamma}$-curves, with $k \geq 1$, that are joined by strictly convex corners (the angles α_i are in $(0,\pi)$). Say $\partial\Omega = \Gamma_1 \cup \Gamma_2 \cup \Gamma_3 \cup \Gamma_4$ counterclockwise oriented with $\Gamma_i = \gamma_i([0,1])$ and the corners will be at $\gamma_i(1) = \gamma_{i+1}(0)$ (i mod 4). Moreover assume $\varphi, \psi : \partial\Omega \to I\!R^2$ satisfy $\varphi_{|\Gamma_i}, \psi_{|\Gamma_i} \in C^{k,\gamma}$,*

$$\begin{aligned} \varphi_\tau > 0 \quad &\text{and} \quad \psi_\tau = 0 \quad \text{on } \Gamma_1, \\ \varphi_\tau = 0 \quad &\text{and} \quad \psi_\tau > 0 \quad \text{on } \Gamma_2, \\ \varphi_\tau < 0 \quad &\text{and} \quad \psi_\tau = 0 \quad \text{on } \Gamma_3, \\ \varphi_\tau = 0 \quad &\text{and} \quad \psi_\tau < 0 \quad \text{on } \Gamma_4, \end{aligned}$$

and that $(\varphi,\psi)(\partial\Omega) = \partial R$ for some open rectangle R. Then there is an extension Φ of (ϕ,ψ) such that

$$\Phi \in C^{k,\gamma}\left(\bar{\Omega}; \bar{R}\right), \tag{3.1}$$

$$\Phi : \bar{\Omega} \to \bar{R} \text{ is a bijection} \tag{3.2}$$

and

$$\det\begin{pmatrix} \Phi_{1,x_1} & \Phi_{1,x_2} \\ \Phi_{2,x_1} & \Phi_{2,x_2} \end{pmatrix} > 0 \quad \text{on } \bar{\Omega}. \tag{3.3}$$

Remark 4.1 If $\Gamma_i = \gamma_i([0,1])$ we mean by $\psi_\tau > 0$ on Γ_i that

$$\begin{cases} \psi_{\tau+} > 0 & \text{on } \gamma_i([0,1)), \\ \psi_{\tau-} > 0 & \text{on } \gamma_i((0,1]), \end{cases}$$

where τ^+ (τ^-) is the upper (lower) counterclockwise tangential direction.

Proof. We start with a series of rather technical transformations and fix a corner at $(0,0)$ with $\gamma'_{i+1}(0) = (1,0)$. By $\vartheta = \vartheta(x^2+y^2)$ we denote a positive C^∞-function with $\vartheta(r) = 1$ for $r < \varepsilon$ and $\vartheta(r) = 0$ for $r > 2\varepsilon$ where ε is small enough. It will be used to construct transformations that act locally involving just one corner. We restrict ourselves to the corner at $(0,0) = \gamma_1(1) = \gamma_2(0)$.

i. By transformations of the type $(x,y) \mapsto (x + c\vartheta\ y, y)$ one may enlarge or reduce any strictly convex $(0 < \alpha < \pi)$ corner at 0 in a regular way to a corner with angle $\frac{1}{2}\pi$. Hence we may suppose that at the corner $\gamma'_1(1) = (0,-1)$ and $\gamma'_2(0) = (1,0)$.
Higher order derivatives of the second component $\left(\gamma_2^{(m)}(0)\right)_2$, $2 \le m \le k$, can be set to 0 by $(x,y) \mapsto (x + c\vartheta\ y^2 P_1(y), y)$, where P_1 is a polynomial. In a similar way one takes care of $\left(\gamma_1^{(m)}(1)\right)_1$.

ii. By a transformation $(x,y) \mapsto ((1 + c_1\ \vartheta)\, x, y)$ we may assume that $\psi_{\tau+} = 1$ at $(0,0)$, and $(x,y) \mapsto ((1 + \vartheta\ x P_2(x))\, x, y)$ is used to obtain $\left(\frac{\partial}{\partial\tau+}\right)^m \psi = 0$ at $(0,0)$. Similarly one takes care of φ.

iii. The transformation $(x,y) \mapsto (1-\vartheta)(x,y) + \vartheta\ (x^2 - y^2, 2xy)$ stretches the corner such that in the new parameterization we find $\gamma'_1(1) = \gamma'_2(0) = (1,0)$. This mapping is no longer regular but its singularity is explicit. As a result we find that the transformed Ω (let's call it Ω^*) has a $C^{k,\gamma}$ boundary and moreover, if ℓ denotes the parametrization by arclength of $\partial\Omega^*$ with $\ell(0) = (0,0)$, we find

$$\begin{aligned} \varphi(\ell(t)) - \varphi(0,0) &= -\left(-t + \mathcal{O}\left(|t|^{k+\gamma}\right)\right)^2 && \text{for } t \le 0, \\ \psi(\ell(t)) - \psi(0,0) &= \left(t + \mathcal{O}\left(t^{k+\gamma}\right)\right)^2 && \text{for } t \ge 0. \end{aligned} \tag{3.4}$$

Writing T_Ω for these combined transformations we find that the determinant satisfies in the new coordinates

$$\det(\nabla T_\Omega) = \sqrt{x^2+y^2}\ \left(4 + \mathcal{O}(1)\right).$$

In a similar way we transform the rectangle R to a $C^{k,\gamma}$-domain (even C^∞) R^*. For the rectangle the 'boundary' conditions one starts with are $(\varphi_R, \psi_R) = Id$. After stretching we obtain a formula as in (3.4) for φ_R, ψ_R near $(0,0)$.

We continue as in the proof of Proposition 1. The Kellogg-Warschawski extension of the Riemann Mapping Theorem yields the existence of $f_{\Omega^*} \in C^{k,\gamma}\left(\bar{\mathbf{D}}; \bar{\Omega}^*\right)$ that is conformal inside $\mathbf{D}$. See (2.9) for $\mathbf{D}$. Similar results hold for f_{R^*}. In order to show that the function α determined by $\tilde{h} : \mathbf{T} \to \mathbf{T}$ is $C^{k,\gamma}$ and still satisfies $0 < \alpha' < \infty$ we use (3.4) both for (φ, ψ) and (φ_R, ψ_R). From $\alpha \in C^{k,\gamma}$ it follows that $\hat{\tilde{\Phi}} \in C^{k,\gamma}$. Finally we have

$$\Phi = T_R \circ f_{R^*} \circ \hat{\tilde{\Phi}} \circ f_{\Omega^*}^{inv} \circ T_\Omega^{inv}.$$

Since the singularities in the determinants of T_R and T_Ω^{inv} cancel, we find that

$$0 < \det(\nabla\Phi) < \infty \quad \text{on } \bar{\Omega}.$$

□

Theorem 5 *Let Ω be a simply connected domain in $I\!R^2$, with Lipschitz boundary consisting of finitely many C^2-curves and finitely many corners. Let Σ be convex and let $\Phi \in C^3(\bar{\Omega}; I\!R^2)$ satisfy* (2.10) *and* (2.11). *Again set $(\varphi, \psi) = \Phi_{|\partial\Omega}$.*
Then problem (2.4) *possesses exactly one solution $u, v \in C^{1,\gamma}(\bar{\Omega}) \cap C^{2,\gamma}(\Omega)$ and* (2.12), (2.13) *hold.*

Remark 5.1 The boundary is $C^{1,\gamma}$ except for finitely points, and Lipschitz. The assumption implies that Ω is similar to a convex domain in the sense of Kadlec [11]. Due to a result of Kadlec [11] solving (2.4) on a domain with C^2-smooth 'convex' corners one still has $C^{1,\gamma}(\bar{\Omega})$-solutions. To apply his result we have to assume that $\Phi \in C^3(\bar{\Omega}; I\!R^2)$. Assuming $\Phi \in C^2(\bar{\Omega}; I\!R^2)$ gives $u, v \in C^{0,\gamma}(\bar{\Omega}) \cap C^{2,\gamma}(\Omega)$. Near a 'concave' corner the derivatives of a solution will become unbounded in general.

Remark 5.2 A similar remark as in Remark 3.3 can be made. If $\partial\Omega$ is non smooth (having finitely many corners) the condition on for example Γ_1 has to be replaced by

$$\begin{cases} \varphi_{\tau^-} \geq 0, \quad \psi_{\tau^-} \geq 0 \quad \text{on } \gamma_1(0,1], \\ \varphi_{\tau^+} \geq 0, \quad \psi_{\tau^+} \geq 0 \quad \text{on } \gamma_1[0,1), \end{cases} \tag{3.5}$$

where Γ_1 is parameterized by $\gamma_1 : [0,1] \to I\!R^2$, and $\tau^\pm$ are the upper/lower counter-clockwise tangential direction.

Remark 5.3 The conditions in (2.14-2.15) do not imply convexity of Σ. For example (see Figure 2.) the boundary of the square $[-1,1]^2$ is mapped in a non regular way by the following boundary conditions φ and ψ:

$$\begin{pmatrix} \varphi \\ \psi \end{pmatrix} = \begin{pmatrix} -yx^2 + 2x \\ yx^2 + 2x \end{pmatrix} \quad \text{for } (x,y) \in \partial\left([-1,1]^2\right).$$

A straightforward computation shows that the conditions in (2.14-2.15) in the sense of (3.5) however are satisfied.

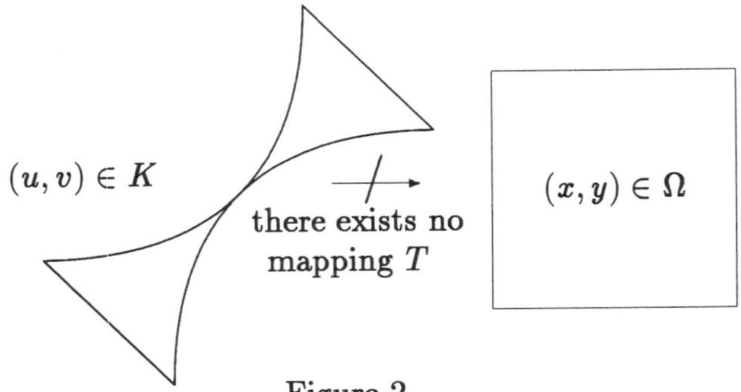

Figure 2.

Remark 5.4 An approach that takes care of the corners in cases II, III and IV is the following. Define an explicit intermediate mapping Ψ on Ω that stretches the concave corners (and possible some of the convex corners) and apply the mapping defined by the differential equations on $\Psi(\Omega)$. The singularities of the mapping $F \circ \Psi : \Omega \to \Sigma$ will only come from the explicitly known singularities of Ψ.

An intermediate mapping Ψ from Ω onto a square is used by Hagmeijer in [8]. His motivation is based upon the availability of such a mapping within grid adaptation problems and considering adaptation as a modification of an existing mapping instead of constructing a new mapping. Getting rid of possible singularities in solving the differential equations gives a second motivation. Although the approach in [8] uses different boundary conditions, no singularities appear in the solution of the differential equation. See [2].

The proof of Theorem 5: We will only mention the parts that differ from the proof of Theorem 2.

i. *Existence:* By Theorem 6.24 of [6] there are solutions $u, v \in C^0(\bar{\Omega}) \cap C^{1,\gamma}(\Omega)$. Denoting the corner points of Ω by K it can be used to find $u, v \in C^{1,\gamma}(\bar{\Omega}\backslash K)$. See also [5]. Since the corners are convex a result of Kadlec [11] implies that $u - \Phi_1$, $\frac{\partial}{\partial x_1}(u - \Phi_1)$ and $\frac{\partial}{\partial x_2}(u - \Phi_1)$ are in $W^{2,2}(\Omega)$. Since $W^{3,2}(\Omega)$ is imbedded in $C^{1,\gamma}(\bar{\Omega})$ for $C^{0,1}$-domains (see page 144 of [1]) we find $u, v \in C^{1,\gamma}(\bar{\Omega})$.

iv. *The Jacobian is positive on the boundary of Ω:* For all boundary points on the smooth part of $\partial\Omega$ one shows the result as before. Let x^* be a boundary point where $\partial\Omega$ has an angle and let τ^-, τ^+ denote the upper and lower tangential direction of the boundary at x^*. Since $\tau^- \neq \tau^+$ and since $\Phi_{1,\tau^\pm} = u_{\tau^\pm}$ respectively $\Phi_{2,\tau^\pm} = v_{\tau^\pm}$ we have $\mathcal{J}_F = \mathcal{J}_\Phi$ at x^*.

v. *The global degrees related with the gradients are zero:* Similarly as before one makes the homotopy for the smooth part. At a corner point $\nabla u = \nabla \Phi_1$. □

4 Appendix, some auxiliary results

Theorem A (see Theorem 8 of [4]) *Let $A \subset I\!R^n$ be convex. Then for every $y \in \partial A$ there is a closed half space S, with $y \in \partial S$ and $A \subset S$.*

Theorem B (see Theorem 11 of [4]) *Let $B \subset I\!R^n$ be bounded. Then* co$(\bar{B})$, *the convex hull of the closure of B, is the intersection of all the closed half-spaces that contain B.*

Theorem C (Corollary of Schulz's [18] version of a theorem by Carleman-Hartman-Wintner, see [2]) *Let $\phi \in W^{2,p}(\Omega)$, with $p > 2$, satisfy*

$$\left(a_{11}\left(\frac{\partial}{\partial x_1}\right)^2 + a_{12}\frac{\partial^2}{\partial x_1 \partial x_2} + a_{22}\left(\frac{\partial}{\partial x_2}\right)^2 + b_1\frac{\partial}{\partial x_1} + b_2\frac{\partial}{\partial x_2}\right)\phi = 0 \quad in\ \Omega,$$

with $a_{ij}, b_i \in C^{0,1}(\Omega)$ and such that for some $c > 0$ we have $\sum a_{ij}\xi_i\xi_j \geq c|\xi|^2$ on Ω for all $\xi \in I\!R^2$. If $x^* \in \Omega$ is such that $\nabla\phi(x^*) = 0$, then there exists $r > 0$ such that $\overline{B_r(x^*)} \subset \Omega$ and either

$$\nabla\phi \equiv 0 \quad on\ \overline{B_r(x^*)}$$

or

$$\begin{cases} \nabla\phi \neq 0 & on\ \overline{B_r(x^*)} \setminus \{x^*\}, \\ \deg(\nabla\phi, B_r(x^*)) < 0. \end{cases}$$

$B_r(x^*) = \left\{x \in I\!R^2; \|x - x^*\| < r\right\}.$

Remark: There is no equivalent of Theorem C in dimensions $n \geq 3$. As a consequence one cannot generalize the proofs in this paper in order to obtain a version of Theorem 1 in higher dimensions.

Before stating the last theorem we will recall a definition.

Definition *A set Ω in $I\!R^n$ is called a Jordan domain if there exists $h : \overline{B_1(0)} \subset I\!R^n \to I\!R^n$ that is continuous, injective and such that $\Omega = h(B_1(0))$.*

Remark: A Jordan domain Ω is open, connected and $\partial\Omega = h(\partial B_1(0))$. For $n = 2$ it is known that the inside of a closed Jordan curve is a Jordan domain (see page 81 of [7]).

Theorem D *Let $\Omega, \Sigma \subset I\!R^n$ both be Jordan domains and let $F \in C\left(\bar{\Omega}; I\!R^n\right)$. Suppose that $F : \partial\Omega \to \partial\Sigma$ is bijective and that $F(\Omega) \subset \Sigma$. Then*

i. $F : \Omega \to \Sigma$ is surjective.

ii. If moreover $F : \Omega \to \Sigma$ is a locally injective, then $F : \Omega \to \Sigma$ is injective.

Proof of Theorem D *i.*: Let h_Ω (resp. h_Σ) be an injective continuous map from $\overline{B_1(0)}$ to $\bar{\Omega}$ (resp. $\bar{\Sigma}$) such as in the definition of a Jordan domain. We consider the continuous map

$$\tilde{F} := h_\Sigma^{inv} \circ F \circ h_\Omega : \overline{B_1(0)} \to \overline{B_1(0)}.$$

The map $\tilde{F} : \partial B_1(0) \to \partial B_1(0)$ is a bijection and $\tilde{F}(B_1(0)) \subset B_1(0)$. It remains to show that $\tilde{F}(B_1(0)) \supset B_1(0)$. To prove the last inclusion it is sufficient that for every $z \in B_1(0)$ the Brouwer degree $\deg\left(\tilde{F}(\cdot) - z, B_1(0)\right)$ is well defined and not equal zero.

Since $\left\|\tilde{F}(x)\right\| = 1$ for $x \in \partial B_1(0)$ we find that $\tilde{F} - z \neq 0$ on $\partial B_1(0)$, hence $\deg\left(\tilde{F}(\cdot) - z, B_1(0)\right)$ is well defined. By considering the homotopy

$$H(t,\cdot) = \tilde{F}(\cdot) - t\,z \quad \text{for } t \in [0,1]$$

we obtain that

$$\deg\left(\tilde{F}(\cdot) - z, B_1(0)\right) = \deg\left(\tilde{F}(\cdot), B_1(0)\right).$$

Since $\tilde{F} : \partial B_1(0) \to \partial B_1(0)$ is a bijection , it follows from the multiplicative property of the degree that

$$\deg\left(\tilde{F}(\cdot), B_1(0)\right) = \pm 1. \tag{4.1}$$

Indeed, for an open bounded set $A \ni 0$ with a continuous injection $J : \bar{A} \to J\left(\bar{A}\right)$ the multiplicative property (see Theorem 5.1 on page 24 of [3]) shows

$$1 = \deg(Id, A) = \deg\left(J^{inv} \circ J, A\right) = \deg\left(J^{inv}, J(A)\right) \deg(J, A) \tag{4.2}$$

which implies $\deg(J, A) = \pm 1$. Since J defined by

$$J(x) = \begin{cases} \|x\|\ \tilde{F}\left(\dfrac{x}{\|x\|}\right) & \text{for } x \neq 0, \\ 0 & \text{for } x = 0, \end{cases}$$

is a continuous bijection from $\overline{B_1(0)}$ onto itself, with $J = \tilde{F}$ on $\partial B_1(0)$, we find (by property d6 on page 17 of [3]) that (4.1) holds.

D *ii.*: Now suppose that $F : \Omega \to \Sigma$, and hence $\tilde{F} : B_1(0) \to B_1(0)$, is locally injective. Then the local degree of $\tilde{F}(\cdot) - \tilde{F}(x)$ near x, for $x \in B_1(0)$, is well defined by

$$d(x) = \lim_{\varepsilon \downarrow 0} \deg\left(\tilde{F}(\cdot) - \tilde{F}(x), B_\varepsilon(x)\right).$$

By (4.2) one finds that it equals ± 1. We will show that $d(\cdot)$ is locally constant on $B_1(0)$ and hence constant. Indeed, if $x \in B_1(0)$ there exists $\delta > 0$ such that $\tilde{F}(\cdot)$ is locally injective on $B_{3\delta}(x^*)$. If $|x - y| < \delta$ we find by homotopy that

$$d(x) = \deg\left(\tilde{F}(\cdot) - \tilde{F}(x), B_{2\delta}(x)\right) = \deg\left(\tilde{F}(\cdot) - \tilde{F}(y), B_{2\delta}(x)\right).$$

Since for all $\varepsilon \in (0, \delta]$ we have that $B_{2\delta}(x)$ contains $B_\varepsilon(y)$ and $\tilde{F}(\cdot) - \tilde{F}(y) \neq 0$ on $B_{2\delta}(x) \backslash B_\varepsilon(y)$, it follows that

$$\deg\left(\tilde{F}(\cdot) - \tilde{F}(y), B_{2\delta}(x)\right) = \deg\left(\tilde{F}(\cdot) - \tilde{F}(y), B_\varepsilon(y)\right) = d(y).$$

Using (4.1) and the additivity property of the degree we find that $\tilde{F}(\cdot) = z$ has at most one solution for all $z \in B_1(0)$. □

Acknowledgement: We thank J. M. Aarts for discussions on Theorem D, E. Coplakova for making the text of [11] accessible to us, and A. Zegeling for reference [13].

References

[1] R. A. Adams, Sobolev spaces, Academic Press, New York-San Francisco-London, 1975.

[2] Ph. Clément, R. Hagmeijer and G. Sweers, On the invertibility of mappings arising in 2D grid generation problems, Report 94-20, Faculty of TWI, T.U.Delft 1994.

[3] K. Deimling, Nonlinear Functional Analysis, Springer-Verlag, Berlin, 1985.

[4] H. G. Eggleston, Convexity, Cambridge Tracts in Mathematics and Mathematical Physics 47, Cambridge University Press, Cambridge, 1958.

[5] D. Gilbarg and L. Hörmander, Intermediate Schauder Estimates, Arch. Rational Mech. Anal. 74 (1980), 297-318.

[6] D. Gilbarg and N. S. Trudinger, Elliptic partial differential equations of second order, 2nd edition, Springer-Verlag, Berlin Heidelberg New York Tokyo, 1983.

[7] M. Greenberg, Lectures on Algebraic Topology, W.A. Benjamin Inc, New York Amsterdam, 1967.

[8] R. Hagmeijer, Grid adaption based on modified anisotropic diffusion equations formulated in the parametric domain, to appear in J.Comp.Phys.

[9] R. Hagmeijer and de Cock, Grid adaptation in computational aerodynamics, Multiblock Grid Generation, Notes on Numerical Fluid Mechanics, Vieweg, 1993.

[10] R. Hagmeijer and de Cock, Grid adaptation for problems in computational fluid dynamics, Proceedings First European Fluid Dynamics Conference, Brussels, 1992.

[11] J. Kadlec, On the regularity of the solution of the Poisson problem for a domain which locally resembles the boundary of a convex domain, Czech. Mat. J. 14 (1964), 363-393 (in Russian).

[12] O. D. Kellogg, Foundations of potential theory, Springer-Verlag, Berlin Heidelberg New York, 1929 (reprint 1967).

[13] H. Liu and G. Liao, A note on harmonic maps, preprint 1994.

[14] C.W. Mastin and J. F. Thompson, Elliptic systems and numerical transformations, J. Math. Anal. Appl. 62 (1978), 52-62.

[15] C. W. Mastin and J. F. Thompson, Transformation of three-dimensional regions onto rectangular regions by elliptic systems, Numerische Mathematik 29 (1978), 397-407.

[16] Ch. Pommerenke, Boundary behaviour of conformal maps, Springer-Verlag, Berlin-Heidelberg 1992.

[17] M. H. Protter, H. F. Weinberger, Maximum principles in differential equations, Prentice Hall, Englewood Cliffs N.J., 1967.

[18] F. Schulz, Regularity theory for quasilinear elliptic systems and Monge-Ampère Equations in two dimensions, Springer Lecture Notes 1445, 1990.

[19] J. F. Thompson and Z. U. A. Warsi, Three-dimensional grid generation from elliptic systems, AIAA 83-1905

[20] Z. U. A Warsi, Basic differential models for coordinate generation, Numerical grid generation, ed. J.F. Thompson, NorthHolland, 1982.

[21] A. Winslow, Adaptive mesh zoning by the equipotential method, UCID-19062, Lawrence Livermore National Laboratories, University of California, 1981.

Addresses:

Ph. Clément, Department of Pure Mathematics, Delft University of Technology, P.O.box 5031, 2600 GA Delft, The Netherlands.

R. Hagmeijer, Department of Theoretical Aerodynamics, National Aerospace Laboratory N.L.R., P.O.box 90502, 1006 BM Amsterdam, The Netherlands.

G. Sweers, Department of Pure Mathematics, Delft University of Technology, P.O.box 5031, 2600 GA Delft, The Netherlands.

B DACOROGNA

On rank one convex functions which are homogeneous of degree one

ABSTRACT

Let $f\colon \mathbb{R}^{2\times 2} \to \mathbb{R}$ be positively homogeneous of degree one. If, in addition, f is rotationally invariant and rank one convex, then it is necessarily convex.

1. INTRODUCTION

Let

$$I(u) = \int_{\Omega} f(\nabla u(x))dx \tag{1.1}$$

where :

i) $\Omega \subset \mathbb{R}^2$ is a bounded open set

ii) $u\colon \Omega \to \mathbb{R}^2$ and hence $\nabla u \in \mathbb{R}^{2\times 2}$

iii) $f\colon \mathbb{R}^{2\times 2} \to \mathbb{R}$ is continuous.

As it is well known (c.f. Morrey [8], Ball [2] or Dacorogna [3]), in order to study minimization problems involving (1.1) one has to impose some convexity hypotheses on f. Besides the usual convexity notion, one introduces the following conditions.

Definitions:

1) $f: \mathrm{IR}^{2\times 2} \to \mathrm{IR}$ is <u>rank one convex</u> if

$$f(\lambda A + (1-\lambda)B) \le \lambda f(A) + (1-\lambda) f(B) \tag{1.2}$$

for every $\lambda \in [0,1]$, $A, B \in \mathrm{IR}^{2\times 2}$ with $\det(A-B) = 0$.

2) $f: \mathrm{IR}^{2\times 2} \to \mathrm{IR}$ is <u>quasiconvex</u> if

$$\int_\Omega f(A + \nabla\varphi(x))dx \ge f(A)\, meas\Omega \tag{1.3}$$

for every $A \in \mathrm{IR}^{2\times 2}$ and every $\varphi \in C_o^\infty(\Omega; \mathrm{IR}^2)$.

3) $f: \mathrm{IR}^{2\times 2} \to \mathrm{IR}$ is <u>polyconvex</u> if there exists $g: \mathrm{IR}^{2\times 2} \times \mathrm{IR} \to \mathrm{IR}$ convex such that for every $A \in \mathrm{IR}^{2\times 2}$

$$f(A) = g(A, \det A). \tag{1.4}$$

In general one has

$$f \text{ convex} \Rightarrow f \text{ polyconvex} \Rightarrow f \text{ quasiconvex} \Rightarrow f \text{ rank one convex}.$$

Recently a considerable effort was made to extend the results of Goffman and Serrin [7] to quasiconvex integrands (c.f. for example Fonseca-Müller [6] and their bibliography). One then needs in particular to know if there are positively homogeneous functions of degree one that are quasivoncex but not convex. Such functions where shown to exist by Müller [9], following earlier work of Sverak [10] and Zhang [12]. Müller shows his result using weak type estimates for Cauchy-Riemann operators. So his example is in some sense implicit.

We show here that if one assumes in addition that f is rotationally invariant then Müller's result does not hold. More precisely.

Theorem 1:

Let $f\colon \mathbb{R}^{2\times 2} \to \mathbb{R}$ be continuous and satisfying

i) $f(tA) = tf(A)$ for every $t \geq 0$ and every $A \in \mathbb{R}^{2\times 2}$

ii) f is rotationally invariant, i.e. $f(RAR') = f(A)$
for every $A \in \mathbb{R}^{2\times 2}$, $R, R' \in O^+ = \{RR^t = I \text{ and } \det R = 1\}$

then

$$f \text{ rank one convex} \Leftrightarrow f \text{ convex}.$$

Therefore such a result rules out all functions $f(A) = g(|A|, \det A)$ where $|A|^2 = \sum_{i,j=1}^{2} A_{ij}^2$. In terms of elasticity our theorem shows that no isotropic function can be homogeneous of degree one and quasiconvex but not convex.

We will also show that Theorem 1 can be slightly generalized.

2. THE CASE OF ROTATIONALLY INVARIANT FUNCTIONS

We start with the key lemma.

Lemma 2:

Let $g\colon \mathbb{R}^2 \to \mathbb{R}$ be such that

i) $g(tx, ty) = tg(x, y)$ for every $t \geq 0$ and $x, y \in \mathbb{R}$

ii) g is separately convex (i.e. $g(x, \cdot)$ and $g(\cdot, y)$ are convex for fixed x and fixed y)

then g is convex.

Remark:

For related results on separately convex functions see Tartar [11].

Proof:

The proof is of an elementary nature. We have to show that given $\lambda_1, \lambda_2 \geq 0$ with $\lambda_1 + \lambda_2 = 1$, $x_1, y_1, x_2, y_2 \in \mathrm{IR}$ then

$$g(\lambda_1 x_1 + \lambda_2 x_2, \lambda_1 y_1 + \lambda_2 y_2) \leq \lambda_1 g(x_1, y_1) + \lambda_2 g(x_2, y_2). \tag{2.1}$$

In order to prove this fact we have to consider serveral cases. The idea is however essentialy included in the first case.

Case 1:

Assume that either $\underline{x_1 x_2 \geq 0 \text{ or } y_1 y_2 \geq 0}$. Since g is continuous (the fact that g is separately convex implies that g is continuous) it is enough to prove the result for $x_1 x_2 > 0$ or $y_1 y_2 > 0$. We consider the case where $x_1 x_2 > 0$, otherwise we interchange the roles of x and y. We then let

$$\sigma = \frac{\lambda_1 x_1 + \lambda_2 x_2}{|\lambda_1 x_1 + \lambda_2 x_2|} = \frac{x_1}{|x_1|} = \frac{y_1}{|y_1|}$$

(since $x_1 x_2 > 0$, then σ is well defined and equal + 1 or - 1). Using the properties of g we get

$$\begin{aligned}
&g(\lambda_1 x_1 + \lambda_2 x_2, \lambda_1 y_1 + \lambda_2 y_2) \\
&= |\lambda_1 x_1 + \lambda_2 x_2|\, g\left(\sigma, \frac{\lambda_1 |x_1|}{|\lambda_1 x_1 + \lambda_2 x_2|} \frac{y_1}{|x_1|} + \frac{\lambda_2 |x_2|}{|\lambda_1 x_1 + \lambda_2 x_2|} \frac{y_2}{|x_2|}\right) \\
&\leq |\lambda_1 x_1 + \lambda_2 x_2| \left[\frac{\lambda_1 |x_1|}{|\lambda_1 x_1 + \lambda_2 x_2|}\, g\left(\sigma, \frac{y_1}{|x_1|}\right) + \frac{\lambda_2 |x_2|}{|\lambda_1 x_1 + \lambda_2 x_2|}\, g\left(\sigma, \frac{y_2}{|x_2|}\right)\right].
\end{aligned}$$

Using again the homogeneity of g we find

$$g(\lambda_1 x_1 + \lambda_2 x_2, \lambda_1 y_1 + \lambda_2 y_2) \le \lambda_1 g(\sigma|x_1|, y_1) + \lambda_2 g(\sigma|x_2|, y_2)$$

which is nothing else than (2.1).

Case 2:

We now assume that $x_1 x_2 < 0$ and $y_1 y_2 < 0$ and that there exists $i \in \{1,2\}$ such that $x_i(\lambda_1 x_1 + \lambda_2 x_2) \ge 0$ and $y_i(\lambda_1 y_1 + \lambda_2 y_2) \le 0$. Using again the continuity of g we may assume without loss of generality that $\lambda_1 x_1 + \lambda_2 x_2 \ne 0$ and $\lambda_1 y_1 + \lambda_2 y_2 \ne 0$.

We adopt the notations that $\lambda_3 = \lambda_1$, $x_3 = x_1$ and $y_3 = y_1$ so that $\lambda_i + \lambda_{i+1} = \lambda_1 + \lambda_2 = 1$. We then define

$$a = \frac{x_i\, y_{i+1} - x_{i+1}\, y_i}{y_{i+1} - y_i} \tag{2.2}$$

$$\mu_i = \frac{y_i}{\lambda_{i+1}(y_i - y_{i+1})} = \left[\lambda_{i+1}(1 - \frac{y_{i+1}}{y_i})\right]^{-1}. \tag{2.3}$$

$$\nu_{i+1} = \frac{\lambda_i\, y_i + \lambda_{i+1}\, y_{i+1}}{y_{i+1}} \tag{2.4}$$

Note that since $y_i y_{i+1} < 0$, then $\mu_i \ge 0$. Furthermore $\mu_i \le 1$, since

$$1 - \mu_i = \frac{-(1-\lambda_{i+1})y_i - \lambda_{i+1}\, y_{i+1}}{\lambda_{i+1}(y_i - y_{i+1})} = \frac{-(\lambda_i\, y_i + \lambda_{i+1}\, y_{i+1})}{\lambda_{i+1}\, y_i(1 - \frac{y_{i+1}}{y_i})} \ge 0 \tag{2.5}$$

since $y_i(\lambda_i\, y_i + \lambda_{i+1}\, y_{i+1}) < 0$ and $y_i y_{i+1} < 0$. Therefore

$$0 \le \mu_i \le 1. \tag{2.6}$$

Since $y_i y_{i+1} < 0$ and $y_i(\lambda_i y_i + \lambda_{i+1}\, y_{i+1}) < 0$, we deduce that $\nu_{i+1} > 0$. We now show that $\nu_{i+1} \le 1$, indeed

$$1 - \nu_{i+1} = \frac{y_{i+1}(1-\lambda_{i+1}) - \lambda_i y_i}{y_{i+1}} = \lambda_i \left[1 - \frac{y_i}{y_{i+1}} \right] \geq 0. \tag{2.7}$$

Hence we deduce that

$$0 \leq \nu_{i+1} \leq 1. \tag{2.8}$$

Observe that

$$\left[\begin{array}{ll} \mu_i(\lambda_i x_i + \lambda_{i+1} x_{i+1}) + (1-\mu_i) x_i = a & \quad (2.9) \\ \mu_i(\lambda_i y_i + \lambda_{i+1} y_{i+1}) + (1-\mu_i) y_i = 0. & \quad (2.10) \end{array} \right.$$

Let us check (2.9),

$$\mu_i(\lambda_i x_i + \lambda_{i+1} x_{i+1}) + (1-\mu_i) x_i = x_i[1 - \mu_i(1-\lambda_i)] + \lambda_{i+1}\mu_i x_{i+1}$$

$$= x_i[1 - \lambda_{i+1}\mu_i] + \lambda_{i+1}\mu_i x_{i+1} = x_i \left[1 - \frac{y_i}{y_i - y_{i+1}} \right] + x_{i+1} \cdot \frac{y_i}{y_i - y_{i+1}}$$

$$= \frac{x_{i+1} y_i - x_i y_{i+1}}{y_i - y_{i+1}} = a.$$

(2.10) is verified similarly.

We also note that

$$\left[\begin{array}{ll} \nu_{i+1} x_{i+1} + (1-\nu_{i+1}) a = \lambda_i x_i + \lambda_{i+1} x_{i+1} & \quad (2.11) \\ \nu_{i+1} y_{i+1} + (1-\nu_{i+1}) 0 = \nu_{i+1} y_{i+1} = \lambda_i y_i + \lambda_{i+1} y_{i+1}. & \quad (2.12) \end{array} \right.$$

(2.12) is just a reformulation of (2.4). We now verify (2.11)

$$\nu_{i+1} x_{i+1} + (1-\nu_{i+1}) a = \frac{\lambda_i y_i x_{i+1} + \lambda_{i+1} y_{i+1} x_{i+1} + \lambda_i (y_{i+1} - y_i) a}{y_{i+1}}$$

$$= \frac{y_{i+1}(\lambda_i x_i + \lambda_{i+1} x_{i+1})}{y_{i+1}} = \lambda_i x_i + \lambda_{i+1} x_{i+1}.$$

We are now in a position to conclude to the convexity of g.

By (2.11) and (2.12) we have

$$\begin{aligned} &g(\lambda_1 x_1 + \lambda_2 x_2, \lambda_1 y_1 + \lambda_2 y_2) = g(\nu_{i+1} x_{i+1} + (1-\nu_{i+1})a, \nu_{i+1} y_{i+1} + (1-\nu_{i+1})0) \\ &\le \nu_{i+1}\, g(x_{i+1}, y_{i+1}) + (1-\nu_{i+1}) g(a,0), \end{aligned} \tag{2.13}$$

the inequality coming from the fact that $y_{i+1}\cdot 0 \ge 0$ and from Case 1. We then use (2.9) and (2.10) to write.

$$\begin{aligned} &g(a,0) = g(\mu_i(\lambda_1 x_1 + \lambda_2 x_2) + (1-\mu_i)x_i, \mu_i(\lambda_1 y_1 + \lambda_2 y_2) + (1-\mu_i) y_i) \\ &\le \mu_i\, g(\lambda_1 x_1 + \lambda_2 x_2, \lambda_1 y_1 + \lambda_2 y_2) + (1-\mu_i) g(x_i, y_i), \end{aligned} \tag{2.14}$$

the inequality coming from Case 1 and the fact that $x_i(\lambda_1 x_1 + \lambda_2 x_2) \ge 0$ (by hypothesis of Case 2).

Combining (2.13) and (2.14) we get

$$\begin{aligned} &\left[1-\mu_i(1-\nu_{i+1})\right] g(\lambda_1 x_1 + \lambda_2 x_2, \lambda_1 y_1 + \lambda_2 y_2) \\ &\le \nu_{i+1}\, g(x_{i+1}, y_{i+1}) + (1-\nu_{i+1})(1-\mu_i) g(x_i, y_i). \end{aligned} \tag{2.15}$$

To conclude to (2.1) i.e. the convexity of g we only need to show that

$$\left[\begin{aligned} &\frac{\nu_{i+1}}{1-\mu_i(1-\nu_{i+1})} = \lambda_{i+1} && (2.16) \\ &\frac{(1-\nu_{i+1})(1-\mu_i)}{1-\mu_i(1-\nu_{i+1})} = \lambda_i. && (2.17) \end{aligned}\right.$$

We verify the first one (using (2.3),(2.4),(2.7))

$$\frac{\nu_{i+1}}{1-\mu_i(1-\nu_{i+1})} = \frac{\lambda_i y_i + \lambda_{i+1} y_{i+1}}{y_{i+1} - \dfrac{y_i}{\lambda_{i+1}(y_i - y_{i+1})}\left[\lambda_i(y_{i+1} - y_i)\right]} = \lambda_{i+1}$$

and similarly for (2.17) since

$$\frac{\nu_{i+1}}{1-\mu_i(1-\nu_{i+1})}+\frac{(1-\nu_{i+1})(1-\mu_i)}{1-\mu_i(1-\nu_{i+1})}=1=\lambda_i+\lambda_{i+1}.$$

This concludes Case 2.

Case 3:

We now assume that $\underline{x_1 x_2 < 0}$ and $\underline{y_1 y_2 < 0}$ and that there exists $i \in \{1,2\}$ with $\underline{x_i(\lambda_1 x_1 + \lambda_2 x_2) < 0}$ and $\underline{y_i(\lambda_1 y_1 + \lambda_2 y_2) < 0}$.

Note first that Cases 1, 2 and 3 cover all the possibilities, since by inverting the roles of i and $i+1$, taking into account that $x_i x_{i+1} < 0$, $y_i y_{i+1} < 0$, the convexity inequality holds by Case 2 also for $x_i(\lambda_1 x_1 + \lambda_2 x_2) \le 0$ and $y_i(\lambda_1 y_1 + \lambda_2 y_2) \ge 0$. In particular if $y_i(\lambda_1 y_1 + \lambda_2 y_2) = 0$, then the convexity inequality holds independently of the sign of $x_i(\lambda_1 x_1 + \lambda_2 x_2)$.

Therefore to establish (2.1) when Case 3 holds, we proceed exactly as in Case 2, the only point to be checked is that (2.14) holds also under the hypotheses of Case 3. Letting

$$\begin{aligned} X_1 &= \lambda_i x_i + \lambda_{i+1} x_{i+1}, \qquad & X_2 &= x_i \\ Y_1 &= \lambda_i y_i + \lambda_{i+1} y_{i+1}, \qquad & Y_2 &= y_i. \end{aligned}$$

We have by hypothesis that $X_1 X_2 < 0, Y_1 Y_2 < 0$ and by (2.10)

$$Y_1(\mu_i Y_1 + (1-\mu_i) Y_2) = 0.$$

Using the above remark we deduce from Case 2, (2.2) and (2.9), that

$$\begin{aligned} g(a,0) &= g(\mu_i X_1 + (1-\mu_i) X_2,\ \mu_i Y_1 + (1-\mu_i) Y_2) \\ &\le \mu_i g(X_1, Y_1) + (1-\mu_i) g(X_2, Y_2) \end{aligned}$$

which is nothing else than (2.14). The remaining part of the proof is then exactly the same as that of Case 2. Therefore the lemma is established. ◆

We are now in a position to prove Theorem 1.

Proof:

i) $\underline{f \text{ convex} \Rightarrow f \text{ rank one convex}}$. This is true for any function.

ii) $\underline{f \text{ rank one convex} \Rightarrow f \text{ convex}}$: Let $g\colon \mathrm{IR}^2 \to \mathrm{IR}$ be defined by

$$g(x,y) = f\begin{pmatrix} x & 0 \\ 0 & y \end{pmatrix}.$$

Since f is positively homogeneous of degree one and rank one convex, we deduce that g is positively homogeneous of degree one and separately convex. Applying Lemma 2 to g, we deduce that g is convex and so that f is convex when restricted to diagonal matrices. Using then a result of Dacorogna-Koshigoe [4], we conclude that f is convex on the whole of $\mathrm{IR}^{2\times 2}$. ◆

We now turn to an example

Proposition 3:

Let

$$f(A) = \begin{cases} |A| + \alpha \dfrac{\det A}{|A|} & \text{if } \quad A \neq 0 \\ 0 & \text{if } \quad A = 0 \end{cases}$$

then

$$f \text{ rank one convex} \Leftrightarrow f \text{ convex} \Leftrightarrow |\alpha| \leq \frac{2}{3}.$$

Remark:

It is interesting to contrast this example with that of Dacorogna-Marcellini [5] (see also Alibert-Dacorogna [1])

$$f(A) = |A|^4 - 2\gamma|A|^2 \det A.$$

In this case the notions of convexity, polyconvexity and rank one convexity are different (the exact values for which these notions hold are $|\gamma| \leq \frac{2}{3}\sqrt{2}, 1, \frac{2}{\sqrt{3}}$ respectively). Note also that if one looks for the simplest example of homogeneous function of degree two, one has

$$f(A) = |A|^2 - \gamma \det A.$$

In this case we have that f is rank one convex (and polyconvex) for every $\gamma \in \mathrm{IR}$, while it is convex if and only if $|\gamma| \leq 2$. Therefore a natural attempt to find a rank one convex (but not convex) and homogeneous of degree one function is the one of Proposition 3. However in view of Theorem 1 this fails to give the expected example.

Proof:

The proof is elementary. Using Theorem 1 and again the result of Dacorogna-Koshigoe [4], the only thing to be checked is that

$$g(x,y) = \begin{cases} \sqrt{x^2+y^2} + \alpha \dfrac{xy}{\sqrt{x^2+y^2}} & \text{if } x^2+y^2 \neq 0 \\ 0 & \text{if } x = y = 0 \end{cases}$$

is separately convex if and only if $|\alpha| \leq \frac{2}{3}$. Since g is symmetric in x and y it is enough to show that

$$\frac{\partial^2}{\partial x^2} g(x,y) \geq 0 \text{ for every } (x,y) \neq (0,0) \Leftrightarrow |\alpha| \leq \frac{2}{3}.$$

This is true since (provided $(x,y) \neq (0,0)$)

$$\frac{\partial^2}{\partial x^2} g(x,y) = \frac{y^2}{(x^2+y^2)^{5/2}} (x^2 - 3\alpha xy + y^2).$$

Obviously the right hand side is positive if and only if $|\alpha| \leq \frac{2}{3}$. ◆

3. SOME GENERALIZATIONS

We now show that Theorem 1 holds also for some non rotationally invariant functions. We first fix some notations.

Notations :

We here identify $\mathbb{R}^{2\times 2}$ with $\mathbb{R}^4$. For $X \in \mathbb{R}^4$, $X = (x_1, x_2, x_3, x_4)$ we let

$$|X|^2 = x_1^2 + x_2^2 + x_3^2 + x_4^2$$
$$\det X = x_1 x_4 - x_2 x_3$$
$$\operatorname{scal} X = x_1 x_2 + x_3 x_4.$$

We now assume that f satisfy the following.

Hypothesis (H) :

Let $f : \mathbb{R}^4 \cong \mathbb{R}^{2\times 2} \to \mathbb{R}$ be continuous and positively homogeneous of degree one. Assume that there exists $S \in \mathbb{R}^{4\times 4}$ invertible and $0 \neq A \in \mathbb{R}^4$ such that $\det A = \det(SA) = 0$. Finally let for $X \in \mathbb{R}^4$, $g: \mathbb{R}^4 \to \mathbb{R}$

$$g(X) \equiv f(SX)$$

and assume that g is rotationally invariant.

Theorem 4 :

Let f satisfy (H) then

$$f \text{ rank one convex} \Leftrightarrow f \text{ convex} \Leftrightarrow g \text{ convex} \Leftrightarrow g \text{ rank one convex}.$$

We immediately give examples of such functions.

Corollary 5 :

Let $f\colon \mathrm{IR}^4 \to \mathrm{IR}$ be positively homogeneous of degree one and having one of the following forms

1) $f(X) = \varphi(|X|, \det X)$

2) $f(X) = \varphi(|X|, \operatorname{scal} X)$

3) $f(X) = \varphi(|X|, \alpha \det X + \beta \operatorname{scal} X)$ for some $\alpha, \beta \in \mathrm{IR}\backslash\{0\}$

4) $f(X) = \varphi(\sqrt{x_1^2 + x_3^2}, \sqrt{x_2^2 + x_4^2})$

for some $\varphi\colon \mathrm{IR}^2 \to \mathrm{IR}$. Then

$$f \text{ rank one convex} \Leftrightarrow f \text{ convex}.$$

Remark :

The first result is a consequence of Theorem 1. However all the three other cases give rise to functions which are not rotationally invariant. One can also imagine several other examples where Theorem 4 apply, such as (for $\alpha, \beta \in \mathrm{IR}$)

$$f(x_1, x_2, x_3 x_4) = (x_1^2 + x_3^2 + \beta^2(x_2^2 + x_4^2))^{1/2} + \alpha \frac{x_1 x_2 + x_3 x_4}{((x_1^2 + x_3^2) + \beta^2(x_2^2 + x_4^2))^{1/2}}.$$

Proof of Corollary 5 :

1) C.f. Theorem 1, since f is then rotationally invariant.

2) Let

$$S = \begin{pmatrix} 1 & 0 & 0 & 0 \\ 0 & 0 & 0 & 1 \\ 0 & 0 & 1 & 0 \\ 0 & -1 & 0 & 0 \end{pmatrix},$$

choose $A = (1,0,0,0)$. Then S and A are as in (H) and

$$g(X) = f(SX) = \varphi(|SX|, \operatorname{scal}(SX)) = \varphi(|X|, \det X).$$

Applying Theorem 4 and the first part of the Corollary we get the result.

3) Let

$$S = \begin{pmatrix} 1 & 0 & 0 & 0 \\ 0 & \frac{-\alpha}{\sqrt{\alpha^2+\beta^2}} & 0 & \frac{\beta}{\sqrt{\alpha^2+\beta^2}} \\ 0 & 0 & -1 & 0 \\ 0 & \frac{\beta}{\sqrt{\alpha^2+\beta^2}} & 0 & \frac{\alpha}{\sqrt{\alpha^2+\beta^2}} \end{pmatrix}.$$

Choose as before $A = (1,0,0,0)$ and observe that for every $X \in \mathbb{R}^4$

$$\begin{cases} |X| = |SX| \\ \alpha \det(SX) + \beta \operatorname{scal}(SX) = \sqrt{\alpha^2+\beta^2} \det X \end{cases}.$$

In view of the above identities, we find

$$g(X) = f(SX) = \varphi(|X|, \sqrt{\alpha^2+\beta^2} \det X).$$

We then apply Theorem 1 and Theorem 4 to obtain the claimed result. ◆

Remark :

If we combine Proposition 3 and Corollary 5 we find that if

$$f(X) = \begin{cases} |X| + \dfrac{\alpha \det X + \beta \operatorname{scal} X}{|X|} & \text{if} \quad |X| \neq 0 \\ 0 & \text{if} \quad X = 0 \end{cases}$$

then

$$f \text{ convex} \Leftrightarrow f \text{ rank one convex} \Leftrightarrow \alpha^2 + \beta^2 \leq \frac{4}{9}.$$

Proof of Theorem 4 :

Observe that since S is invertible we have trivially

$$f \text{ convex} \Leftrightarrow g \text{ convex}. \tag{3.1}$$

From Theorem 1 we deduce that

$$g \text{ convex} \Leftrightarrow g \text{ rank one convex}. \tag{3.2}$$

So if we show that

$$f \text{ rank one convex} \Rightarrow g \text{ rank one convex}, \tag{3.3}$$

we will have proved the theorem (since trivially f convex $\Rightarrow f$ rank one convex).

To prove that g is rank one convex, we have to show that given $X, Y \in \mathbb{R}^{2\times 2} \cong \mathbb{R}^4$ with $Y \neq 0$ and $\det Y = 0$, then the function $\varphi : \mathbb{R} \to \mathbb{R}$ defined by

$$\varphi(t) \equiv g(X + tY) \tag{3.4}$$

is convex.

Since g is rotationally invariant, there is no loss of generality in assuming that $Y = A$ (where A is the matrix given in (H)). Indeed there exist $R, R^{'}, Q, Q^{'} \in O^{+}$ such that

$$Y = R\begin{pmatrix} |Y| & 0 \\ 0 & 0 \end{pmatrix} R^{'}$$

$$A = Q\begin{pmatrix} |A| & 0 \\ 0 & 0 \end{pmatrix} Q^{'}.$$

i.e.

$$Y = \frac{|Y|}{|A|} RQ^{t}\ AQ^{'t}R^{'}. \tag{3.5}$$

Therefore we in fact only need to show that given any $X \in \mathrm{IR}^4$ then the function

$$\varphi(t) = g(X + tA) \tag{3.6}$$

is convex for every $t \in \mathrm{IR}$. However, we have, by definition,

$$\varphi(t) = g(X + tA) = f(SX + tSA). \tag{3.7}$$

Since f is assumed to be rank one convex (c.f. (3.3)) and that by hypothesis $\det SA = 0$, we deduce immediately that φ is convex and thus g is rank one convex. This concludes the proof of the theorem. ◆

Acknowledgments:

It would like to thank A. Curnier for stimulating discussions on this problem.

References :

[1] J.J. Alibert, B. Dacorogna: An example of a quasiconvex function that is not polyconvex in two dimensions; Arch. Rational Mech. Anal. 117 (1992), 155-166.

[2] J.M. Ball: Convexity conditions and existence theorems in nonlinear elasticity; Arch. Rational Mech. Anal. 64 (1977), 337-403.

[3] B. Dacorogna: *"Direct methods in the calculus of variations"*; Springer-Verlag, Berlin (1989).

[4] B. Dacorogna, H. Koshigoe: On the different notions of convexity for rotationally invariant functions; Annales Fac. Sciences de Toulouse (1993), 163-184.

[5] B . Dacorogna, P. Marcellini: A counterexample in the vectorial calculus of variations; in *"Material instabilities in continuum mechanics"*; ed. by J.M. Ball, Oxford University Press (1988), 77-83.

[6] I. Fonseca, S. Müller: Relaxation of quasiconvex functions in BV $(\Omega;\mathbb{R}^p)$; Arch. Ration. Mech. Anal. 123(1993), 1-49.

[7] C. Goffman, J. Serrin: Sublinear functions of measures and variational integrals, Duke Math. J. (1964), 159-178.

[8] C.B. Morrey: *"Multiple integrals in the calculus of variations"*; Springer-Verlag (1966).

[9] S. Müller: On quasiconvex functions which are homogeneous of degree one; Indiana Univ. Math. J. 41 (1992), 295-300.

[10] V. Sverak: Quasiconvex functions with subquadratic growth; Proc. Roy. Soc. London 433A (1991), 723-725.

[11] L. Tartar: Some remarks on separately convex functions; preprint.

[12] K. Zhang: A construction of quasiconvex functions with linear growth at infinity; preprint.

Bernard DACOROGNA
Ecole Polytechnique Fédérale de Lausanne
Department of Mathematics
CH 1015 LAUSANNE

K DECKELNICK AND G DZIUK

On the approximation of the curve shortening flow

1 Introduction

Analysis and numerical approximation of mean curvature flow has been treated by many authors during the last time. The corresponding onedimensional problem is the so called curve shortening flow ([6], [7]). It is now well known that this flow shrinks embedded plane curves to round points in finite time. Curves which are not embedded may exhibit cusps while shrinking. The curve shortening problem for space curves has been considered in [1], [2] .
In this note we shall propose a numerical scheme for the curve shortening flow in possibly higher codimension and prove convergence for a spatial discretization if the continuous problem has a smooth solution on some time interval. Numerical test computations will show that the numerical algorithm will continue past singularities although there is no convergence proof in this case.

First we shall recall an algorithm together with a convergence result for the curve shortening flow which was proved in [5] by the second author and which also provides a convergence estimate for the curvature. The method of proof for that result was the same as the proof for asymptotic convergence for a finite element approximation of the mean curvature flow problem for twodimensional graphs in [3]. The algorithm in [5] was based on a mathematical formulation of the mean curvature flow problem which used the Laplace–Beltrami operator on the moving surface. A discretization of this formulation was proposed in [4] and shown to work until singularities of the surface appear. For curves the mathematical problem is described by the following nonlinear degenerate system of parabolic equations

$$x_t - \frac{1}{|x_s|}\Big(\frac{x_s}{|x_s|}\Big)_s = 0 \tag{1}$$

on $I \times (0,T), I = [0,2\pi]$ for $x = x(s,t)$. It is important to mention that here s does not denote the arc length parameter. x is assumed to be periodic in s, $|x_s| > 0$ and

$$x(\cdot, 0) = x_0. \tag{2}$$

If we denote by κ the curvature of $\Gamma(t) = x(I, t)$ and by ν the principal normal to $\Gamma(t)$ then the above equations say that Γ is moved in normal direction with velocity equal to minus curvature.

$$(x_t, \nu) = -\kappa. \tag{3}$$

Throughout this note we shall treat closed curves and thus shall assume all functions to be periodic with respect to the parameter s. The problem with a convergence proof for (1), (2) is that the equation does not provide estimates for the tangential part of the derivative x_s. In [5] this problem has been overcome by differentiating the equation (1) with respect to time and then proving error estimates for this new equation.

Here we shall use a variant of (1), (2) which replaces equation (1), which is

$$x_t - \frac{1}{|x_s|^2}\Big(x_{ss} - (x_{ss}, \frac{x_s}{|x_s|})\frac{x_s}{|x_s|}\Big) = 0 \tag{4}$$

by the system

$$x_t - \frac{1}{|x_s|^2} x_{ss} = 0. \tag{5}$$

This only changes the solution Γ tangentially, i.e. it only changes the parametrization. The shape of the solution remains unchanged.

In [8] numerical methods for free boundary value problems are used to solve the equation for the curvature which numerically is equivalent to a porous media type equation. This method can handle plane curves with $\kappa > 0$ and breaks down at cusp singularities. In [9] a line method for plane curves is formulated.

2 Convergence for purely normal motion

We recall the results from [5] and supply the reader with some informations about the mechanics of the numerical algorithm. The numerical scheme presented in [4] for surfaces leads to the following onedimensional scheme for (1), (2). A weak formulation for (1) is

$$\int_I x_t |x_s| \phi + \int_I \frac{x_s}{|x_s|}\phi_s = 0 \tag{6}$$

for every test function ϕ. If

$$X_h = span\{\phi_1, .., \phi_n\}$$

is the space of piecewise linear continuous functions on a grid $s_j = jh (j = 1, \ldots, n)$ in I with grid size h, then a spatial discretization of (6) is

$$x_h(s,t) = \sum_{j=1}^{n} x_j(t)\phi_j(s),$$

$$\int_I x_{ht}|x_{hs}|\phi_j + \int_I \frac{x_{hs}}{|x_{hs}|}\phi_{js} = 0 \ (j = 1, .., n) \tag{7}$$

$$x_h(\cdot, 0) = x_{h0} \tag{8}$$

where x_{h0} is some approximation of x_0 in X_h. The lumped version of this finite element formulation leads to the diffference scheme

$$\frac{1}{2}(h_j + h_{j+1})\dot{x}_j = \frac{x_{j+1} - x_j}{h_{j+1}} - \frac{x_j - x_{j-1}}{h_j}, \tag{9}$$

$$h_j = |x_j - x_{j-1}|.$$

with initial values $x_j(0) =: x_{0j}$ $(j = 1, \ldots, n)$. The main result of [5] then is the following theorem which uses high regularity of the continuous solution. This is no problem as long as there are no cusps developing. For numerical tests see section 6 in [5] and section 4 of this paper.

2.1 Theorem *Let $x = x(s,t), (s,t) \in I\!R/2\pi \times [0,T]$, be a smooth solution of the curve shortening flow (1) with initial data x_0 and $|x_s| \geq c_0 > 0$. Then there exists an $h_0 > 0$ depending on x and T such that for every $0 < h \leq h_0$ there exists a unique solution $x_h(s,t) = \sum_{j=1}^{n} x_j(t)\phi_j(s)$ of the difference scheme (9) with initial data $x_{h0}(s_j) = x_0(s_j)(j = 1, \ldots, n)$ and*

$$\max_{t\in[0,T]} \|x - x_h\|_{L^2(I)} + (\int_0^T \|x_s - x_{hs}\|^2_{L^2(I)} dt)^{1/2} \leq ch, \tag{10}$$

$$\max_{t\in[0,T]} \|x_t - x_{ht}\|_{L^2(I)} + (\int_0^T \|x_{ts} - x_{hts}\|^2_{L^2(I)} dt)^{1/2} \leq ch,$$

where c depends on x and T.

3 Convergence for the reparametrized problem.

Let $x \in C^{2,1}(I\!R/2\pi \times [0,T])$ be a solution of

$$\begin{aligned} x_t &= \frac{1}{|x_s|^2} x_{ss} && \text{in } I\!R/2\pi \times [0,T] \\ x(\cdot, 0) &= x_0 && \text{in } I\!R/2\pi. \end{aligned}$$

In addition we assume that

$$x_t \in L^\infty((0,T), H^1(I\!R/2\pi)) \cap L^2((0,T), H^2(I\!R/2\pi))$$

and

$$|x_s| \geq c_0 > 0 \qquad \text{in } I\!R/2\pi \times [0,T]. \tag{11}$$

A weak formulation of the above system is given by

$$\int_I |x_s|^2 x_t \phi + \int_I x_s \phi_s = 0, \qquad \phi \in H^1(I\!R/2\pi), 0 \leq t \leq T. \tag{12}$$

3.1 Theorem *Assume that x satisfies the above conditions. Then there exists an $h_0 > 0$ depending on x and T such that for every $0 < h \leq h_0$ there is a unique solution $x_h \in H^{1,2}((0,T), X_h)$ of the semi-discrete problem*

$$\int_I |x_{hs}|^2 x_{ht} \phi_h + \int_I x_{hs} \phi_{hs} = 0, \qquad \phi_h \in X_h,\ 0 \leq t \leq T$$

and

$$\max_{t\in[0,T]} \|(x - x_h)(t)\|_{H^1(I)} + \left(\int_0^T \|x_t - x_{ht}\|^2_{L^2(I)} dt\right)^{\frac{1}{2}} \leq ch.$$

The constant c depends on the continuous solution x and on T.

Proof. Consider the Banach space $Z_h = C^0([0,T], X_h)$ equipped with the norm

$$\|x_h\|_{Z_h} := \sup_{t\in[0,T]} \|x_h(t)\|_{L^2(I)}$$

and define

$$B_h := \{x_h \in Z_h | \sup_{t\in[0,T]} e^{-\alpha t} \|(x_s - x_{hs})(t)\|^2_{L^2(I)} \leq K^2 h^2\}$$

where the constants $\alpha > 0, K \geq 1$ are to be chosen later. Clearly B_h is a nonempty (take e.g. $x_h(t) = I_h x(t)$, I_h being the piecewise linear interpolation operator) closed and convex subset of Z_h. Interpolation and inverse inequalities yield for $x_h \in B_h, 0 \leq t \leq T$

$$\begin{aligned}
\|(x_s - x_{hs})(t)\|_{L^\infty} &\leq \|(x_s - (I_h x)_s)(t)\|_{L^\infty} + \|((I_h x)_s - x_{hs})(t)\|_{L^\infty} \\
&\leq ch + ch^{-\frac{1}{2}} \|((I_h x)_s - x_{hs})(t)\|_{L^2} \leq ch \\
&\quad + ch^{-\frac{1}{2}} (\|(x_s - (I_h x)_s)(t)\|_{L^2} + \|(x_s - x_{hs})(t)\|_{L^2}) \\
&\leq ch^{\frac{1}{2}} (1 + e^{\frac{\alpha}{2}T} K)
\end{aligned}$$

so that in view of (11) we may assume that for $x_h \in B_h, h \leq h_0 = h_0(\alpha, K, T, x)$

$$|x_{hs}| \geq \frac{1}{2} c_0, \qquad |x_{hs}| \leq c \qquad \text{in } I\!R/2\pi \times [0,T]. \tag{13}$$

Next, let us define the operator $F : B_h \to Z_h$ as follows: for a given $x_h \in B_h$ we obtain $y_h = F(x_h)$ as the unique solution of the linear problem

$$\int_I |x_{hs}|^2 y_{ht} \phi_h + \int_I y_{hs} \phi_{hs} = 0, \qquad \phi_h \in X_h, 0 \leq t \leq T \tag{14}$$

with the initial condition $y_h(0) = I_h x_0$. We claim that $y_h \in B_h$ provided α and K are chosen in an appropriate way. Taking the difference of (12) and (14) we arrive at

$$\int_I |x_{hs}|^2(x_t - y_{ht})\phi_h + \int_I (x_s - y_{hs})\phi_{hs} = \int_I (|x_{hs}|^2 - |x_s|^2)x_t\phi_h.$$

For $t \in [0, T]$ we use $\phi_h = I_h x_t(t) - y_{ht}(t)$ as a test function and get

$$\begin{aligned} &\int_I |x_{hs}|^2|x_t - y_{ht}|^2 + \int_I (x_s - y_{hs})(x_{ts} - y_{hts}) \\ = \;& \int_I (|x_{hs}|^2 - |x_s|^2)\, x_t(I_h x_t - y_{ht}) + \int_I |x_{hs}|^2(x_t - y_{ht})(x_t - I_h x_t) \\ &+ \int_I (x_s - y_{hs})(x_{ts} - (I_h x_t)_s) \end{aligned}$$

and therefore by (13)

$$\begin{aligned} &\frac{c_0^2}{4}\int_I |x_t - y_{ht}|^2 + \frac{1}{2}\frac{d}{dt}\int_I |x_s - y_{hs}|^2 \\ \leq\;& c\int_I |x_s - x_{hs}|\,|x_t|\,(|x_t - I_h x_t| + |x_t - y_{ht}|) \\ &+ c\int_I |x_t - y_{ht}|\,|x_t - I_h x_t| + \int_I |x_s - y_{hs}|\,|x_{ts} - (I_h x_t)_s| \\ \leq\;& c\|x_t\|_{L^\infty}\|x_s - x_{hs}\|_{L^2}(\|x_t - I_h x_t\|_{L^2} + \|x_t - y_{ht}\|_{L^2}) \\ &+ c\|x_t - y_{ht}\|_{L^2}\|x_t - I_h x_t\|_{L^2} + \|x_s - y_{hs}\|_{L^2}\|x_{ts} - (I_h x_t)_s\|_{L^2}. \end{aligned}$$

In view of the interpolation estimates

$$\|x_t - I_h x_t\|_{L^2} \leq ch\|x_t\|_{H^1}, \qquad \|x_{ts} - (I_h x_t)_s\|_{L^2} \leq ch\|x_t\|_{H^2},$$

the regularity of x and the fact that $x_h \in B_h$ we may further estimate

$$\begin{aligned} &\frac{c_0^2}{4}\|x_t - y_{ht}\|_{L^2}^2 + \frac{1}{2}\frac{d}{dt}\|x_s - y_{hs}\|_{L^2}^2 \\ \leq\;& ce^{\frac{\alpha}{2}t}Kh^2 + ce^{\frac{\alpha}{2}t}Kh\|x_t - y_{ht}\|_{L^2} + ch\|x_t - y_{ht}\|_{L^2} \\ &+ ch\|x_t\|_{H^2}\|x_s - y_{hs}\|_{L^2} \\ \leq\;& \frac{c_0^2}{8}\|x_t - y_{ht}\|_{L^2}^2 + \|x_s - y_{hs}\|_{L^2}^2 + ch^2(e^{\alpha t}K^2 + 1 + \|x_t\|_{H^2}^2). \end{aligned}$$

Integration with respect to time gives

$$\begin{aligned} &\frac{c_0^2}{4}\int_0^t \|x_t - y_{ht}\|_{L^2}^2 dt + \|(x_s - y_{hs})(t)\|_{L^2}^2 \\ \leq\;& \|x_{0s} - (I_h x_0)_s\|_{L^2}^2 + \int_0^t \|x_s - y_{hs}\|_{L^2}^2 ds \\ &+ ch^2(\alpha^{-1}e^{\alpha t}K^2 + T + \int_0^T \|x_t\|_{H^2}^2) \\ \leq\;& ch^2 + c\alpha^{-1}e^{\alpha t}K^2h^2 + \int_0^t \|x_s - y_{hs}\|_{L^2}^2 ds. \end{aligned} \tag{15}$$

With the help of Gronwall's lemma we finally conclude for $t \in [0,T]$

$$\|(x_s - y_{hs})(t)\|_{L^2}^2 \le ch^2 + c\alpha^{-1}e^{\alpha t}K^2h^2$$

and therefore

$$\sup_{t\in[0,T]} e^{-\alpha t}\|(x_s - y_{hs})(t)\|_{L^2}^2 \le ch^2 + c\alpha^{-1}K^2h^2.$$

If we fix $K^2 > 2c$ and take α such that $c\alpha^{-1} \le \frac{1}{2}$ we see that

$$\sup_{t\in[0,T]} e^{-\alpha t}\|(x_s - y_{hs})(t)\|_{L^2}^2 \le K^2h^2$$

or in other words, $y_h = F(x_h) \in B_h$. Next, observe that by (15)

$$\int_0^T \|y_{ht}\|_{L^2}^2 dt \le c$$

which then implies together with the initial condition $y_h(0) = I_h x_0$, that $\|y_h\|_{H^{1,2}((0,T),X_h)} \le c$, the constant c being independent of the particular choice of x_h with $F(x_h) = y_h$. Since the embedding $H^{1,2}(0,T) \hookrightarrow C^0([0,T])$ is compact we may thus infer that $F(B_h)$ is precompact (note that X_h is a finite-dimensional space). Furthermore it is not hard to see that F is continuous.
Schauder's fixed point theorem now implies the existence of some $x_h \in B_h$ satisfying $F(x_h) = x_h$ which means that x_h is a discrete solution of (12). The error estimates then are a consequence of the definition of B_h and (15). A similar argument as above shows that the discrete solution is unique.

4 Numerical results

In this section we present a numerical result for the curve shortening flow with singularities. The purely normal motion of a curve and the tangentially modified motion were described in the previous section. To this we add a suitable time discretization for both methods. The numerical results for both algorithms geometrically are the same.

Let us start with the purely normal motion which was spatially discretized in (9). By the upper index we count the time level. If $\tau > 0$ is a time step then g^m stands for $g(m\tau)$, $m = 0, \ldots, M$ with $M\tau = T$. A fully discrete scheme for (9) then is given by

$$\frac{1}{2}(h_j^{m-1} + h_{j+1}^{m-1})\frac{1}{\tau}x_j^m + \left(-\frac{1}{h_{j+1}^{m-1}}x_{j+1}^m + \left(\frac{1}{h_{j+1}^{m-1}} + \frac{1}{h_j^{m-1}}\right)x_j^m - \frac{1}{h_j^{m-1}}x_{j-1}^m\right)$$
$$= \frac{1}{2}(h_j^{m-1} + h_{j+1}^{m-1})\frac{1}{\tau}x_j^{m-1},$$

and

$$h_j^{m-1} = |x_j^{m-1} - x_{j-1}^{m-1}|.$$

$(j = 1, \ldots, n; m = 1, \ldots, M)$. With this discretization we computed the curve shortening flow for a non embedded curve which exhibits cusps during the evolution. The initial curve is given by

$$x_0(s) = cos(2s)(cos(s), sin(s)),\ s \in [0, 2\pi].$$

The results for some time steps are shown in Fig. 1. The curves were rescaled graphically. The true lengths were 8.2, 4.7, 1.5 and 0.8.

Fig. 1. Four states of the evolution of a non–embedded curve.

In Fig. 2 we also add some information about the development of the parametrization which according to the scheme develops automatically. We show the piecewise constant function $|x_{hs}^m|$ which is the length element. One observes that at cusps this function becomes zero. The second derivatives $|x_{hss}^m|$ are represented as the modulus of the difference quotient of x_{hs}^m. In each of the four states of Fig. 2 we plotted $|x_{hs}|$ above and $|x_{hss}|$ below. The horizontal axis ranges from 0 to 2π. The vertical axis runs from 0. to 0.85, scaled by 1/3 for $|x_{hs}|$ and by 1/60 for $|x_{hss}|$.

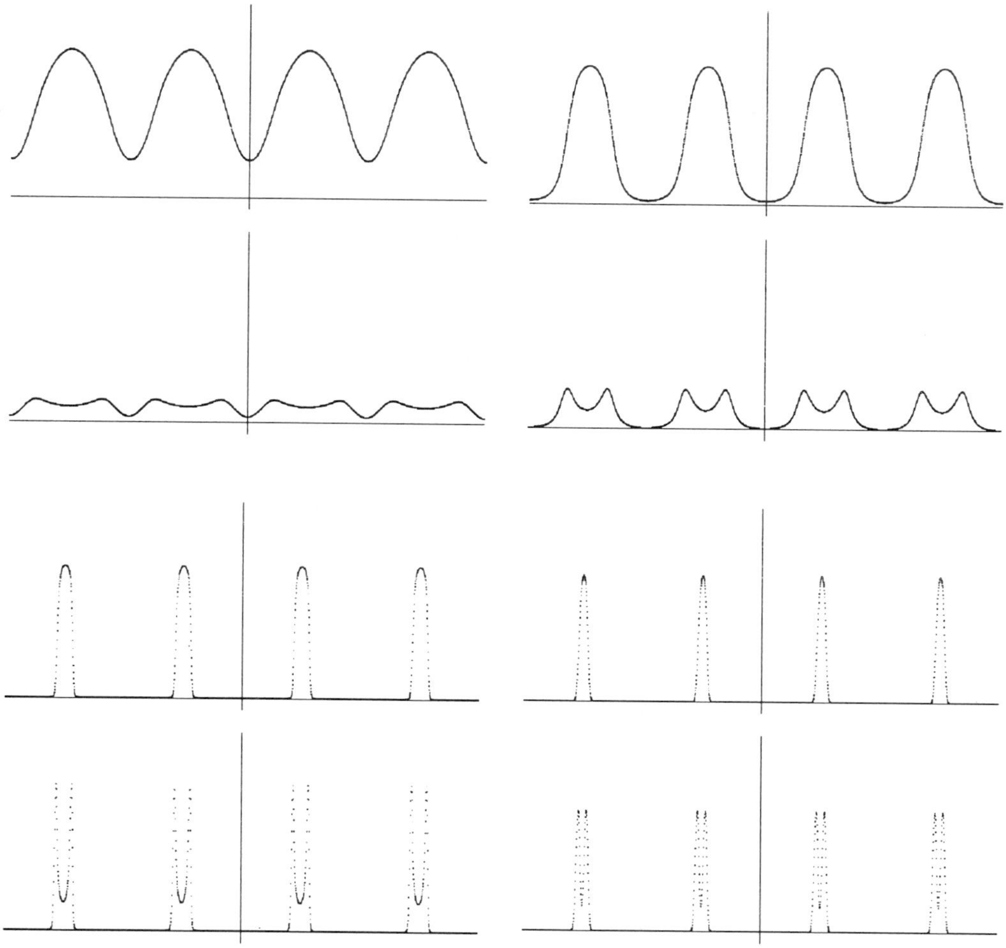

Fig. 2. First and second derivatives for the states shown in Fig. 1.

A fully discrete scheme for the problem (5) is derived from the first equation in 3.1 Theorem in a similar way as the fully discrete scheme for (1). It reads

$$\frac{1}{2}((h_j^{m-1})^2 + (h_{j+1}^{m-1})^2)\frac{1}{\tau}x_j^m + (-x_{j+1}^m + 2x_j^m - x_{j-1}^m)$$
$$= \frac{1}{2}((h_j^{m-1})^2 + (h_{j+1}^{m-1})^2)\frac{1}{\tau}x_j^{m-1}.$$

Last but not least it should be mentioned here that the numerical techniques which are necessary for the computations of the curve shortening flow with one of the two proposed schemes are very simple. The fully discrete schemes are semi implicit schemes and lead to tridiagonal linear systems which can be solved very fast.

References

[1] S. J. Altschuler, *Singularities of the curve shortening flow for space curves*, J. Diff. Geom. 34 (1991) 491-514.

[2] S. J. Altschuler, M. A. Grayson, *Shortening space curves and flow through singularities*, J. Diff. Geom. 35 (1992) 283-298.

[3] K. Deckelnick, G. Dziuk, *Convergence of a Finite Element Method for non-parametric mean curvature flow*, Preprint No. 312 SFB 256 Bonn 1993.

[4] G. Dziuk, *An algorithm for evolutionary surfaces*, Num. Math. 58 (1991) 603-611.

[5] G. Dziuk, *Convergence of a semi–discrete scheme for the curve shortening flow*, Math. Mod. Meth. Appl. Sc. 4 (1994) 589 - 606.

[6] M. A. Grayson, *The heat equation shrinks embedded plane curves to round points*, J. Diff. Geom. 26 (1987) 285 - 314.

[7] M. Gage, R. S. Hamilton, *The heat equation shrinking convex plane curves*, J. Diff. Geom. 23 (1986) 69 - 96.

[8] J. Kacur, K. Mikula, *Numerical solution of the anisotropic curve shortening flow*, Proc. ISNA'92, Prag 1992.

[9] S. Roberts, *A line element algorithm for curve flow problems in the plane*, Research Report CMA-R58-89, Canberra 1989.

Klaus Deckelnick and Gerhard Dziuk
Institut für Angewandte Mathematik
der Universität Freiburg
Hermann-Herder-Str. 10
D-79104 Freiburg i.Br.
Germany

A FASANO AND M PRIMICERIO

Flows through saturated mass exchanging porous media under high pressure gradients

Abstract. Some models of incompressible flows through porous media are examined in various cases exhibiting a relevant interaction between the flow and the solid matrix both of chemical and of mechanical origin.

1. Introduction

In recent years an intense research has been performed about some non-standard filtration phenomena characterized by the presence of a mutual action between the flow and the porous medium.

Such an action can be both of mechanical and of chemical origin. The motivation was provided by illycaffè s.p.a (an Italian Company based in Trieste) who supported a theoretical and experimental research program aimed at producing a mathematical model for the filtration of water through a layer of ground coffee in the operating conditions of an expresso coffee machine (95°C, 9 bars). Under such extreme conditions Darcy's law does no longer apply. Indeed a first series of experiments performed at the laboratories of illycaffè using compact coffee beds and low temperature water (4°C), in order to isolate the mechanical effects, has shown that the volumetric discharge q (amount of water crossing the unit surface p. u. time) through the system exhibits an exponential decay with time to an asymptotic value q_∞ depending on the external constant pressure p_0 applied at the inflow surface. In addition, not even q_∞ satisfy Darcy's law, since it is not proportional to p_0 and can on the contrary have a non-monotone behaviour (typically, the value of q_∞ for $p_0 = 8\,\text{bar}$ is less than the corresponding value for $p_0 = 6\,\text{bar}$) [Baldini, 1992]. Such a behaviour is not explained by any of the nonlinear generalizations of the Darcy's law proposed in the literature (see e.g. [Bear, 1972]).

An experimentally observed fact is that a fine component of the porous medium can be removed by the matrix and be transported by the flux. If the outflow surface is impermeable to solid particles, they accumulate and give rise to a high resistance layer. In [Fasano, Primicerio, Watts, 1994], [Fasano, Primicerio, 1993] a mathematical model has been developed for this complex phenomenon and it has been, shown that this mechanism could explain the anomalous behaviour of the flow.

A concurrent phenomenon is the flow-induced deformation of the porous matrix, leading to a modification of the hydraulic properties of the system. In [Baldini, Petracco, 1991] it is shown that the plasticity of the matrix can also account for the peculiar pressure and time temperature dependence of the flow.

The success of the above mentioned research has encouraged us to thoroughly review the entire question of the filtration of incompressible liquids in the presence of high pressure gradients and of nonlinear phenomena due to various types of interactions between the flow and the porous medium.

Flows through porous media accompanied by chemical reactions, diffusion, crystallization, have been studied in many papers. We just quote here some of them, where additional references can be found. [Pawell, Krannich (1994), Knabner, Van Duijn, Hengst (1994), Van Duijn, Knabner (1991), Chadam, Peirle, Sem (1998), Chadam, Chen, Comparini, Ricci, De Graaf (1987)].

The distinctive feature of the present investigation, which for the moment is confined to isothermal processes, is that a mechanical action of the flow on the medium is considered accompanied by mass exchange, and several effects on the behaviour of the hydraulic properties of the system are taken into account.

On the contrary, diffusion is neglected in view of the time scales fixed by relatively high values of the pressure gradients considered. In addition, our analysis is limited to a one-dimensional geometry.

The research program which is being carried out in Florence is oriented towards two main branches:

(I) Problems with a wetting front penetrating in a dry medium,

(II) Problems of filtration through a saturated layer.

In this paper we will mention briefly the problems of the first class (Part I), while we will deal more extensively with the second class (Part II), presenting a model encompassing various phenomena. Such a model will be shown to be well posed.

PART I. Wetting fronts

1. Introduction

Neglecting capillarity effects and gravity, the simplest version of the wetting front problem is the so-called Green-Ampt model [Green, Ampt (1991)], well known to the soil scientists. The model (in its simple version, corresponding to horizontal flow, where growth plays no role) considers a porous medium occupying the half space $x > 0$ and obeying Darcy's law

$$q = -k\frac{\partial p}{\partial x} \tag{1.1}$$

in the saturated region $0 < x < s(t)$. Here q is the volumetric velocity (already defined), $k > 0$ is a physical constant (the hydraulic conductivity), and p is pressure.

The incompressibility condition $\frac{\partial q}{\partial x} = 0$ yields that p is a linear function of x. If $p_0(t) > 0$ is the pressure in the dry region (and on $x = s(t)$), the pressure gradient is easily found, and the condition that the wetting front moves with the speed of the penetrating liquid, i.e. $\dot{s}(t) = q/\varepsilon$, ε being the porosity, leads immediately to the equation of motion of the wetting front

$$s(t) = \left[\frac{k}{\varepsilon}\int_0^t p_0(\tau)\, d\tau\right]^{1/2} . \tag{1.2}$$

The problem can be adapted to different situations in various ways. For instance, suppose that in the porous medium a substance is distributed initially with some uniform concentration c_0 and that it dissolves in the flow according to

$$\frac{\partial c}{\partial t} = -\lambda c\ , \qquad t > \theta(x), \tag{1.3}$$

with $\theta(x)$ being the inverse function of $x = s(t)$. Assume that $k = k(c)$ is a C^1 function bounded between two positive constants and define

$$f(\tau) = \{\, k(c_0\, e^{-\lambda\tau})\,\}^{-1}. \tag{1.4}$$

Then it is easy to show that the problem of determining the motion of the wetting front $x = s(t)$ is reduced to the solution of the integro-differential equation

$$\dot{s}(t)\left[f(0)\, s(t) + \int_0^t s(\tau) f'(t-\tau)\, d\tau\right] = \frac{1}{\varepsilon}\, p_0, \tag{1.5}$$

with the initial condition

$$s(0) = 0 \ , \tag{1.6}$$

which produces a degeneracy for $t = 0$. It is possible to prove existence and uniqueness.

Other generalizations, like the one briefly described in the next section, lead to much harder problems.

2. The wetting front problem with an elastic response of the porous medium

Leaving aside any chemical interaction, we now suppose that the flow produces an elastic deformation of the medium which results in a dependence of ε and k on q. We assume that $\varepsilon(q)$, $k(q)$ are C^2-functions such that $\varepsilon'(q) \leq 0$, $k'(q) \leq 0$, in agreement with the physical picture, and $\varepsilon''(q) \geq 0$, $k''(q) \geq 0$. Moreover ε and k are a-priori bounded between two positive constants.

The pressure is specified at $x = 0$:

$$p(0,t) = p_0(t) \ , \qquad t > 0 \ , \tag{2.1}$$

where $p_0(t)$ is a positive twice continuously differentiablefunction. We also assume that the pressure vanish at the wetting front $x = s(t)$.

The problem consists in finding a triple (q,p,s) satisfying the system

$$q(x,t) = -k(q)\,\frac{\partial p}{\partial x} \ , \qquad 0 < x < s(t) \ , \ t > 0 \ , \tag{2.2}$$

$$\varepsilon'(q)\frac{\partial q}{\partial t} + \frac{\partial q}{\partial x} = 0 \ , \qquad 0 < x < s(t) \ , \ t > 0 \ , \tag{2.3}$$

$$\dot{s}(t) = \frac{q(s(t),t)}{\varepsilon(q(s(t),t))} \ , \qquad t > 0 \ , \tag{2.4}$$

the second equation expressing the incompressibily of the liquid together with the condition that the medium remains saturated in the wet region.

In the scheme above the cumulative (i.e. macroscopic) deformation is neglected.

This problem has been considered in [Tani, 1996]. It is a rather difficult free boundary problem, because of the nonlinearity of (2.3) and of the interplay between (2.2) and (2.3). Its character is essentially hyperbolic.

By means of a fixed argument it has been shown that a unique classical solution exists when $\varepsilon' < 0$ and $\dot{p}_0 \geq 0$ in a suitable time interval $[0,T]$ (only existence is guaranteed for $\dot{p} \leq 0$).

3. Further developments

An alternative approach to the description of the mechanical action of the flow on the medium, which will be adopted in the case of the saturated layer, is a relaxation law for ε with coefficients dependent on q, so that $\frac{\partial \varepsilon}{\partial t}$ is basically a function of q. The wetting front problem has not been treated in this form and it should be slightly simpler.

Completely open (and interesting) problems arise when to the mechanical action we add

(i) chemical interactions of the type considered in the rest of this paper,

(ii) the thermal interaction of the liquid (injected at a constant temperature, higher than the initial temperature of the medium) and the medium itself, affecting both the mechanical and the chemical properties of the system.

PART II. Filtration in a saturated layer with mass removal.

1. The mathematical model

We will examine this very complex phenomenon under the assumption that the temperature is constant throughout the system. We focus our attention on the coupling between the flow and the consequent mass release from the porous matrix.

This work is in the spirit of [Fasano, Primicerio, Watts (1994)], [Fasano, Primicerio, 1993], where the removal of fine solid particles has been considered. Here we add the possible dissolution of several different species, letting the porosity change, due to the mass loss from the porous matrix and to the deformations (possibly partially elastic) induced by the friction of the flowing incompressible fluid on the solid grains. However we will always neglect the macroscopic deformation and the actual displacement of the grains, considering that a substantial change of the porosity can be produced by micrometric movements of the grains.

Contrary to the above mentioned analysis, in which the transported particles were giving rise to a compact layer of high hydraulic resistance, here we allow all the species to flow out of the system, thus avoiding the difficulties connected to the presence of an internal free boundary.

The ultimate goal is to calculate the concentration of each of the extracted species in the outgoing flow.

We denote by L the thickness of the layer and by $\tilde{x} \in (0,L)$ the coordinate in the direction of the flow. In order to describe the removal of each species from the porous matrix (either solid particles or soluble substances), we introduce the concentrations $\tilde{b}_i$, $\tilde{m}_i$ (grains per unit volume of the whole system) of the i-th species in the porous matrix and in the flow, respectively, and we model the release process as follows

$$\frac{\partial \tilde{b}_i}{\partial \tilde{t}} = -\tilde{F}_i(\tilde{q}, \tilde{\mathbf{b}})\ \tilde{G}_i(\tilde{b}_i - \tilde{\beta}_i(\tilde{q}, \tilde{\mathbf{b}}))\ , \qquad i = 1, \ldots\ldots, n\ , \tag{1.1}$$

where the symbols with $\tilde{}$ denote physical quantities (the tilda will be dropped after suitable rescaling). Here the functions $\tilde{F}_i \geq 0$ express how effective the flow $\tilde{q}$ is in the extraction process, while the functions $\tilde{G}_i$ are such that $\tilde{G}_i(\theta) > 0$ for $\theta > 0$ and $\tilde{G}_i(\theta) = 0$ for $\theta \leq 0$. Accordingly, the functions $\tilde{\beta}_i(\tilde{q}, \tilde{\mathbf{b}})$ represent threshold values for b_i below which the process is switched off. The importance of such terms has been pointed out in the above mentioned papers for the specific case of the removal of solid particles. In case no threshold effect is present one can take $\tilde{\beta}_i = 0$ and $\tilde{G}_i(b_i) = 1$. The dependence of $\tilde{F}_i$ and $\tilde{\beta}_i$ on $\tilde{\mathbf{b}}$ is explained by the influence that the relative amount of all the species can have on the extraction of a single species (for instance the presence of a coating of fat substance can interfere with the dissolution of underlying components).

In (1.1) we have assumed that the release process is not influenced by the concentrations $\tilde{m}_i$ of the dissolved substances, since the total time of permanence of the liquid in the layer is relatively small (for the kind of applications we have described).

We must also remark that which on one hand we consider that porosity changes do affect the flow (mainly through the permeability of the medium), we decided to neglect their influence on the concentrations $\tilde{\mathbf{b}}$ and $\tilde{\mathbf{m}}$.

The transport equation for each species can be written in the form

$$\frac{\partial \tilde{m}_i}{\partial \tilde{t}} + \frac{\partial}{\partial \tilde{x}}\left(\alpha_i \frac{1}{\tilde{\varepsilon}} \tilde{q}\, \tilde{m}_i\right) = -\frac{\partial \tilde{b}_i}{\partial \tilde{t}}\ , \qquad i = 1, \ldots\ldots, n, \tag{1.2}$$

where the porosity $\tilde{\varepsilon}$ is an unknown function of x, t, and the coefficients $\alpha_i \in (0,1]$ are slowing factors. Typically $\alpha_i = 1$ for solutes (i.e. the substance is transported with the fluid velocity) and $\alpha_i \leq 1$ for solid particles.

The combined effect of mass removal and friction of the flow on the porous matrix can be expressed as follows,

$$\frac{\partial\tilde{\varepsilon}}{\partial\tilde{t}} = -\sum_{i=1}^{n}\tilde{\delta}_i\frac{\partial\tilde{b}_i}{\partial\tilde{t}} - \tilde{g}(\tilde{q})\tilde{\xi}\,[\tilde{\varepsilon} - \tilde{\varepsilon}_*(\tilde{q},\tilde{\mathbf{b}})] + \tilde{h}(\tilde{q})\,\tilde{\eta}[\,\tilde{\varepsilon}^*(\tilde{\mathbf{b}}) - \tilde{\varepsilon}\,]\,. \tag{1.3}$$

The three terms of the right hand side of (1.3) contribute to the rate of change of the
porosity $\tilde{\varepsilon}(\tilde{x}, \tilde{t})$ in the following way:

(i) increase of porosity due to the mass losses $\frac{\partial\tilde{b}_i}{\partial\tilde{t}}$, each with a conversion factor $\tilde{\delta}_i \geq 0$,

(ii) decrease of porosity due to flow-induced compression,

(iii) increase of porosity following an elastic response of the medium.

The functions $\tilde{g} \geq 0$, $\tilde{h} \geq 0$ are nondecreasing and nonincreasing, respectively, with $\tilde{g}(0) = 0$. The functions $\tilde{\xi}$, $\tilde{\eta}$ are nondecreasing and vanish for negative values of their argument. The values $\tilde{\varepsilon}_*$ and $\tilde{\varepsilon}^*$ $(\tilde{\varepsilon}_* \leq \tilde{\varepsilon}^*)$ are ideal equilibrium values of porosity for the process of compression and of elastic relaxation, respectively.

Other governing equations are

$$\frac{\partial\tilde{\varepsilon}}{\partial\tilde{t}} + \frac{\partial\tilde{q}}{\partial\tilde{x}} = 0\,, \tag{1.4}$$

expressing both mass conservation and saturation, and Darcy's law

$$\tilde{q}(x,t) = -\tilde{k}(\tilde{\mathbf{b}}, \tilde{\mathbf{m}}, \tilde{\varepsilon})\frac{\partial\tilde{p}}{\partial\tilde{x}}\,, \tag{1.5}$$

in which gravity is neglected.

The model is completed by the initial and boundary conditions

$$\tilde{\mathbf{b}}(\tilde{x},0) = \tilde{\mathbf{b}}_0(\tilde{x})\,, \quad \tilde{\mathbf{m}}(\tilde{x},0) = \tilde{\mathbf{m}}_0(\tilde{x})\,, \qquad 0 < \tilde{x} < L\,, \tag{1.6}$$

$$\tilde{\mathbf{m}}(0,\tilde{t}) = 0\,, \qquad \tilde{t} > 0\,, \tag{1.7}$$

$$\tilde{\varepsilon}(\tilde{x},0) = \tilde{\varepsilon}^*(\tilde{\mathbf{b}}_0) \equiv \tilde{\varepsilon}_0(\tilde{x})\,, \qquad 0 < \tilde{x} < L\,, \tag{1.8}$$

$$\tilde{p}(0,\tilde{t}) = \tilde{p}_0(\tilde{t})\,, \quad \tilde{p}(L,\tilde{t}) = 0\,, \qquad \tilde{t} > 0\,. \tag{1.9}$$

2. Formulation of the problem in dimensionless variables. Statement of the results.

We define

$$x = \tilde{x}/L\,, \qquad t = \tilde{t}/\tilde{t}_0\,,$$

with

$$\tilde{t}_0 = L/\tilde{q}_0 \ , \qquad \tilde{q}_0 = \tilde{k}(\bar{\mathbf{b}}_0,\ \bar{\mathbf{m}}_0, \bar{\varepsilon}_0)\ \tilde{p}_0(0)/L \ ,$$

where $\bar{\mathbf{b}}_0$, $\bar{\mathbf{m}}_0$, $\bar{\varepsilon}_0$ are average values and $\tilde{p}_0(0)$ is supposed to be positive (note that all the components $\bar{b}_{i,0}$ have to be positive).

Next we introduce the dimensionless quantities

$$p(x,t) = \tilde{p}(\tilde{x},\tilde{t})/\tilde{p}_0(0) \ , \qquad p_0(t) = \tilde{p}_0(\tilde{t})/\tilde{p}_0(0) \qquad q(x,t) = \tilde{q}(\tilde{x},\tilde{t})/\tilde{q}_0 \ ,$$

$$b_i(x,t) = \tilde{p}_i(\tilde{x},\tilde{t})/\bar{b}_{i,0} \ , \qquad m_i(x,t) = \tilde{m}_i(\tilde{x},\tilde{t})/\bar{b}_{i,0} \ , \qquad i{=}1, \ldots, n \ ,$$

$$\beta_i(q,\mathbf{b}) = \tilde{\beta}_i(\tilde{q},\tilde{\mathbf{b}})/\bar{b}_{i,0} \ , \qquad i = 1, \ldots, n \ ,$$

$$F_i(q,\mathbf{b}) = \tilde{t}_0\,\tilde{F}_i(\tilde{q},\tilde{\mathbf{b}}) \ , \qquad G_i(b_i,\beta_i) = \tilde{G}_i(\tilde{b}_i,\tilde{\beta}_i) \ , \qquad i = 1, \ldots, n \ ,$$

$$\varepsilon(x,t) = \tilde{\varepsilon}(\tilde{x},\tilde{t}) \ , \quad \varepsilon_0(x) = \tilde{\varepsilon}_0(\tilde{x}) \ , \quad \varepsilon_*(q,\mathbf{b}) = \tilde{\varepsilon}_*(\tilde{q},\tilde{\mathbf{b}}) \ , \qquad \varepsilon^*(\mathbf{b}) = \tilde{\varepsilon}^*(\tilde{\mathbf{b}}) \ ,$$

$$\delta_i = \bar{b}_{i,0}\,\tilde{\delta}_i \ , \qquad g(q) = \tilde{t}_0\,\tilde{g}(\tilde{q}) \ , \qquad h(q) = \tilde{t}_0\,\tilde{h}(\tilde{q}) \ , \qquad \xi = \tilde{\xi} \ , \ \ \eta = \tilde{\eta} \ ,$$

$$k(\mathbf{b},\ \mathbf{m},\ \varepsilon) = \tilde{k}(\tilde{\mathbf{b}},\ \tilde{\mathbf{m}},\ \tilde{\varepsilon})/\tilde{k}(\bar{\mathbf{b}}_0,\ \bar{\mathbf{m}}_0,\ \bar{\varepsilon}_0) \ ,$$

$$b_{i,0}(x) = \tilde{b}_{i,0}(\tilde{x})/\bar{b}_{i,0} \ , \qquad m_{i,0}(x) = \tilde{m}_{i,0}(\tilde{x})/\bar{b}_{i,0} \ , \qquad i = 1, \ldots, n.$$

Now we can write down the set of governing equations in dimensionless form

$$\frac{\partial b_i}{\partial t} = -F_i(q,\ \mathbf{b})\ G_i[b_i - \beta_i(q,\ \mathbf{b})] \ , \qquad i = 1, \ldots, n \ , \tag{2.1}$$

$$\frac{\partial m_i}{\partial t} + \frac{\partial}{\partial x}\Big(\alpha_i \frac{q}{\varepsilon} m_i\Big) = -\frac{\partial b_i}{\partial t} \ , \qquad i = 1, \ldots, n \ , \tag{2.2}$$

$$\frac{\partial \varepsilon}{\partial t} = -\sum_{i=1}^{n} \delta_i \frac{\partial b_i}{\partial t} - g(q)\,\xi[\,\varepsilon - \varepsilon_*(q,\ \mathbf{b})] + h(q)\,\eta[\,\varepsilon^*(\mathbf{b}) - \varepsilon] \ , \tag{2.3}$$

$$\frac{\partial \varepsilon}{\partial t} + \frac{\partial q}{\partial x} = 0 \ , \tag{2.4}$$

$$q = -k(\mathbf{b},\ \mathbf{m},\ \varepsilon)\frac{\partial p}{\partial x} \ , \tag{2.5}$$

to be satisfied in the classical sense in $(0,1)\times(0,T)$, for some given $T > 0$, together with the initial and boundary conditions

$$\mathbf{b}(x,0) = \mathbf{b}_0(x) \ , \qquad \mathbf{m}(x,0) = \mathbf{m}_0(x) \ , \qquad 0 < x < L \ , \tag{2.6}$$

$$\mathbf{m}(0,t) = 0 \ , \qquad 0 < t < T \ , \tag{2.7}$$

$$\varepsilon(x,0) = \varepsilon_0(x) = \varepsilon^*(\mathbf{b}_0) \ , \qquad 0 < x < L \ , \tag{2.8}$$

$$p(0,t) = p_0(t) \ , \qquad p(1,t) = 0 \ , \qquad 0 < t < T \ . \tag{2.9}$$

We assume that all the data in (2.6)-(2.9) *and all the coefficients in the system* (2.1)-(2.5) *are* C^1 *functions and their first derivatives are Lipschitz continuous.*

It is convenient to take the compatibility condition

$$\mathbf{m}_0(0) = 0, \tag{2.10}$$

and to assume, according to the physics of the problem, that there exist constants ε_m, ε_M such that

$$0 < \varepsilon_m \leq \varepsilon_* \leq \varepsilon_M < 1\ , \qquad 0 < \varepsilon_m \leq \varepsilon^* \leq \varepsilon_M < 1\ , \quad \forall\ q,\ \mathbf{b}\ , \tag{2.11}$$

and two constants k_m, k_M such that for all $\mathbf{b}$, $\mathbf{m}$, ε

$$0 < k_m \leq k(\mathbf{b},\ \mathbf{m},\ \varepsilon) \leq k_M\ . \tag{2.12}$$

Moreover the meaning of F_i and g requires

$$F_i(0,\mathbf{b}) = g(0) = 0\ , \qquad i = 1,\ ....,\ n. \tag{2.13}$$

For q variable in a given interval $[q_m,q_M]$, $q_m > 0$, let us denote by g_0, h_0, the respective maxima of the functions g, h, set $G_0 = \max\limits_{i=1,...n} G_i(\|b_{i,0}\|)$ and denote by F_0 (resp. β_0) the maximum of the sup of $F_i(q,\mathbf{b})$ (resp. of $\beta_i(q,\mathbf{b})$) for $q \in [q_m,q_M]$ and $0 < b_i \leq \|b_{i,0}\|$, for $i = 1,\ ....,\ n$, where here and in the sequel $\|\cdot\|$ denotes the sup-norm. In addition we impose

$$-E_* \leq \frac{\partial \varepsilon_*}{\partial q} \leq 0 \tag{2.14}$$

in the same set, for some positive constant E_*, and

$$\|\varepsilon_0\| + \sum_{i=1}^{n} \delta_i\ \|b_{i,0}\| \leq \varepsilon_M\ . \tag{2.15}$$

We also suppose that

$$\frac{\partial F_i}{\partial q} \geq 0\ , \qquad \frac{\partial \beta_i}{\partial q} \leq 0\ , \qquad g'(p) \geq 0, \tag{2.16}$$

and we denote by F_0', g_0', h_0', β_0', G_0' the sup of the absolute values of the first derivatives of the respective functions in the set of values of q and $\mathbf{b}$ considered above. It is not restrictive to assume that these bounds hold for the whole range of q and $\mathbf{b}$. When necessary we specify F_{0q}' for a bound of the q-derivative of F_i, etc.

For the functions $\xi(\theta)$, $\eta(\theta)$ we take

$$0 \le \xi \le \xi_0\,, \qquad 0 \le \eta \le \eta_0\,, \qquad 0 \le \xi' \le \xi_0', \quad 0 \le \eta' \le \eta_0' \tag{2.17}$$

for some positive constants ξ_0, η_0, ξ_0', η_0' and $|\theta| \le \varepsilon_M$.

Concerning $p_0(t)$ we assume that it belongs to $C^1([0,T])$ and that

$$k_m p_0(t) - F_0 G_0 \sum_{i=1}^{n} \delta_i - h_0 \eta (\varepsilon_M - \varepsilon_m) \ge \rho > 0\,, \qquad t \in [0,T] \tag{2.18}$$

for some positive constant ρ, possibly depending on T (nonincreasing).

Finally the following assumption is also needed

$$F_0' \sum_{i=1}^{n} \delta_i \parallel b_{i,0} \parallel + F_0 \sum_{i=1}^{n} \delta_i \beta_0' + (g_0' + h_0')(\varepsilon_M - \varepsilon_m) + E_* g_0 \le \left(1 + \frac{k_M}{k_m}\right)^{-1}, \tag{2.19}$$

in other words the variation with q of the various functions entering the model has to be sufficiently slow. The latter condition appears in a natural way in the proof of the existence and uniqueness and does not seem to be purely technical.

In the rest of the paper we will prove the following theorem

Theorem 2.1. *Under the assumptions listed above problem* (2.1)-(2.9) *has a unique classical solution for any* $T > 0$.

The proof goes through several steps and is based upon a fixed point argument.

3. Determination of the initial flux

Combining (2.1), (2.3), (2.4) we can write

$$\frac{\partial q}{\partial x} = \Gamma(q,\varepsilon,\mathbf{b}) \tag{3.1}$$

with

$$\Gamma(q,\varepsilon,\mathbf{b}) = -\sum_{i=1}^{n} \delta_i F_i(q)\, G_i[b_i - \beta_i(q,\mathbf{b})] + g(q)\,\xi\,[\varepsilon - \varepsilon_*(q,\mathbf{b})] +$$

$$+ h(q)\,\eta\,[\varepsilon^*(\mathbf{b}) - \varepsilon]\,. \tag{3.2}$$

For $t = 0$, ε and $\mathbf{b}$ are known, so that (3.2) is a prescribed function of q and of x and (3.1) is a nonlinear o.d.e. for the unknown $q(x,0)$. Because of the assumptions made on the functions appearing on the r.h.s. of (3.2) we can say that there exists one and only one integral $q(x,0;q_0^*)$ defined in $(0,1)$ for any constant $q_0^* > 0$ and such that

$$q(0,0;q_0^*) = q_0^*\,. \tag{3.3}$$

Thus the problem of determining $q(x,0)$ is reduced to finding q_0^*. The condition to be imposed is the validity of (2.5) ot $t = 0$. Integrating both sides and recalling $p(0,0) = p_0(0) = 1$, $p(1,0) = 0$, we have

$$\int_0^1 \frac{q(x,0;q_0^*)}{k(\mathbf{b}_0, \mathbf{m}_0, \varepsilon_0)} \, dx = 1. \tag{3.4}$$

Lemma 3.1. *Equation* (3.4) *determines* $q_0^* > 0$ *uniquely.*

Proof. First we note that $\lim_{q_0^* \to +\infty} q(x,0;q_0^*) = +\infty$ uniformly in $(0,1)$ and that $q(x,0;0) \equiv 0$, owing to (2.8) and to (2.13). Next we show $q(x,0;q_0^*)$ depends monotonically on q_0^*. To this end we consider the function

$$Q(x,q_0^*) = \frac{\partial}{\partial q_0^*} q(x,0;q_0^*) ,$$

which can be easily calculated from (3.1), (3.2) noting that $Q(0,q_0^*) = 1$

$$Q(x,q_0^*) = \exp\left[\int_0^x \frac{\partial}{\partial q}\Gamma(q(\xi,0;q_0^*),\varepsilon_0(\xi),\mathbf{b}_0(\xi)) \, d\xi\right] > 0 .$$

Thus equation (3.4) has exactly one positive solution.

□

We conclude that $q(x,0)$ can be calculated and that its C^1-norm is easily estimated in terms of the data.

4. The fixed point argument

We remark that equations (2.1), (2.2), (2.3) with the corresponding data (2.6), (2.7), (2.8) provide $\mathbf{b}$, $\mathbf{m}$, ε when $q(x,t)$ is known. Then we construct a mapping on q such that any possible fixed point of it is associated to a solution of our problem. This procedure goes through the following steps:

(i) we compute the initial flux $q_0(x)$, using the method illustrated in Sect. 3,

(ii) we define $R = (0,1) \times (0,T)$ and we introduce the set

$$\Theta = \{q \in C(\bar{R}) \mid q(x,0){=}q_0(x),\ 0 < \rho(T) \le q \le q_M(T),$$

$$\sup \frac{|q(x',t) - q(x'',t)|}{|x' - x''|} \le A_x , \quad \sup \frac{|q(x,t') - q(x,t'')|}{|t' - t''|} \le A_t \ \} \tag{4.1}$$

where "sup" refers to distinct points in R and A_x, A_t are positive constants to be specified later. The constant ρ is the one appearing in (2.18), and

$$q_M(T) = k_M \| p_0 \|_T + g_0 \eta(\varepsilon_M - \varepsilon_m) , \qquad \| p_0 \|_T = \max_{0 \le t \le T} p_0(t) . \tag{4.2}$$

(iii) We take any $q \in \Theta$ and we insert it in the o.d.e.'s system (2.1), then we integrate with the appropriate initial condition thus obtaining the vector function $\mathbf{b}(x,t)$.

(iv) With the same specification of q and with the above determination of $\mathbf{b}$ we obtain $\varepsilon(x,t)$ integrating equation (2.3), which we write in the form

$$\frac{\partial \varepsilon}{\partial t} = -\Gamma(q,\varepsilon,\mathbf{b}) . \tag{4.3}$$

(v) Computing the correspondent vector $\mathbf{m}$ is more complicated, since equations (2.2) involve $\frac{\partial \varepsilon}{\partial x}$. Therefore we use the modified system

$$\frac{\partial m_i}{\partial t} + \alpha_i \frac{q}{\varepsilon} \frac{\partial m_i}{\partial x} + \alpha_i \frac{1}{\varepsilon} \Gamma(q,\varepsilon,\mathbf{b}) - \alpha_i \frac{q}{\varepsilon^2} E m_i =$$

$$= F_i(q)\, G_i[b_i - \beta_i(q,\mathbf{b})] , \qquad i = 1, \ldots, n , \tag{4.4}$$

which is derived from (2.2) replacing $\frac{\partial q}{\partial x}$ by $\Gamma(q,\varepsilon,\mathbf{b})$ and $\frac{\partial \varepsilon}{\partial x}$ by the function $E(x,t)$, which is obtained by integrating

$$\frac{\partial E}{\partial t} = -\Gamma \Gamma_q - E \Gamma_\varepsilon - \nabla_{\mathbf{b}} \Gamma \cdot \mathbf{B} , \tag{4.5}$$

$$E(x,0) = \varepsilon_0'(x) , \qquad 0 \le x \le 1 , \tag{4.6}$$

with $\mathbf{B}$ playing the role of $\frac{\partial \mathbf{b}}{\partial x}$ and being the solution of the system

$$\frac{\partial B_i}{\partial t} = \{ -\Gamma(q,\varepsilon,\mathbf{b}) \frac{\partial F_i}{\partial q} + \nabla_{\mathbf{b}} F_i \cdot \mathbf{B} \}\, G_i[b_i - \beta_i(q,\mathbf{b})]$$

$$+ F_i(q,\mathbf{b}) \{ B_i - \Gamma(q,\varepsilon,\mathbf{b}) \frac{\partial p_i}{\partial q} - \nabla_{\mathbf{b}} \beta_i \cdot \mathbf{B} \}\, G_i'[b_i - \beta_i(q,\mathbf{b})] , \tag{4.7}$$

$$B_i(x,0) = b_{i,0}'(x) , \qquad i = 1, \ldots, n . \tag{4.8}$$

Once the function $E(x,t)$ has been found, we can integrate (4.4) with the appropriate initial and boundary conditions.

(vi) The last step is to determine a function $q^*(x,t)$ associated to $q(x,t)$ and such that when $q^* = q$ it provides a solution of the problem.
Using the selected function $q \in \Theta$ and the functions ε, $\mathbf{b}$ found above, we obtain a known expression for $\Gamma(q,\varepsilon,\mathbf{b})$ and recalling (3.1) we define

$$\hat{q}(x,t,\chi(t)) = \chi(t) + \int_0^x \Gamma(q,\varepsilon,\mathbf{b})\, d\xi \,, \tag{4.9}$$

for each $t \in (0,T)$, where $\chi(t)$ has to be chosen in such a way to make (4.9) consistent with Darcy's law. We proceed as we did for $t = 0$ imposing for each $t \in (0,T]$ the condition

$$\int_0^1 \frac{\hat{q}(x,t;\chi(t))}{k(\mathbf{b},\, \mathbf{m},\, \varepsilon)}\, dx = p_0(t) \,, \tag{4.10}$$

where $(\mathbf{b}, \mathbf{m}, \varepsilon)$ is the triple calculated in the previous steps.
The same argument used in Sect. 3 ensures that (4.10) possesses a unique solution $\chi^*(t)$. At this point we can assert that the operator

$$\mathcal{T} q = q^* \,, \tag{4.11}$$

with

$$q^*(x,t) = \hat{q}(x,t;\chi^*(t)) \tag{4.12}$$

is well defined for any q in Θ.

5. Auxiliary estimates

We begin with some estimates which will allow us to choose the constants in the set Θ so that $\mathcal{T}(\Theta) \subset \Theta$.

Clearly the functions b_i are non-increasing w.r.t. t and

$$0 \le b_i(x,t) \le \| b_{i,0} \| \,, \qquad i = 1, \ldots, n \,, \tag{5.1}$$

yielding a uniform bound for $\frac{\partial b_i}{\partial t}$:

$$0 \le -\frac{\partial b_i}{\partial t} \le F_0 G_0 \,, \qquad i = 1, \ldots, n \,. \tag{5.2}$$

Owing to (2.11) and (2.15) and to the structure of the functions ξ and η in (2.3), we can assert that

$$0 < \varepsilon_m \le \varepsilon(x,t) \le \varepsilon_M < 1 \tag{5.3}$$

and that

$$-g_0\,\xi\,(\varepsilon_M-\varepsilon_m)\le\frac{\partial\varepsilon}{\partial t}\le\delta F_0\,G_0+h_0\,\eta\,(\varepsilon_M-\varepsilon_m)\ , \tag{5.4}$$

where $\delta=\sum_{i=1}^{n}\delta_i$. Because of (4.9) we have the reverse inequality for $\frac{\partial q^*}{\partial x}$

$$-g_0\,\xi\,(\varepsilon_M-\varepsilon_m)\le\frac{\partial q^*}{\partial x}\le g_0\,\xi\,(\varepsilon_M-\varepsilon_m)\ . \tag{5.5}$$

Now we note that from (4.10) we have an estimate of the average $\bar{q}(t)$ of $q^*(x,t)$ over $0<x<1$:

$$k_m p_0(t)\le\bar{q}(t)\le k_M p_0(t)\ . \tag{5.6}$$

Therefore, recalling (2.18) and making use of (5.5), we get

$$0<\rho\le q^*(x,t)\le k_M\,p_0(t)+g_0\,\xi\,(\varepsilon_M-\varepsilon_m)\ . \tag{5.7}$$

Thus we are led to identify $q_M(T)$ in (4.1) with

$$q_M(T)=k_M\,\|\,p_0\,\|_T+g_0\,\xi\,(\varepsilon_M-\varepsilon_m)\ . \tag{5.8}$$

We pass to estimating the Lipschitz norm of b_i w.r.t. x.

For any pair $0<x'<x''<1$ from (2.1) we deduce that the difference

$$\varphi_i(t)=b_i(x',t)-b_i(x'',t) \tag{5.9}$$

satisfies

$$\begin{aligned}-\frac{d\varphi_i}{dt}=&\left\{\frac{\partial F}{\partial q}\Big|_{x=x_1(t)}G_i(b_i-\beta_i)|_{x=x'}+\right.\\&\left.+F_i(q(x'',t),\,\mathbf{b}(x'',t))\ G_i'(b_i-\beta_i)|_{x=x_2(t)}\right\}[q(x',t)-q(x'',t)]\\&+\Big\{\nabla_{\mathbf{b}}F_i|_{x=x_3(t)}\ G_i(b_i-\beta_i)|_{x=x'}+\\&+F_i(q(x'',t),\,\mathbf{b}(x'',t))\ G_i'(b_i-\beta_i)|_{x=x_4(t)}\nabla_{\mathbf{b}}\beta_i\Big\}\cdot\varphi\ ,\end{aligned} \tag{5.10}$$

where $x'\le x_j(t)\le x''$ for $t\in(0,T)$ and $j=1,\,....,\,4$. For $t=0$ we have

$$\varphi(0)=\mathbf{b}_0(x')-\mathbf{b}(x'') \tag{5.11}$$

and thanks to the assumption made in Section 2 we conclude that there exist two nondecreasing functions $C_1(T)$, $C_2(T)$ such that

$$\|\,\mathbf{b}(x',\cdot)-\mathbf{b}(x'',\cdot)\,\|_T\le[C_1(T)+C_2(T)\,A_x]\,|x'-x''|\ . \tag{5.12}$$

A similar procedure applied to (4.3) leads to a parallel estimate for ε

$$\| \varepsilon(x',\cdot) - \varepsilon(x'',\cdot) \|_T \le [C_1(T) + C_2(T)\, A_x]\, |x' - x''| \ . \tag{5.13}$$

Recalling that the data and the coefficients in the governing equations have Lipschitz continuous first derivatives, we can prove a similar inequality for the vector $\mathbf{B}(x,t)$ proceeding as above (see (4.7), (4.8)) and eventually conclude that also the function $E(x,t)$ satisfies the same inequality (see (4.5), (4.6)).

Now we are in position to study the system (4.4) for $\mathbf{m}(x,t)$. For each $i = 1, \ldots,$ n there is a set of characteristic curves which are obtained by integrating

$$\frac{dx}{d\tau} = \alpha_i \frac{q(x,\tau)}{\varepsilon(x,\tau)} \tag{5.14}$$

and setting

$$t = \theta + \tau \ , \qquad \text{with } \theta \ge 0 \text{ and } x|_{\tau=0} = 0 \tag{5.15}$$

or

$$t = \tau \ , \qquad \text{and } x|_{\tau=0} = x_0 \in (0,1) \ . \tag{5.15'}$$

Thus we have a one-parameter family or regular curves whose slope is uniformly bounded for all $q \in \Theta$:

$$0 < \alpha_i \frac{\rho}{\varepsilon_M} \le \alpha_i \frac{q}{\varepsilon} \le \alpha_i \frac{q_M(T)}{\varepsilon_m} \ . \tag{5.16}$$

For a given characteristic curve γ the function m_i satisfies the linear o.d.e.

$$D_\tau m_i - \alpha_i \frac{q}{\varepsilon^2} E\, m_i = -\alpha_i \frac{1}{\varepsilon}\, \Gamma(q,\varepsilon,\, \mathbf{b}) + F_i(q,\mathbf{b})\, G_i[b_i - \beta_i(q,\mathbf{b})] \equiv \Xi_\gamma(t) \ , \tag{5.17}$$

where all the functions are evaluated on γ and D_τ denotes the tangential derivative, i.e. $D_\tau = \left(\frac{\partial}{\partial t} + \frac{dx}{d\tau}\frac{\partial}{\partial x}\right)_{x=x_\gamma(\tau)}$, $x = x_\gamma(\tau)$ being the equation of γ.

The initial value for m_i on γ is either 0 or $m_{i,0}(x_0)$, we call it $m_{i,\gamma}$. Thus we have an explicit representation of $m_i(x_\gamma(t),t)$:

$$m_i(x_\gamma(t),t) = m_{i,\gamma} \exp\left[\int_0^{t-\theta} \alpha_i \frac{q(x_\gamma(\tau),\tau)}{\varepsilon^2(x_\gamma(\tau),\tau)}\, E(x_\gamma(\tau),\tau)\, d\tau \right] +$$

$$+ \int_0^{t-\theta} \Xi_\gamma(\tau) \exp\left[\int_\tau^{t-\tau} \alpha_i \frac{q(x_\gamma(\tau'),\tau')}{\varepsilon^2(x_\gamma(\tau'),\tau')}\, E(x_\gamma(\tau'),\tau')\, d\tau' \right] d\tau \ . \tag{5.18}$$

In order to evaluate the Lipschitz norm of m_i w.r.t. x, we take two points (x',t),

(x'',t) and the corresponding characteristics passing through them γ', γ''. Next we evaluate the difference $m_i(x_{\gamma'}(t),t) - m_i(x_{\gamma''}(t),t)$ by means of (5.18). We can easily realize that for all $\tau \in (0,t-\theta)$ we have

$$|x_{\gamma'}(\tau) - x_{\gamma''}(\tau)| \leq \omega\,|x' - x''| \,, \tag{5.19}$$

the constant ω depending on the bounds in (5.16) and on the Lipschitz coefficients w.r.t. x of q and of ε, i.e. ultimately on the constant A_x (see (4.1).

This information, together with the assumptions made on the chain of data for m_i (including (2.10)), allows us to infer from (5.18) than an equality like (5.12) is valid for $\mathbf{m}(x,t)$ too

$$\| \mathbf{m}(x',t) - \mathbf{m}(x'',t) \|_T \leq [C_1(T) + C_2(T)\,A_x]\,|x' - x''| \,. \tag{5.20}$$

What we have seen so far proves most of the following statement:

Lemma 5.1. *The Lipschitz norms of* $\mathbf{b}$, $\mathbf{m}$, ε *for any given* $q \in \Theta$ *are estimated in terms of the data and of* A_x, *but are independent of* A_t.

Proof. It is enough to recall the estimates (5.2), (5.4), (5.12), (5.13), (5.20) and to deduce a bound for $|\mathbf{m}(x,t') - \mathbf{m}(x,t'')| \,/\, |t' - t''|$ directly from (4.4), having established a simple uniform bound for the function E.

□

The last auxiliary result we need before passing to the existence proof is the estimation of the derivative $\dfrac{\partial q^*}{\partial t}$.

We recall that

$$q^*(x,t) = \chi^*(t) + \int_0^x \Gamma(q,\varepsilon,\mathbf{b})\,d\xi \,, \tag{5.21}$$

with $q \in \Theta$ and $\chi^*(t)$ being the unique solution of (4.10). Thus

$$\frac{\partial q^*}{\partial t} = \dot{\chi}^* + \int_0^x \frac{\partial \Gamma}{\partial t}\,d\xi \,. \tag{5.22}$$

Differentiating (4.10) we get

$$\int_0^1 \frac{1}{k(\mathbf{b},\mathbf{m},\varepsilon)} \left[\dot{\chi}^*(t) + \int_0^x \frac{\partial \Gamma}{\partial t}\,d\xi \right] dx =$$

$$\dot{p}_0(t) + \int_0^1 \frac{q^*(x,t)}{k^2(\mathbf{b},\mathbf{m},\varepsilon)} \left[\nabla_{\mathbf{b}} k \cdot \frac{\partial \mathbf{b}}{\partial t} + \nabla_{\mathbf{m}} k \cdot \frac{\partial \mathbf{m}}{\partial t} + \frac{\partial k}{\partial \varepsilon}\frac{\partial \varepsilon}{\partial t} \right] dx \; . \tag{5.23}$$

Since $q \in \Theta$ is Lipschitz, $\frac{\partial \Gamma}{\partial t}$ is defined a.e. in R:

$$\begin{aligned} \frac{\partial \Gamma}{\partial t} = \Big\{ &- \sum_{i=1}^{n} \delta_i \left[(\nabla_{\mathbf{b}} F_i)\, G_i(b_i - \beta_i) + F_i (1 - \nabla_{\mathbf{b}} \beta_i)\, G_i'(b_i - \beta_i) \right] \\ &- [h(q)\, \eta'(\varepsilon^* - \varepsilon)\, \nabla_{\mathbf{b}} \varepsilon^* + g(q)\, \xi'(\varepsilon - \varepsilon_*)\, \nabla_{\mathbf{b}} \varepsilon_*] \Big\} \cdot \frac{\partial \mathbf{b}}{\partial t} + \\ &+ [h(q)\, \eta'(\varepsilon^* - \varepsilon) + g(q)\, \xi'(\varepsilon - \varepsilon_*)]\Big\} \cdot \frac{\partial \varepsilon}{\partial t} + \\ \frac{\partial q}{\partial t} \Big\{ &- \sum_{i=1}^{n} \delta_i \left(\frac{\partial F_i}{\partial q} G_i(b_i - \beta_i) - F_i\, G_i'(b_i - \beta_i) \frac{\partial \beta_i}{\partial q} \right) + \\ &+ g'(q)\, \xi(\varepsilon - \varepsilon_*) - g(q)\, \xi'(\varepsilon - \varepsilon_*)\, \frac{\partial \varepsilon_*}{\partial q} - h'(q)\, \eta(\varepsilon^* - \varepsilon) \Big\} . \end{aligned} \tag{5.24}$$

On the basis of the assumptions listed in Section 2 and of Lemma 5.1, we have

$$\| \frac{\partial \Gamma}{\partial t} \|_{L_\infty} \le C_3 + C_4 A_t \; , \tag{5.25}$$

with

$$C_4 = \delta\, (F'_{0q}\, G_0 + F_0\, G'_{0q}\, \beta'_{0q}) + g'_0\, \xi(\varepsilon_M - \varepsilon_m) + g_0\, \xi'_0\, \xi_* + h'_0\, \eta(\varepsilon_M - \varepsilon_m) \; . \tag{5.26}$$

Note that each term in (5.23) contains the sup of a derivative w.r.t. q.

Going back to (5.23) and using again Lemma 5.1, we obtain the basic estimate

$$\| \dot{\chi}^* \|_T \le k_M \left\{ \| \dot{p}_0 \|_T + \frac{1}{k_m} (C_3 + C_4 A_t) + C_5 + C_6 A_x \right\} . \tag{5.27}$$

6. Proof of Theorem 2.1

At this point we are in position for proving the following

Proposition 6.1. *If we take* $q_M(T)$ *as specified in* (5.8) *and*

$$A_x = \max\, [q_0\, \xi(\varepsilon_M - \varepsilon_m),\ \delta F_0 G_0 + h_0\, \eta(\varepsilon_M - \varepsilon_m)] \; , \tag{6.1}$$

and if the constant C_4 *defined by* (5.26) *is such that*

$$C_4 \le \left(\frac{k_M}{k_m} + 1 \right)^{-1} , \tag{6.2}$$

then we can select A_t *in terms of the data only so that the operator* $\mathcal{T}$ *maps* Θ *into itself.*

Proof. Concerning the choice of $q_M(T)$ and of A_x it is enough to recall (5.7) and (5.5). Next we recall (5.22), (5.25) and (5.27) and we conclude that

$$\|\frac{\partial q^*}{\partial t}\| \le k_M[\,\|\dot{p}_0\|_T + C_5 + C_6 A_x] + \left(\frac{k_M}{k_m} + 1\right)(C_3 + C_4 A_t)\ , \tag{6.3}$$

where A_x is now specified by (6.1).

Therefore, if (6.2) is true we can make the r.h.s. of (6.3) less than A_t taking

$$A_t \ge \left\{1 - C_4\left(1 + \frac{k_M}{k_m}\right)\right\}^{-1} \left\{k_M[\,\|\dot{p}_0\|_T + C_5 + C_6 A_x] + C_3\left(1 + \frac{k_M}{k_m}\right)\right\}. \tag{6.4}$$

□

Next, we prove the continuity of the operator $\mathcal{T}$:

Proposition 6.2. *For* $q_1, q_2 \in \Theta$

$$\|q_1^* - q_2^*\| \le C\,\|q_1 - q_2\| \tag{6.5}$$

for some positive constant C *depending on the data.*

Proof. We denote by $\mathbf{b}^{(j)}$, $\mathbf{m}^{(j)}$, ε_j, χ_j^* the quantities calculated for $q = q_j$, $j = 1, 2$.

Recalling (5.21) we have

$$\|q_1^* - q_2^*\| \le \|\chi_1^* - \chi_2^*\|_T + \|\Gamma(q_1,\varepsilon_1,\mathbf{b}^{(1)}) - \Gamma(q_2,\varepsilon_2,\mathbf{b}^{(2)})\|\ . \tag{6.6}$$

The argument for estimating the differences $\|\mathbf{b}^{(1)} - \mathbf{b}^{(2)}\|$ and $\|\varepsilon_1 - \varepsilon_2\|$ follows the same pattern as the proof of (5.12), (5.13) and leads to

$$\|\mathbf{b}^{(1)} - \mathbf{b}^{(2)}\| \le M_{\mathbf{b}} T\,\|q_1 - q_2\|\,, \tag{6.7}$$

$$\|\varepsilon_1 - \varepsilon_2\| \le M_\varepsilon T\,\|q_1 - q_2\|\,, \tag{6.8}$$

with $M_{\mathbf{b}}$, M_ε depending on the data. Remembering (5.26), it is not difficult to realize that

$$\|\Gamma(q_1,\varepsilon_1,\mathbf{b}^{(1)}) - \Gamma(q_2,\varepsilon_2,\mathbf{b}^{(2)})\| \le (C_4 + M_\Gamma)\,\|q_1 - q_2\|\,, \tag{6.9}$$

where M_Γ is easily constructed in terms of $M_{\mathbf{b}}$, M_ε and of $\|\nabla_{\mathbf{b}}\Gamma\|$, $\|\frac{\partial\Gamma}{\partial\varepsilon}\|$.

It remains to estimate $\|\chi_1^* - \chi_2^*\|_T$. From (4.10) we have

$$\int_0^1 \left[\frac{\chi_1^* + \int_0^x \Gamma_1 d\xi}{k(\mathbf{b}^{(1)},\mathbf{m}^{(1)},\varepsilon_1)} - \frac{\chi_2^* + \int_0^x \Gamma_2 d\xi}{k(\mathbf{b}^{(2)},\mathbf{m}^{(2)},\varepsilon_2)} \right] dx = 0 \;, \tag{6.10}$$

with a clear meaning of Γ_1 and Γ_2. Hence

$$\frac{1}{k_M} \| \chi_1^* - \chi_2^* \|_T \le \frac{1}{k_m} \| \Gamma_1 - \Gamma_2 \| + q_M(T) \frac{1}{k_m^2} \| k_1 - k_2 \| , \tag{6.11}$$

k_1, k_2 denoting the denominators in (6.10).

The quantity to be estimated for bounding $\| k_1 - k_2 \|$ is the difference $\| \mathbf{m}^{(1)} - \mathbf{m}^{(2)} \|$.

Let us consider a given pair $(m_i^{(1)}, m_i^{(2)})$ and for the same index i the characteristics γ_1, γ_2 passing through a previously chosen point (x,t) and corresponding to q_1 and q_2, respectively.

Integrating the equations

$$\dot{x}_{\gamma_j}(\tau) = \alpha_i \, q_j(x_{\gamma_j},\tau) \, / \, \varepsilon_j(x_{\gamma_j},\tau) \;, \qquad j = 1, 2$$

backward in time with the conditions $x_{\gamma_1}(t) = x_{\gamma_2}(t) = x$, we can estimate the difference

$$\| x_{\gamma_1} - x_{\gamma_2} \|_t \le Mt \, \| q_1 - q_2 \| \tag{6.12}$$

for some computable constant M, thanks to (6.8). It can also be seen from (4.5)-(4.8) that the functions E_1, E_2 corresponding to q_1, q_2 satisfy inequality of the same type as (6.8), i.e.

$$\| E_1 - E_2 \| \le M_E T \, \| q_1 - q_2 \| . \tag{6.13}$$

Thus from the explicit formula (5.18) we obtain the desired inequality

$$\| \mathbf{m}^{(1)} - \mathbf{m}^{(2)} \| \le M_m T \, \| q_1 - q_2 \| , \tag{6.14}$$

whence

$$\| k_1 - k_2 \| \le M_k T \, \| q_1 - q_2 \| . \tag{6.15}$$

From (6.9) and (6.15) we obtain

$$\| \chi_1^* - \chi_2^* \|_T \le \frac{k_M}{k_m} (C_4 + M_\chi T) \, \| q_1 - q_2 \| , \tag{6.16}$$

with M_χ (like M_m, M_k) depending in a known way on the data.

Thus we can go back to (6.6) and we write

$$\| q_1^* - q_2^* \| \le \left\{ C_4\left(\frac{k_M}{k_m}+1\right)+M^*T \right\} \| q_1 - q_2 \| , \tag{6.17}$$
with M^* depending on the data.

□

Since the set Θ is closed, bounded, convex and compact in the selected topology, the existence of at least one fixed point follows from Schauder's theorem.

At this point we remember condition (6.2) on the smallness of the constant C_4 and we take T sufficiently small in order to make

$$C_4\left(\frac{k_M}{k_m}+1\right)+M^*T < 1 , \tag{6.18}$$

thus (6.17) ensures that the operator $\mathcal{T}$ is contractive and consequently proves uniqueness.

The latter result is extended with no limitation in time by a simple iteration.

We conclude this section with the following positivity result

Theorem 6.3. *All the components of the vectors* **b** *and* **m** *are nonnegative.*

Proof. The proof is trivial for **b**. Concerning **m**, for each index i, we take any two characteristics $x = x_{\gamma'}(\tau)$, $x = x_{\gamma''}(\tau)$, $x_{\gamma'} < x_{\gamma''}$, and integrate (2.2) in the region between them bounded by $\tau = t$ and by the axis. Due to the non-negativity of the source term, we obtain

$$\int_{x_{\gamma'}(t)}^{x_{\gamma''}(t)} m_i(x,t)\, dx \ge \int_{I(\gamma',\gamma'')} m_{i,0}(x)\, dx \ge 0 , \tag{6.19}$$

where $I(\gamma',\gamma'')$ is the intersection (possibly void) of the closure of the region considered with $x = 0$. The non-negativity of m_i follows trivially.

□

References

Baldini, G. (1992) Filtrazione non linare di un fluido attraverso un mezzo poroso deformabile, Thesis, University of Florence, Italy.

Baldini, G. and Petracco, M. (1994) Models for water percolation during the penetration of espresso coffee, Proc. ECMI 93, A. Fasano, M. Primicerio eds., 131-138, Teubner, Stuttgart.

Bear, J. (1972) Dynamics of Fluids in Porous Media, American Elsevier, New York.

Fasano, A. and Primicerio, M. (1993) Mathematical models for filtration through porous media interacting with the flow. In "Nonlinear Mathematical problems in Industry, I", M. Kawarada, N. Kenmochi, N. Yanagihara eds., Math. Sci. & Appl. 1, 61-85, Gakkotosho, Tokyo.

Fasano, A., Primicerio, M. and Watts, A. (1994) On a filtration problem with flow-induced displacement of free particles. In "Boundary Control and Boundary Variations", 7th IFIP Conference, J.P. Zolesio ed., to appear.

Green, W.H. and Ampt, C.A. (1911) Studies on soil physics 1. The flow of air and water through soils, J. Agric. Sci. 4, 1-24.

Tani, P. (1994) Fronti di saturazione in mezzi porosi con caratteristiche idrauliche variabili, Thesis, University of Florence, Italy.

Pawell, A. and Krannich (1994), Dissolution effects arising in transport in porous media which affect a chemical equilibrium, Math. Meth. in Appl. Sci., to appear.

Knabner, P., Van Duijn, C.J. and Nengst, S. (1994) An analysis of crystal dissolution fronts in flows through porous media, preprint IAAS, n. 90, Berlin.

Chadam, J., Peirle, A. and Sem, A. (1988) Weakly nonlinear stability of reaction-infiltration interfaces, SIAM J. Appl. Math. 48, 1362-1378.

Chadam, J., Chen, X., Comparini, E., and Ricci, R. (1994) Travelling wave solutions of a reaction-infiltration problem and a related free boundary problem, EJAM, to appear.

Van Duijn, C.J. and Knabner, P. (1991) Solute transport in porous media with equilibrium and non-equilibrium multiplicity adsorption in travelling waves, J. Reine Angew. Math., 1-49.

De Graaf, J.M. (1987) Nonlinear diffusion problems in hydrology and biology, Ph.D. Thesis, Univ. Leiden.

Authors' address: Dipartimento di Matematica "U.Dini", viale Morgagni 67/a, 50134 FIRENZE (Italy)

O KAVIAN AND B RAO

Existence of nontrivial solutions to the Marguerre–von Kármán equations

Abstract We prove that there exist an infinite family of shallow shells configurations for which the corresponding Marguerre-von Kármán equations, with the homogeneous Dirichlet boundary condition, have at least two nontrivial solutions.

1. Introduction

Let $\omega \subset \mathbb{R}^2$ be a bounded domain with smooth boundary $\partial\omega$. Given one smooth function $\theta : \omega \to \mathbb{R}$, we define the middle surface $\widetilde{\omega} \subset \mathbb{R}^3$ of a shallow shell as follows :

$$\widetilde{\omega} = \{(x_1, x_2, \theta(x_1, x_2)) \in \mathbb{R}^3 \ ; \quad \forall x = (x_1, x_2) \in \omega\}$$

We assume that the clamped shell is subjected to a applied vertical body force and that the load is subsequently removed. We want to know if the shell will return to its original state or will find another stable defomed state. In other word, we want to study the existence of nontrivial solutions to the homogeneous Marguerre-von Kármán equations involving a parameter $\varepsilon > 0$, which characterizes the thickness of the shell :

$$(1.1) \qquad \begin{cases} \varepsilon^2 \Delta^2 \zeta = [\varphi, \zeta] + [\varphi, \theta] & \text{in } \omega, \\ \Delta^2 \varphi = -[\zeta, \zeta] - 2[\zeta, \theta] & \text{in } \omega, \\ \zeta = \partial_\nu \zeta = 0 & \text{on } \partial\omega, \\ \varphi = \partial_\nu \varphi = 0 & \text{on } \partial\omega, \end{cases}$$

where the unknowns $\zeta, \varphi : \omega \to R$ are respectively the vertical displacement and the Airy's stress function of the shell, and where for smooth functions η and ζ, the bracket $[\cdot\ ,\ \cdot]$ is defined as follows :

$$(1.2) \qquad [\eta, \zeta] = \frac{\partial^2 \eta}{\partial x_1^2} \frac{\partial^2 \zeta}{\partial x_2^2} - 2 \frac{\partial^2 \eta}{\partial x_1 \partial x_2} \frac{\partial^2 \zeta}{\partial x_1 \partial x_2} + \frac{\partial^2 \eta}{\partial x_2^2} \frac{\partial^2 \zeta}{\partial x_1^2} \ .$$

For the details of the obtention of these equations by the asymptotic expansion method, we refer to Ciarlet-Paumier [1986].

In the case of a simply supported shell, Rupprecht [1981] proved that these equations admit at least two nontrivial solutions, if the thickness $\varepsilon > 0$ is sufficiently small. In the present case of a clamped shell, Kesavan-Srikanth [1983] proved that the equations (1.1) have only trival solution if $\varepsilon > 0$ is large enough. It is conjectured by Rupprecht [1981] that the trivial solution is the unique solution to the equations (1.1) for all parameter $\varepsilon > 0$. In this work (the results of which were announced in Kavian–Rao [1993]), we will prove that this conjecture is not true.

For general discussions regarding the question of bifurcation and numerical simulation of equations (1.1), we refer to Ciarlet-Rabier [1980], Paumier-Rao [1989] and Rao [1993].

Let us first introduce in the Sobolev space $H_0^2(\omega)$ the following inner product :

$$(1.3) \qquad \langle u, v\rangle = \int_\omega \Delta u \Delta v dx, \qquad \|u\|_2^2 = \Big(\int_\omega |\Delta u|^2 dx\Big), \qquad \forall u, v \in H_0^2(\omega).$$

Next, we define the operateur B as follows :

$$(1.4) \qquad \forall\ (\eta, \zeta) \in H_0^2(\omega) \times H_0^2(\omega), \qquad \begin{cases} \Delta^2 B(\eta,\zeta) = [\eta,\zeta] & \text{in } \omega, \\ B(\eta,\zeta) \in H_0^2(\omega), \end{cases}$$

Now using the operator B, we can write the equations (1.1) into the following form :

$$(1.5) \qquad \begin{cases} \varepsilon^2\zeta = B(\zeta+\theta, \varphi)\ , \\ \varphi = -B(\zeta,\zeta) - 2B(\zeta,\theta)\ . \end{cases}$$

Substituting the second equation into the first one, we obtain

$$(1.6) \qquad \varepsilon^2\zeta - B(\zeta+\theta, B(\theta,\theta)) + B(B(\zeta+\theta,\zeta+\theta),\zeta+\theta) = 0\ .$$

Once the equation (1.6) is solved, the stress function φ can be calculayed explicitely from the second equation of (1.5). Hence it is sufficient to consider the equation (1.6). In the next section, we will prove that there exist an infinity of functions $\theta \in H_0^2(\omega)$ for which the equation (1.6) admits at least two nontrivial solutions. We next prove that for each deformed configuration of the shell, there exist a neighborhood $V \subset H_0^2(\omega)$ and a mapping $h : \eta \in V \longrightarrow H_0^2(\omega)$, such that for each function $\theta := h(\eta)$ the equations (1.6) admit at least one non-trivial solution $\zeta := \eta - h(\eta)$.

2. Existence of Nontrivial Solutions

We first give the following fondamental proprities of the operator B whose proof can be found in Berger [1967, 1977] and Ciarlet-Rabier [1980] .

Proposition 1. The operator B defined in (1.4) is bilinear, symmetric, continuous from the space $H_0^2(\omega) \times H_0^2(\omega)$ into the space $H_0^2(\omega)$ such that
(i) for any $\eta \in H_0^2(\omega), B(\eta,\eta) = 0, \quad \Longrightarrow \eta = 0,$

(ii) for any $\xi, \eta, \zeta \in H_0^2(\omega)$ we have : $\langle B(\xi,\eta),\zeta\rangle = \langle B(\xi,\zeta),\eta\rangle,$

(iii) for any $(\eta,\zeta) \in (H^2(\omega))$ we have : $\|B(\eta,\zeta)\|_2 \leq C(\omega)\|\eta\|_{W^{1,4}(\omega)}\|\zeta\|_{W^{1,4}(\omega)}$.

Proposition 2. For any $\theta \in H_0^2(\omega)$, the operator $u \to B(\theta, B(\theta,u))$ is compact and self-adjoint in the space $H_0^2(\omega)$.

Proof. The compacteness is a direct consequence of (iii) of Proposition 1. Now given any $u, v \in H_0^2(\omega)$, using (ii) of Proposition 1, we have :

$$\langle B(\theta, B(\theta,u)), v\rangle = \langle B(\theta,v), B(\theta,u)\rangle = \langle u, B(\theta, B(\theta,v))\rangle \ .$$

The proof is thus complete.

Theorem 3. (Kesavan-Srikanth [1983]) Let $\lambda_1 > 0$ be the largest eigenvalue of the self-adjoint operator $B(\theta, B(\theta,\cdot))$. Then the homogeneous equation (1.6) has only trivial solution if the thickness ε is large enough.

Proof. Let $\zeta \in H_0^2(\omega)$ be a solution of the equation (1.6). Then we have :

$$\varepsilon^2\|\zeta\|^2 - \frac{1}{4}\|B(\theta,\zeta)\|^2 + \|B(\zeta,\zeta) + \frac{3}{2}B(\theta,\zeta)\|^2 = 0. \tag{2.1}$$

By the definition of $\lambda_1 > 0$ and by neglecting the positive term in the equality (2.1), it follows that

$$\|B(\theta,\zeta)\|^2 \leq \lambda_1\|\zeta\|^2 \quad \text{and} \quad \varepsilon^2\|\zeta\|^2 \leq \frac{1}{4}\|B(\theta,\zeta)\|^2 \ .$$

We deduce that $\zeta \equiv 0$ if $4\varepsilon^2 > \lambda_1$. The proof is complete.

This result means that thick shells snap back to their original states. In the next result, we will show that there exist an infinity of functions $\theta \in H_0^2(\omega)$, for which the equation (4.1) admits at least two nontrivial solutions. To do this end, we first recall the following classical result of critical points theory (*cf.* Palais [1970]) :

Theorem 4. Let J, F be two functionals of class $C^1(H_0^2(\omega)\,;\mathbb{R})$. We define the constraint $S := \{\theta \in H_0^2(\omega)\,; F(\theta) = 1\}$. Assume that J is even, lower bounded, non constant and satisfies the Palais-Smale condition on the constraint S. For all integer $j \geq 1$, setting

$$E_j := \{g(S^{j-1})\ ;\ g \in C(S^{j-1}, S),\ \text{old}\ \} \quad \text{and} \quad c_j = \inf_{A \in E_j} \max_{\theta \in A} J(\theta)\ .$$

Then for all $j \geq 1, E_j \neq \emptyset$ and c_j is a critical value of J on S. Moreover, we have :

$$c_j \leq c_{j+1} \quad \text{and} \quad \lim_{j \to +\infty} c_j = +\infty\ .$$

Now let us define the functionals :

$$J(\theta) = \|\theta\|^2, \quad F(\theta) = \|B(\theta, \theta)\|^2\ , \qquad \forall \theta \in H_0^2(\omega), \tag{2.2}$$

and the smooth infinite dimensional manifold :

$$S = \{\theta \in H_0^2(\omega)\ : F(\theta) = 1\}. \tag{2.3}$$

Proposition 5. There exist a infinite sequence of critical values c_j :

$$0 < c_j \leq c_{j+1}, \qquad \lim_{j \to +\infty} c_j = +\infty, \tag{2.4}$$

and a corresponding sequence of critical points $\theta_j \in S$ such that the pair (c_j, θ_j) satisfies the following nonlinear eigenvalue problem :

$$\theta_j = c_j B(\theta_j, B(\theta_j, \theta_j)), \quad \forall j \geq 1\ . \tag{2.5}$$

Proof. We first verify that the functional J satisfies the Palais-Smale condition on the manifold S. Let $(\theta_n, \lambda_n) \in S \times \mathbb{R}$ be a sequence satisfying

$$\varepsilon_n := J'(\theta_n) - \lambda_n F'(\theta) \to 0\ , \qquad \text{in } H_0^2(\omega)\ , \tag{2.6}$$

$$J(\theta_n) \to c \in \mathbb{R}\ . \tag{2.7}$$

Then using (ii) of Proposition 1, one straightforward computation gives that

$$\varepsilon_n = 2\theta_n - 4\lambda_n B(\theta_n, B(\theta_n, \theta_n)) \to 0 , \qquad \text{in } H_0^2(\omega) . \tag{2.8}$$

This implies that

$$\langle \varepsilon_n, \theta_n \rangle = 2J(\theta_n) - 4\lambda_n \to 0 , \qquad \text{in } H_0^2(\omega) , \tag{2.9}$$

which, together with (2.7) shows that $\lambda_n \to c/2$.

On the other hand, using (iii) of Proposition 1, we see that B is compact. Hence since θ_n is bounded, and since λ_n converges to $c/2$, we deduce from the condition (2.8) that there exist a subsequence denoted still by θ_n converging to $\theta \in H_0^2(\omega)$. We have thus proved that J satisfies the Palais-Smale condition on S.

Now applying the Theorem 4 to the function J with the constraint S defined in (2.2)-(2.3), we deduce that there exist an infinite sequence of critical values $c_j > 0$ and a corresponding sequence of critial points $\theta_j \in S$ suth that for all integer $j \geq 1$ the pair (c_j, θ_j) is a solution of the nonlinear value problem (2.5).

Theorem 6. There exist an infinity of functions $\theta \in H_0^2(\omega)$ for which the equation (1.6) admits at least two nontrivial solutions.

Proof. We search for a function $\theta \in H_0^2(\omega)$ and a real $\alpha \neq 0$ such that $-\alpha\theta$ is a nontrivial solution of the equation (1.6). If this is true, then the function θ must satisfy the equation :

$$\theta = f(\alpha) B(\theta, B(\theta, \theta)), \qquad f(\alpha) := \frac{1}{\varepsilon^2}(\alpha - 1)(2 - \alpha). \tag{2.10}$$

Now let the pair (c_j, θ_j) be a solution of the nonlinear eigenvalue problem (2.5). For $1 < \alpha < 3/2$, setting

$$\theta_{\alpha,j} = \sqrt{\frac{c_j}{f(\alpha)}}\, \theta_j , \tag{2.11}$$

we verify easily that $\theta_{\alpha,j}$ satisfies the equation (2.10). It follows that $\zeta_{\alpha,j} := -\alpha\theta_{\alpha,j}$ is a nontrivial solution of the equation (1.6). On the other hand, we see that $f(\alpha)$ is invariant if α is replaced by $(3 - \alpha)$. Hence we find that $\widetilde{\zeta}_{\alpha,j} = (\alpha - 3)\theta_{\alpha,j}$ gives another nontrivail solution of the equation (1.6). The proof is thus complete.

Let us denote by $G(\theta, \zeta)$ the homogeneous equation (1.6) :

$$G(\theta, \zeta) := \varepsilon^2 \zeta - B(\zeta + \theta, B(\theta, \theta)) + B(B(\zeta + \theta, \zeta + \theta), \zeta + \theta) . \tag{2.12}$$

Theorem 7. Let $1 < \alpha < 3/2$ such that $\varepsilon^2(2-\alpha)/2c_j$ is not an eigenvalue of the compact and self-adjoint operator $B(\theta_j, B(\theta_j, \cdot))$. Then there exist a neighborhood $V \subset H_0^2(\omega)$ and one mapping $h \; : V \to H_0^2(\omega)$ such that

$$G(h(\eta), \eta - h(\eta)) = 0, \qquad \forall \eta \in V. \tag{2.13}$$

Proof. We first define :

$$H(\theta, \eta) = G(\theta, \eta - \theta), \qquad \eta_{\alpha,j} = \zeta_{\alpha,j} + \theta_{\alpha,j} \, . \tag{2.14}$$

Then a direct computation gives that

$$H(\theta_{\alpha,j}, \eta_{\alpha,j}) = G(\theta_{\alpha,j}, \zeta_{\alpha,j}) = 0 \, , \tag{2.15}$$

$$\begin{aligned} D_\theta H(\theta_{\alpha,j}, \eta_{\alpha,j}) &= D_\theta G(\theta_{\alpha,j}, \zeta_{\alpha,j}) - D_\zeta G(\theta_{\alpha,j}, \zeta_{\alpha,j}) \\ &= -\varepsilon^2 I + \frac{2}{2-\alpha} c_j B(\theta_j, B(\theta_j, \cdot)) \, . \end{aligned} \tag{2.16}$$

Since $\varepsilon^2(2-\alpha)/2c_j$ is not an eigenvalue of the compact, self-adjoint operator $B(\theta_j, B(\theta_j, \cdot))$, then we deduce that the derivative $D_\theta H(\theta_{\alpha,j}, \eta_{\alpha,j})$ invertible in the space $H_0^2(\omega)$. Applying the implicit function theorem to the function $H(\theta, \eta)$ in a neighborhood of the point $(\theta_{\alpha,j}, \eta_{\alpha,j})$, we deduce that there exist a neighborhood $V \subset H_0^2(\omega)$ of $\eta_{\alpha,j}$ and a mapping $h \; : V \to H_0^2(\omega)$ such that

$$h(\eta_{\alpha,j}) = \theta_{\alpha,j}, \qquad H(h(\eta), \eta) = G(h(\eta), \eta - h(\eta)) = 0, \quad \forall \eta \in V \, .$$

Remark. From (2.13), we see that for each configuration function $\theta \; := h(\eta)$ there exists at least one nontrivial solution $\zeta \; := \eta - h(\eta)$, which is not necessarly a homothetic function of θ as that one given by (2.12).

Remark. The idea of the proof of Theorem 7 consists in perturbing the homogeneous equation (1.6) in a neighborhood of the deformed configuration $\eta_{\alpha,j}$. In fact, by one straightforward computation, we find the derivative of the function $G(\theta, \zeta)$ with respect to the variable ζ at the point $(\theta_{\alpha,j}, \zeta_{\alpha,j})$:

$$D_\zeta G(\theta_{\alpha,j}, \zeta_{\alpha,j}) = \varepsilon^2 I + \frac{2(\alpha-1)}{2-\alpha} c_j B(\theta_j, B(\theta_j, \cdot)) + \frac{\alpha}{1-\alpha} c_j B(B(\theta_j, \theta_j), \cdot) \, .$$

We don't know if this derivative is invertible in the space $H_0^2(\omega)$. Thus it would be difficul to apply directly the implicit function theorem to function $G(\theta,\zeta)$.

On the other hand, we have the derivative of the function $G(\theta,\zeta)$ with respect to the variable ζ at the point $(\theta_{\alpha,j},\zeta_{\alpha,j})$:

$$D_\theta G(\theta_{\alpha,j},\zeta_{\alpha,j}) = \frac{2\alpha}{2-\alpha} c_j B(\theta_j, B(\theta_j,\cdot)) + \frac{\alpha}{1-\alpha} c_j B(B(\theta_j,\theta_j),\cdot) \ ,$$

which is compact, therefore has non bounded inverse in $H_0^2(\omega)$. But we hope a good combination of the two derivatives is easily inverted. To do this end, we have introduced the auxiliary function $H(\theta,\zeta)$. The calculation of (2.16) shows that $D_\theta H(\theta_{\alpha,j},\zeta_{\alpha,j})$ is the difference of $D_\theta G(\theta_{\alpha,j},\zeta_{\alpha,j})$ and $D_\zeta G(\theta_{\alpha,j},\zeta_{\alpha,j})$, which is invertible in the space $H_0^2(\omega)$ by means of the Fredholm's alternative.

Similarly, since the function $f(\alpha)$ is invariant if α is replaced by $(3-\alpha)$, we have also the following perturbation result :

Theorem 8. Let $1 < \alpha < 3/2$ such that $\varepsilon^2(\alpha-1)/2c_j$ is not an eigenvalue of the compact and self-adjoint operator $B(\theta_j, B(\theta_j,\cdot))$. Then there exist a neighborhood $\widetilde{V} \subset H_0^2(\omega)$ and one mapping $\widetilde{h} \ : \widetilde{V} \to H_0^2(\omega)$ such that

$$G(\widetilde{h}(\eta), \eta - \widetilde{h}(\eta)) = 0, \quad \forall \eta \in \widetilde{V} \ . \tag{2.17}$$

Proof. It is sufficient to replace α by $(3-\alpha)$ in the proof of Theorem 7.

This work is part of the Project "Junctions in Elastic Multi–Structures" of the Program "SCIENCE" of the Commission of the European Communities (No.SC1*0473–C(EDB)).

3. References

BERGER, M. S. [1967]: *On von Kármán's equations and the buckling of a thin elastic plate,* I. the Clamped Plate, Comm. Pure Appl. Math., 20, 687–719.

BERGER, M. S. [1977]: *Nonlinearity and Functional Analysis,* Academic Press, New York.

CIARLET, P. G. ; RABIER, P. [1980]: *Les Equations de von Kármán*, Lecture Notes in Mathematics, Vol. 826, Springer-Verlag, Berlin.

CIARLET, P. G. ; PAUMIER, J. C. [1986]: *A justification of the Marguerre-von Kármán equations*, Computational Mechanics 1, 177–202.

KAVIAN, O. ; RAO, B. P. [1993]: *Une remarque sur l'existence de solutions non nulles pour les équations de Marguerre-von Kármán*, C. R. Acad. Sci. Paris, 317, Série I, 1137–1142.

KESAVAN, S; SRIKANTH, P. N. [1983]: *On the Dirichlet problem for the Marguerre equations*, Nonlinear Analysis, Theory, Methods and Applications, 7, 209–216.

PAUMIER, J. C. ; RAO, B. P. [1989]: *Qualitative and quantitative analysis of buckling of shallow shells*, Eur. J. Mech. A/Solids, 8, 461–489.

RAO, B. P. [1993]: *Marguerre-von Kármán equations and membrane model*, to appear in Nonlinear Analysis, Theory, Methods and Applications.

RUPPRECHT, G. [1981]: *A singular perturbation approach to nonlinear shell theory*, Rocky Mountain J. of Math., 11, 75–98.

Otared KAVIAN   Bopeng RAO

Université de Nancy I, URA CNRS 750, Département de Mathématiques,
B. P. 239, 54506 Vandœuvre-lès-Nancy, France

H LE DRET AND A RAOULT

Quasiconvex envelopes of stored energy densities that are convex with respect to the strain tensor

In [4], Pipkin gave a simple formula for computing the quasiconvex envelope of stored energy functions defined on the set of real 3×2 matrices that are left $O(3)$-invariant and convex with respect to the strain tensor. We prove here that this formula is also valid in the $n \times m$ case, with $m \leq n$. Note that Pipkin's argument does not work for square matrices. Our generalized result applies in particular for $m = n = 3$ to the Saint Venant-Kirchhoff stored energy function, whose quasiconvex envelope was previously determined in Le Dret and Raoult [2] by an altogether entirely different method. We also show that Pipkin's formula is false for $m > n$ by means of a counterexample.

1. Introduction

Elastic membrane stored energy functions, *i.e.*, functions $W : M_{3,2} \to \mathbb{R}$ where $M_{3,2}$ is the space of 3×2 matrices are studied in detail in a series of papers by Pipkin [1], [2], [3] and [4]. In particular, Pipkin shows how tension field theory can be properly incorporated into the theory of elastic membranes by introducing a suitable relaxed energy. He also analyses the mathematical properties of membrane energy functions together with their mechanical interpretation.

In the last paper of this series, Pipkin gives a nice formula for computing the quasiconvex envelope of membrane stored energy functions that are left $O(3)$-invariant—and can therefore be expressed as functions $\tilde{W}$ of the strain tensor $C = F^T F$—and convex with respect to C on the set $\mathbb{S}_2^+$ of real 2×2 symmetric positive semidefinite matrices, *i.e.*, the function $\tilde{W}$ is convex. He shows that, under the above hypotheses,

$$QW(F) = \inf_{S \in \mathbb{S}_2^+} \tilde{W}(F^T F + S). \tag{1}$$

The above formula is much simpler than Dacorogna's general formula for the quasiconvex envelope of a function W defined on the set $M_{n,m}$ of real $n \times m$ matrices, which reads, see Dacorogna [1],

$$QW(F) = \inf \left\{ \frac{1}{\operatorname{meas} D} \int_D W(F + \nabla\varphi(x))\, dx \;; \varphi \in W_0^{1,\infty}(D; \mathbb{R}^n) \right\}, \tag{2}$$

for all $F \in M_{n,m}$, where D is any bounded open subset of $\mathbb{R}^m$. See section 2 for the definition of quasiconvexity and quasiconvex envelopes. Our main purpose here is to extend formula (1) to functions W defined on $M_{n,m}$ with $m \leq n$ that are convex with respect to $C = F^T F$.

Quasiconvex envelopes of functions defined on $M_{3,2}$ arise naturally when nonlinear elastic membrane theory is derived rigorously from three-dimensional elasticity, see Le Dret and Raoult [1]. In this work, we consider a family of three-dimensional cylindrical bodies made of a given hyperelastic material. Using Γ-convergence arguments, we pass to the limit in the sequence of variational minimization problems describing the equilibrium of the bodies under the action of external loads when the thickness of the cylinders goes to zero. We thus obtain a limit minimization problem that involves a membrane stored energy function. The computation of the limit stored energy function, which acts on gradients of mappings from a subdomain of $\mathbb{R}^2$ into $\mathbb{R}^3$, that is 3×2 matrices, from the three-dimensional stored energy function W, which acts on 3×3 matrices, requires two steps: First, we define a function $\overline{W} : M_{3,2} \mapsto \mathbb{R}$ by $\overline{W}(\bar{F}) = \inf_{z \in \mathbb{R}^3} W((\bar{F}|z))$, where $(\bar{F}|z)$ denotes the 3×3 matrix whose first two column-vectors are those of $\bar{F}$ and third column-vector is z. Second, we take the quasiconvex envelope $Q\overline{W}$ of $\overline{W}$. Several properties of $Q\overline{W}$ are established in Le Dret and Raoult [1], [3]. It should be emphasized that the quasiconvexification step cannot be avoided. Unfortunately, no general algorithm for computing quasiconvex envelopes is known.

Nonetheless, in Le Dret and Raoult [1] we give an explicit formula for $Q\overline{W}$ when W is the Saint Venant-Kirchhoff density. Following the same ideas, we obtain another explicit formula for the quasiconvex envelope of the three-dimensional Saint Venant-Kirchhoff density itself in Le Dret and Raoult [2]. In this case, both W and $\overline{W}$ are convex functions of the strain tensors, and it is the fact that Pipkin's formula yields the right quasiconvex envelope for the three-dimensional Saint Venant-Kirchhoff density that prompted us to prove it in all generality. Let us emphasize again that Pipkin's argument does not apply directly to square matrices.

2. The main result

Following Pipkin's ideas, the main ingredient in the proof is Lemma 1 below on rank 1 convex, left $O(n)$-invariant functions. Let us first recall some definitions and basic results. We denote by $\mathbb{S}_m$ the set of real $m \times m$ symmetric matrices, by $\mathbb{S}_m^+$ the set of real $m \times m$ symmetric, positive, semidefinite matrices and $\mathbb{S}_m^>$ the set of real $m \times m$ symmetric, positive, definite matrices. A mapping $Y : M_{n,m} \to \mathbb{R}$ is said to be left $O(n)$-invariant if $Y(QF) = Y(F)$ for all F in $M_{n,m}$ and all Q in $O(n)$. It is well known that such a mapping can be expressed as $Y(F) = \tilde{Y}(F^T F)$ where $\tilde{Y} : S_m^+ \to \mathbb{R}$ (when $m < n$, left $SO(n)$-invariance is sufficient for such a factorization to exist).

A mapping $Y : M_{n,m} \to \mathbb{R}$ is rank 1 convex if it is convex on any segment whose endpoints differ by a rank 1 matrix, which means that the function $t \in \mathbb{R} \mapsto Y(F + t\, u \otimes v)$ is convex for all F in $M_{n,m}$ and for all u in $\mathbb{R}^n$ and v in $\mathbb{R}^m$. A mapping Y is quasiconvex if

$$\int_D Y(F + \nabla\varphi(x))\, dx \geq (\text{meas}\, D) Y(F)$$

for all $F \in M_{n,m}$, $\varphi \in W_0^{1,\infty}(D;\mathbb{R}^n)$, where D is any bounded open subset of $\mathbb{R}^m$, see Morrey [1]. If Y is not quasiconvex, such minimization problems as

$$\begin{cases} \text{Find } \phi \in \Phi = \{\psi \in W^{1,p}(\Omega\,;\mathbb{R}^n) \text{ with boundary conditions}\}, \\ \text{such that} \\ I(\phi) = \inf_{\Phi} I(\psi),\ I(\psi) = \int_\Omega Y(\nabla\psi)\,dx - \int_\Omega f\cdot\psi\,dx, \end{cases}$$

where Ω is a domain of $\mathbb{R}^m$, may fail to have mimimizers and one is led to introduce a relaxed problem in which Y is replaced by its quasiconvex envelope QY, see Dacorogna [1]. The quasiconvex envelope QY of Y is the largest quasiconvex function that lies below Y, *i.e.*,

$$QY = \sup\{Z : M_{n,m} \to \mathbb{R}\,;\ Z \text{ quasiconvex},\ Z \le Y\}.$$

It is also given by formula (2). Quasiconvexity implies rank 1 convexity.

Lemma 1. *Let $m \le n$ and $Y : M_{n,m} \to \mathbb{R}$ be a left $O(n)$-invariant, rank 1 convex mapping. Then the mapping $\tilde{Y} : \mathbb{S}_m^+ \to \mathbb{R}$ such that $Y(F) = \tilde{Y}(F^TF)$ for all F in $M_{n,m}$ satisfies*

$$\tilde{Y}(C) \le \tilde{Y}(C+S) \text{ for all } C,\ S \in \mathbb{S}_m^+. \tag{3}$$

Comment. In the case when $m < n$, this result is due to Pipkin [4] and the proof follows Pipkin [2], [3]. However, Pipkin's argument does not apply to square matrices. In Pipkin's terminology, $\tilde{Y}$ is said to be increasing.

Proof. Following Pipkin [2], we first remark that proving (3) amounts to proving that

$$\tilde{Y}(C) \le \tilde{Y}(C+\mu v\otimes v) \text{ for all } C \in \mathbb{S}_m^+ \text{ and for all } \mu \ge 0,\ v \in \mathbb{R}^m \setminus \{0\}. \tag{4}$$

Indeed, (3) clearly implies (4). Conversely, any S in $\mathbb{S}_m^+$ admits a spectral decomposition $S = \sum_{i=1,m} \mu_i v_i \otimes v_i$ where $\mu_i \ge 0$, and v_i, $i = 1,\cdots,m$ are orthonormal eigenvectors of S. Applying inequality (4) m times, we obtain (3).

Let us now prove (4). Let $C \in \mathbb{S}_m^+$ and $v \in \mathbb{R}^m \setminus \{0\}$ be given. Without loss of generality, we assume $\|v\|^2 := v^Tv = 1$.

We first consider the case when either $m < n$ or $m = n$ and C is not invertible. In both cases, C can be written as $C = F^TF$ where F^T is a noninjective $m \times n$ matrix. Therefore, there exists u in $\ker F^T$ with $\|u\| = 1$. From the rank 1 convexity of Y, we know that the function $y : t \in \mathbb{R} \mapsto y(t) = Y(F + t\,u\otimes v) \in \mathbb{R}$ is convex. Moreover, since $F^Tu = 0$, $y(t) = \tilde{Y}(C + t^2 v\otimes v)$. Therefore, y is even. It follows that y is monotone increasing on $\mathbb{R}^+$; in particular, $y(0) \le y(t)$ for all $t \ge 0$. Choosing $t = \sqrt{\mu}$, we obtain (4).

We now turn to the case when $m = n$ and C is invertible. For all $\mu \geq 0$, the matrix $C_\mu = C + \mu\, v \otimes v$ is symmetric, positive, definite. Hence, $F_\mu = C_\mu^{1/2}$ is invertible. We define a function h on $\mathbb{R}$ by

$$h : t \in \mathbb{R} \mapsto h(t) = Y(F_\mu + t\, F_\mu^{-1} v \otimes v).$$

It follows from the rank 1 convexity of Y that h is convex. Moreover,

$$h(t) = \tilde{Y}(C + (\mu + 2\,t)\, v \otimes v + t^2 v \otimes v (C + \mu v \otimes v)^{-1} v \otimes v).$$

An easy computation shows that the function

$$t \in \mathbb{R} \mapsto 2\,t\, v \otimes v + t^2 v \otimes v (C + \mu\, v \otimes v)^{-1} v \otimes v$$

is symmetric with respect to $\bar{t} = -(v^T\,(C + \mu\, v \otimes v)^{-1}\, v)^{-1} < 0$. The function h, in turn, is symmetric with respect to $\bar{t}$. Therefore, h attains its minimum at $\bar{t}$ and is monotone increasing on $[\bar{t}, +\infty[$. Obviously, $h(0) = \tilde{Y}(C + \mu\, v \otimes v)$. If we can find t such that $\bar{t} \leq t \leq 0$ and

$$h(t) = \tilde{Y}(C), \tag{5}$$

then inequality (4) is proved. A sufficient condition for a real number t to solve (5) is

$$t^2 v^T (C + \mu v \otimes v)^{-1} v + 2\,t + \mu = 0. \tag{6}$$

The discriminant of equation (6) is positive if and only if

$$\mu v^T (C + \mu v \otimes v)^{-1} v \leq 1. \tag{7}$$

Let us check that this is indeed the case. Let $z = \mu^{1/2} C^{-1/2} v$. Then we have

$$\mu v^T (C + \mu v \otimes v)^{-1} v = z^T (I + z \otimes z)^{-1} z = \frac{\|z\|^2}{1 + \|z\|^2} \leq 1,$$

hence the roots of equation (6) are real. Moreover, they are negative and symmetric with respect to $\bar{t}$. The largest root thus satisfies $\bar{t} \leq t \leq 0$ and (5), which proves our claim. □

With this lemma at hand, we can now state our main result.

Theorem 2. *Let $W : F \in M_{n,m} \mapsto \mathbb{R}$ be a left $O(n)$-invariant, bounded from below, stored energy function such that the associated function $\tilde{W} : C \mapsto \tilde{W}(C)$ is convex on $\mathbb{S}_m^+$. Then,*

$$QW(F) = \inf_{S \in S_m^+} \tilde{W}(F^T F + S). \tag{8}$$

Proof. Since $\tilde{W}$ is bounded from below, we can define, with Pipkin's notation, a function $\tilde{W}_r$ on $\mathbb{S}_m^+$ by $\tilde{W}_r(C) = \inf_{S \in \mathbb{S}_m^+} \tilde{W}(C+S)$. It is easy to check that $\tilde{W}_r(C)$ is convex. Indeed, given C_1 and $C_2 \in \mathbb{S}_m^+$ and an arbitrary $\varepsilon > 0$, there exists S_1 and S_2 in $\mathbb{S}_m^+$ such that $\tilde{W}_r(C_i) \leq \tilde{W}(C_i + S_i)$ and $\tilde{W}(C_i + S_i) \leq \tilde{W}_r(C_i) + \varepsilon$ for $i = 1, 2$. Let $t \in [0,1]$ and $S = t\,S_1 + (1-t)S_2$. Then,

$$
\begin{aligned}
\tilde{W}_r(t\,C_1 + (1-t)C_2) &\leq \tilde{W}(t\,C_1 + (1-t)\,C_2 + S) \\
&\leq t\,\tilde{W}(C_1 + S_1) + (1-t)\,\tilde{W}(C_2 + S_2) \\
&\leq t\,\tilde{W}_r(C_1) + (1-t)\tilde{W}_r(C_2) + \varepsilon.
\end{aligned}
$$

The convexity of $\tilde{W}_r$ follows at once.

Let us now remark that $\tilde{W}_r$ obviously satisfies $\tilde{W}_r(C) \leq \tilde{W}_r(C+S)$ for all C and S in $\mathbb{S}_m^+$. This implies that the function $Z := F \in M_{n,m} \mapsto \tilde{W}_r(F^T F)$ is convex. Indeed, for all F and G in $M_{n,m}$ and for all $t \in [0,1]$,

$$
\begin{aligned}
(tF + (1-t)G)^T(tF + (1-t)G) = t\,F^T F + (1-t)G^T G \\
- t(1-t)(F-G)^T(F-G).
\end{aligned}
$$

Therefore, since $t(1-t)(F-G)^T(F-G)$ is positive semidefinite,

$$
Z(t\,F + (1-t)G) \leq \tilde{W}_r(t\,F^T F + (1-t)G^T G) \leq t\,Z(F) + (1-t)Z(G),
$$

by the convexity of $\tilde{W}_r$. Consequently, since Z is convex and below W, we see that $Z \leq QW$.

The reverse inequality is obtained as follows. From Le Dret and Raoult [1], we know that QW is also left $O(n)$-invariant. Applying Lemma 1 to $Y = QW$, which is rank 1 convex, we obtain

$$
QW(F) = \widetilde{QW}(F^T F) \leq \widetilde{QW}(F^T F + S) = QW\left((F^T F + S)^{1/2}\right) \leq \tilde{W}(F^T F + S)
$$

for all $S \in \mathbb{S}_m^+$. Therefore, $QW(F) \leq Z(F)$ and the proof is complete. □

Comments. i) It follows clearly from the proof that the quasiconvex envelope is also in this case the convex and rank 1 convex envelope of the stored energy function.

ii) Assume, for simplicity, that $m = n$. If Y is left $O(n)$-invariant, rank 1 convex, and in addition right $O(n)$-invariant, a simpler proof of Lemma 1 is available. Let us denote by $\lambda_i(C), i = 1, \cdots n$, the eigenvalues of a semidefinite positive matrix C, arranged in increasing order. Since Y is left and right $O(n)$-invariant, there exists a symmetric mapping $\vartheta : \mathbb{R}_+^n \mapsto \mathbb{R}$, see Ciarlet [1], Marsden and Hughes[1], such that $Y(F) = \tilde{Y}(F^T F) = \vartheta(\lambda_1(F^T F), \cdots, \lambda_n(F^T F))$ for all F in $M_{n,n}$. In other words,

$$
Y(F) = Y\left(\operatorname{diag}(\lambda_1(F^T F)^{1/2}, \cdots, \lambda_n(F^T F)^{1/2})\right).
$$

By the Rayleigh quotient formula for the eigenvalues of semidefinite positive matrices, it is clear that $\lambda_i(C) \leq \lambda_i(C+S)$ for all C and S in $\mathbb{S}_n^+$ and for all $i = 1, \cdots n$. It is also an easy matter to show that

$$Y(\text{diag}(\mu_1, \cdots, \mu_n)) \leq Y(\text{diag}(\nu_1, \cdots, \nu_n)) \tag{9}$$

as soon as $0 \leq \mu_i \leq \nu_i$ for all $i = 1, \cdots n$. Indeed, if $k(t) := Y(\text{diag}(t, \mu_2 \cdots, \mu_n))$, then k is a convex and even function in the real variable t. It follows that k is monotone increasing on $\mathbb{R}^+$. Therefore, $Y(\text{diag}(\mu_1, \cdots, \mu_n)) \leq Y(\text{diag}(\nu_1, \mu_2, \cdots, \mu_n))$. Repeating the argument for the other indices, we obtain (9).

iii) We now proceed to show by means of a simple counterexample that Pipkin's formula fails for $m > n$. Let $m = 2$, $n = 1$. Consider the function $W : M_{1,2} \to \mathbb{R}$, $W(F) = \text{tr}\left((F^T F - I_2)^2\right) = \|F^T F - I_2\|^2$. We thus let $\tilde{W}(C) = \|C - I_2\|^2$. This function is clearly convex with respect to C. If we denote $F = (z_1, z_2)$ with $z_i \in \mathbb{R}$, then we have

$$\begin{aligned} W(F) &= (z_1^2 - 1)^2 + (z_2^2 - 1)^2 + 2z_1^2 z_2^2 \\ &= \left(\|F\|^2 - 1\right)^2 + 1. \end{aligned}$$

It is well known that then

$$QW(F) = CW(F) = \begin{cases} W(F) & \text{if } \|F\| \geq 1, \\ 1 & \text{if } \|F\| \leq 1, \end{cases}$$

see Dacorogna [1].

Let us now take $F = (1, -1)$ so that $C = \begin{pmatrix} 1 & -1 \\ -1 & 1 \end{pmatrix}$ and $\tilde{W}(C) = 2$. With the choice $S = \frac{1}{2}\begin{pmatrix} 1 & 1 \\ 1 & 1 \end{pmatrix}$, we obtain $\tilde{W}_r(C) \leq \tilde{W}(C+S) = 1 < 2 = \tilde{W}(C) = W(F) = QW(F)$.

3. Application

As mentioned earlier, the authors' attention was originally drawn to Pipkin's formula (1) because they had previously computed the quasiconvex envelope of the Saint Venant-Kirchhoff density defined on $M_{3,3}$, see Le Dret and Raoult [2] (and its associated membrane density which, by construction, is a quasiconvex envelope, see Le Dret and Raoult [1]).

The Saint Venant-Kirchhoff stored energy function is defined by

$$W(F) = \frac{\mu}{4}\|F^T F - I_3\|^2 + \frac{\lambda}{8}(\|F\|^2 - 3)^2 \text{ for all } F \in M_{3,3}, \tag{10}$$

where $\mu > 0$ and $\lambda \geq 0$ are the Lamé moduli of the material, and $\|F\|^2 = \text{tr}(F^T F)$, see Ciarlet [1], Truesdell and Noll [1]. Equivalently, $W(F) = \tilde{W}(F^T F)$ where

$$\tilde{W}(C) = \frac{\mu}{4}\|C - I_3\|^2 + \frac{\lambda}{8}(\text{tr}\,\text{C} - 3)^2 \tag{11}$$

for all C in $\mathbb{S}_3^+$. The mapping $\tilde{W}$ is clearly convex with respect to C.

From Le Dret and Raoult [1], we know that the membrane problem associated with the Saint Venant-Kirchhoff stored three-dimensional energy involves the quasiconvex envelope of the function $\overline{W} : M_{3,2} \to \mathbb{R}$ defined by

$$\overline{W}(\bar{F}) = \frac{\mu}{4}\mathrm{tr}\,(\bar{F}^T\bar{F}-I_2)^2+\frac{\lambda\mu}{4(\lambda+2\mu)}w(\bar{F})^2+\frac{1}{8(\lambda+2\mu)}\left([\lambda w(\bar{F})-(\lambda+2\mu)]_+\right)^2, \quad (12)$$

where $w(\bar{F}) = \mathrm{tr}\,(\bar{F}^T\bar{F} - I_2)$. Equivalently, $\overline{W}(\bar{F}) = \widetilde{\overline{W}}(\bar{F}^T\bar{F})$ where

$$\widetilde{\overline{W}}(\bar{C}) = \frac{\mu}{4}\mathrm{tr}\,(\bar{C} - I_2)^2 + \frac{\lambda\mu}{4(\lambda+2\mu)}\tilde{w}(\bar{C})^2 + \frac{1}{8(\lambda+2\mu)}\left([\lambda\tilde{w}(\bar{C}) - (\lambda+2\mu)]_+\right)^2 \quad (13)$$

and $\tilde{w}(\bar{C}) = \mathrm{tr}\,(\bar{C} - I_2)$ for all C in $\mathbb{S}_2^+$. Again, $\widetilde{\overline{W}}$ is clearly convex in $\bar{C}$. Therefore, in both cases Theorem 2 applies. Let us briefly show how computations can be organized. We concentrate on the derivation of QW.

For any C in $\mathbb{S}_3^+$, let $J_C : S \in \mathbb{S}_3^+ \mapsto \tilde{W}(C + S) \in \mathbb{R}$. This is a strictly convex, coercive mapping. Consequently, J_C admits one and only one minimizer on $\mathbb{S}_3^+$. By (8), we have to evaluate $\inf\limits_{S\in\mathbb{S}_3^+} J_C(S) = \min\limits_{S\in\mathbb{S}_3^+} J_C(S)$. Assume first that C is diagonal. We deduce from (11) that $J_C(S) \geq J_C(\mathrm{diag}\,(s_{11}, s_{22}, s_{33}))$. Minimizing $J_C(S)$ among semidefinite positive matrices thus amounts to minimizing $J_C(S)$ among diagonal positive matrices. Equivalently, we have to minimize on $(\mathbb{R}^+)^3$ the mapping j_C such that

$$j_C(s_1, s_2, s_3) = \frac{\mu}{4}\sum_{i=1}^{3}(c_{ii} - 1 + s_i)^2 + \frac{\lambda}{8}(\sum_{i=1}^{3}(c_{ii} - 1 + s_i))^2.$$

Without loss of generality, we assume that $c_{11} \leq c_{22} \leq c_{33}$. The optimality conditions for j_C on $(\mathbb{R}^+)^3$ read

$$\begin{cases} Dj_C(s_1, s_2, s_3)(t_1, t_2, t_3) \geq 0 \ \text{ for all } (t_1, t_2, t_3) \in (\mathbb{R}^+)^3, \\ Dj_C(s_1, s_2, s_3)(s_1, s_2, s_3) = 0, \ (s_1, s_2, s_3) \in (\mathbb{R}^+)^3. \end{cases}$$

They are equivalent to

$$\begin{cases} \partial_i j_C(s_1, s_2, s_3) \geq 0 \ \text{ for } i = 1, 2, 3, \\ \partial_i j_C(s_1, s_2, s_3)\, s_i = 0, \ s_i \geq 0 \ \text{ for } i = 1, 2, 3, \end{cases}$$

that is to say

$$\begin{cases} (2\,\mu + \lambda)(c_{ii} - 1 + s_i) + \lambda\sum_{k\neq i}(c_{kk} - 1 + s_k) \geq 0, \\ ((2\,\mu + \lambda)(c_{ii} - 1 + s_i) + \lambda\sum_{k\neq i}(c_{kk} - 1 + s_k))\, s_i = 0\,, \ s_i \geq 0. \end{cases}$$

We distinguish four different cases:

1) If $c_{33} \leq 1$, then we can choose $s_i = 1 - c_{ii}$ and $\min\limits_{(\mathbb{R}+)^3} j_C = 0$.

2) If $c_{33} \geq 1$ and $2(\lambda+\mu)\,c_{22} + \lambda\,c_{33} \leq 3\lambda + 2\mu$, then we can choose $s_3 = 0$, $s_j = -c_{jj} - \dfrac{\lambda}{2(\lambda+\mu)}c_{33} + \dfrac{3\lambda+2\mu}{2(\lambda+\mu)} \geq 0,\ j = 1,2$ and

$$\min_{(\mathbb{R}+)^3} j_C = \frac{\mu\,(3\lambda+2\mu)}{8(\lambda+\mu)}(c_{33}-1)^2.$$

3) If $2(\lambda+\mu)\,c_{22} + \lambda\,c_{33} \geq 3\lambda + 2\mu$ and $(\lambda+2\mu)\,c_{11} + \lambda\,(c_{22}+c_{33}) \leq 3\lambda + 2\mu$, then $s_2 = s_3 = 0$, $s_1 = -c_{11} - \dfrac{\lambda}{(\lambda+2\mu)}(c_{22}+c_{33}) + \dfrac{3\lambda+2\mu}{(\lambda+2\mu)} \geq 0$ and

$$\min_{(\mathbb{R}+)^3} j_C = \frac{\mu}{4}((c_{22}-1)^2 + (c_{33}-1)^2) + \frac{\lambda\mu}{4(\lambda+2\mu)}(c_{22}+c_{33}-2)^2.$$

4) If $(\lambda+2\mu)\,c_{11} + \lambda\,(c_{22}+c_{33}) \geq 3\lambda + 2\mu$, then we can choose $s_1 = s_2 = s_3 = 0$ and $\min\limits_{(\mathbb{R}+)^3} j_C = \tilde{W}(C)$.

So far, we have determined $QW(F)$ when $C = F^T F$ is diagonal. To extend the result to arbitrary matrices C in $\mathbb{S}_3^+$, we make use of the right $O(3)$-invariance of the Saint Venant-Kirchhoff density W. As proved in Le Dret and Raoult [2], QW inherits from W this right $O(3)$-invariance. Therefore, $QW(F)$ only depends on the singular values of F. We denote by $v_1(F) \leq v_2(F) \leq v_3(F)$ the singular values arranged in increasing order. We leave it to the reader to replace c_{ii} by $v_i(F)^2$ in the above formulas (cases 1) to 4)) to obtain an explicit expression for $QW(F)$. □

Comments. i) We also leave it to the reader to check that the expressions thus obtained are the same as those obtained in Le Dret and Raoult [2]. Formula (8) thus provides an interesting alternative to compute the quasiconvex envelope of the Saint Venant-Kirchhoff density. Note however that the right $O(3)$-invariance is essential to cover all cases. It is quite likely that if this property was missing, computations could not be carried out completely.

ii) Similar computations based on (8) can be performed for the membrane energy $Q\overline{W}$ where $\overline{W}$ is defined in (12). The result agrees with that of Le Dret and Raoult [1].

References

P.G. Ciarlet [1], *Mathematical Elasticity. Volume I: Three-Dimensional Elasticity*, North-Holland, Amsterdam, 1988.

B. Dacorogna [1], *Direct Methods in the Calculus of Variations*, Applied Mathematical Sciences, no. 78, Springer-Verlag, Berlin, 1989.

H. Le Dret, A. Raoult [1], The nonlinear membrane model as variational limit of nonlinear three-dimensional elasticity, to appear in *J. Math. Pures Appl.*

H. Le Dret, A. Raoult [2], Enveloppe quasi-convexe de la densité d'énergie de Saint Venant-Kirchhoff, *C. R. Acad. Sci. Paris*, t. 318, Série I, 1994, p. 93–98.

H. Le Dret, A. Raoult [3], Remarks on the quasiconvex envelope of stored energy functions in nonlinear elasticity, to appear in *Communications on Applied Nonlinear Analysis*.

J.E. Marsden, T.J.R. Hughes [1], *Mathematical Foundations of Elasticity*, Prentice-Hall, Englewood Cliffs, 1983.

C.B. Morrey Jr. [1], Quasiconvexity and the semicontinuity of multiple integrals, *Pacific J. Math.*, 2, 1952, p. 25-53.

A.C. Pipkin [1], The relaxed energy density for isotropic elastic membranes, *IMA J. Appl. Math.*, 36, 1986, p. 85-99.

A.C. Pipkin [2], Convexity conditions for strain-dependent energy functions for membranes, *Arch. Rational Mech. Anal.*, 121, 1993, p. 361-376.

A.C. Pipkin [3], Relaxed energy densities for small deformations of membranes, *IMA J. Appl. Math.*, 50, 1993, p. 225-237.

A.C. Pipkin [4], Relaxed energy densities for large deformations of membranes, to appear in *IMA J. Appl. Math.*

C. Truesdell, W. Noll [1], The Nonlinear Field Theories of Mechanics, in *Handbuch der Physik*, Vol. III/3, Springer Verlag, Berlin, 1965.

Hervé LE DRET
Laboratoire d'Analyse Numérique
Université Pierre et Marie Curie
75252 Paris Cedex 05, France

Annie RAOULT
Laboratoire de Modélisation et Calcul
Université Joseph Fourier
BP 53
38041 Grenoble Cedex 9, France

Y NAKANO AND M PRIMICERIO

Application of quasi-steady solutions of soil freezing for geotechnical engineering

1. Introduction

The scientific study of soil freezing and ice segregation began in the early 1900s. By the 1930s researchers (Taber 1930, Beskow 1935) had already found that ice segregation and the resultant frost heave are caused not only by freezing of in-situ water, but also by freezing of water transported toward a freezing front from the unfrozen part of the soil. The understanding gained in the 1930s was largely qualitative. However, the transport of water was already identified as one of major issues in the study of soil freezing. The problem has attracted the attention of many researchers in the past, but we have as yet no consensus on the mechanism of water transport (Nakano 1991).

The main constituents of saturated frozen soils are a solid porous matrix of soil particles and ice, and water in its supercooled liquid phase called unfrozen water. The physical properties of all constituents except unfrozen water are well understood. It is generally understood that the transport of water in frozen soils is caused by the movement of unfrozen water and that unfrozen water exists in small spaces surrounded with surfaces of soil particles and ice. Unfrozen water contents in various soils under isothermal conditions have been measured by several methods including differential thermal analysis and nuclear magnetic resonance (Andersland and Anderson 1978).

The dynamic and thermodynamic properties of liquids have been known to be modified by confinement in very small spaces such as porous media, cell membranes, etc. The problem of confined liquids has attracted the attention of researchers in many disciplines in recent years (Drake and Klafter 1990, Granick 1991) because the problem is fundamental to new developments in diverse fields such as medicine, biotechnology, petroleum engineering, etc. Unfrozen water in frozen soils is one special case of a wide class of confined liquids. The key issue underlying the transport of unfrozen water is deemed to be the dynamic collective behavior of water confined to small spaces in frozen soils, which strongly depends on complex solid-liquid interactions. The effects of solid-liquid interactions on the freezing of liquids have only begun to be understood (Ritter et al. 1988).

It is generally accepted that a thin transitional zone, often referred to as the frozen fringe, exists between the 0°C isotherm (frost front) and the growing surface of an ice layer or frozen soil. Since the properties of all parts except the frozen fringe are understood, the dynamic behavior of the frozen fringe has been one of the major subjects in the study of soil freezing in recent years. Since the 1960s many mathematical models (Kay and Perfect 1988) of a frozen fringe have been proposed on the basis of various hypotheses. With the widespread use of computers, the methods of numerical analysis became very popular. However, because of the paucity of basic knowledge and the complex nature of the problem, these numerical studies have not been effective for the critical evaluation of the multiple hypotheses used. Recently efforts (Fasano and Primicerio 1984, Nakano and Takeda 1993, Ding and Talamucci 1994) have been initiated to study the problem analytically and to evaluate the hypotheses used in the analysis by comparing the behavior of solutions directly with experimental data.

Around 1980 two important semi-empirical models of soil freezing were introduced for engineering applications: the segregation potential (SP) model (Konrad and Morgenstern 1980) and the Takashi model (Takashi et al. 1978). Today the SP model is widely used for engineering in Europe and North America, while the Takashi model is the standard of engineering design in Japan. These two semi-empirical models share a common approach that the freezing characteristics of a given soil are determined empirically under certain quasi-steady conditions where a frost front moves with a constant speed and that the problem of unsteady freezing is approximated by a series of quasi-steady freezing steps.

It has been recognized that there is a serious communication gap between mathematicians and geotechnical engineers. This paper aims to lessen this gap. Following a description of the mathematical nature of the problem, we will demonstrate the usefulness of quasi-steady solutions for geotechnical engineering by using one of the potentially valid models as an example. The semi-empirical models mentioned above require one or more empirically determined parameters that are known to depend on a particular quasi-steady condition specified by given thermal and hydraulic fields. We will show that quasi-steady solutions inherently can provide the functional dependence of empirical parameters on given fields. The predicted dependence can be used to evaluate the model by comparison with empirical data.

2. Balance Equations of Mass and Heat

We will consider the one-directional freezing of soils. Let the freezing process advance from the top down and the coordinate x be positive upward with its origin fixed at some point in the unfrozen part of the soil. We will treat the soil as a mixture of water in the liquid phase B_1, ice B_2 and soil minerals B_3. The bulk density of B_i is denoted by $\rho_i(x,t)$. If d_i is the density of the ith constituent, then the volumetric content $\theta_i(x,t)$ of the ith constituent is given as:

$$\theta_i = \rho_i / d_i \tag{2.1}$$

It is clear that the sum of θ_i should be unity, namely:

$$\theta_1 + \theta_2 + \theta_3 = 1 \tag{2.2}$$

We will assume that the density of each constituent remains constant.

We will assume that the unfrozen part of the soil is kept saturated with water at all times by using an appropriate water supply device. The balance of mass for the ith constituent is given as (Nakano 1990):

$$\frac{\partial}{\partial t}\rho_i = -\frac{\partial}{\partial x}(\rho_i v_i) + \lambda_i \qquad , \quad i = 1,2,3 \tag{2.3}$$

where $v_i(x,t)$ is the velocity of the ith constituent and $\lambda_i(x,t)$ the time rate of supply of mass of the ith constituent per unit volume of the mixture. The summation convention on index i is not in force here, so that $(\rho_i v_i)$ represents only one term. Since none of the constituents is involved in the chemical reaction, we have:

$$\lambda_1 + \lambda_2 = 0 \qquad \text{and} \qquad \lambda_3 = 0 \tag{2.4}$$

We will assume that the constituents are locally in thermal equilibrium with each other and that the heat capacity c_i of the ith constituent and the latent heat of fusion of water L do not depend on the temperature T. If k is the thermal conductivity of the mixture, the conductive heat flux $q(x,t)$ in the mixture is assumed to be given as:

$$q = -k\frac{\partial T}{\partial x} \tag{2.5}$$

Using (2.5), we will obtain the balance equation of heat for the mixture (Nakano 1990) given as:

$$\frac{\partial}{\partial x}q = L(\lambda_2 + z) \tag{2.6}$$

where $z(x,t)$is defined as

$$Lz = -c\frac{\partial T}{\partial t} + (c_1 - c_2)T\lambda_2 - \sum_i \rho_i v_i c_i \frac{\partial T}{\partial x} \tag{2.7}$$

$$c = c_1\rho_1 + c_2\rho_2 + c_3\rho_3 \tag{2.8}$$

We will consider a special case in which a frost front (0°C isotherm) $x = n_o(t)$ moves with a constant speed, namely:

$$-\frac{d}{dt}n_o(t) = -\dot{n}_o = V_o \geq 0 \tag{2.9}$$

In such a case we will introduce a new independent variable ξ defined as:

$$\xi = x - \dot{n}_o t - n_o(0) \tag{2.10}$$

For the sake of convenience we will define new dependent variables $f_1(\xi)$ and $f_2(\xi)$ as:

$$f_1 = \rho_1(v_1 - v_3) \tag{2.11}$$

$$f_2 = \rho_2(v_2 - v_3) \tag{2.12}$$

It is easy to see that $f_i(i = 1,2)$ is the mass flux of either B_1 or B_2 relative to the mass flux of soil particles. Using (2.10), (2.11) and (2.12), we reduce (2.3) to:

$$(\rho_1 V)' = -f_1' - \lambda_2 \tag{2.13}$$

$$(\rho_2 V)' = -f_2' + \lambda_2 \tag{2.14}$$

$$(\rho_3 V)' = 0 \tag{2.15}$$

where primes denote differentiation with respect to ξ and $V(\xi)$ is defined as:

$$V = v_3 - \dot{n}_o \tag{2.16}$$

Similarly we will reduce (2.6) and (2.7) to:

$$q' = -(kT')' = L(\lambda_2 + z) \tag{2.17}$$

$$Lz = -(c_1 f_1 + c_2 f_2 + cV)T' + (c_1 - c_2)\lambda_2 T \tag{2.18}$$

3. Quasi-Steady Problem

A freezing soil may be considered to consist of three parts: the unfrozen part R_o, the frozen fringe R_1 and the frozen part R_2 as shown in Figure 1. We will assume that 1) the dry density of R_o remains constant, 2) the composition is continuous at n_o, 3) the pressure P of water at n remains constant at P_n, 4) f_2 vanishes in R_1, 5) the bulk density ρ_1 is negligibly small in R_2, and 6) that ρ_1 is given in R_1 as:

$$\rho_1 = \rho_3 \nu(T) \tag{3.1}$$

where $\nu(T)$ is an empirically determined and monotonically increasing function of T.

We will seek a solution to the problem in which the boundaries $n(t)$, $n_o(t)$ and $n_1(t)$ move with the same constant speed V_o, namely:

$$V_o = -\dot{n} = -\dot{n}_o = -\dot{n}_1 \tag{3.2}$$

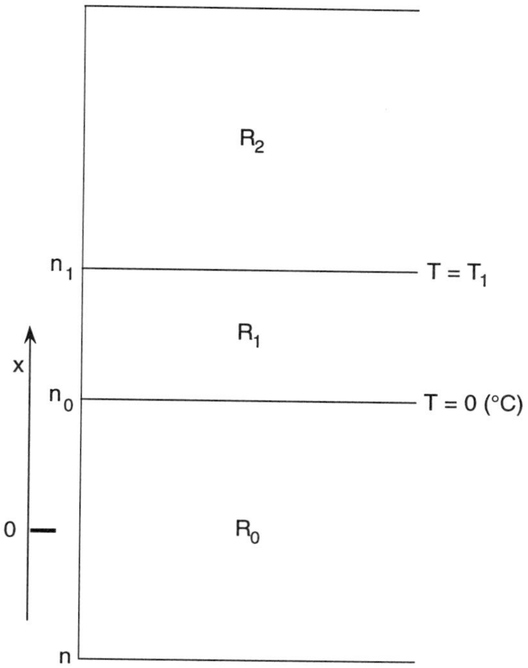

Figure 1. Schematic drawing of a quasi-steady freezing of soil.

From a physical point of view it is difficult to maintain a constant pressure P_n at the moving boundary *n(t)*. However, a solution obtained under such an idealized condition is quite useful for engineering applications as we will show in the later section. If such a solution exists, it must satisfy (2.13)–(2.15), (2.17) and (2.18).

From (2.13), (2.14) and (2.15), we find that the flux of water $f_1(\xi)$ is given in R_1 (Nakano 1994a and b) as:

$$f_1 = f_o + sV_o - d_2(V - V_o)\ , \quad 0 < \xi < \delta \tag{3.3}$$

where $\delta = n_1 - n_o$, f_o is the constant flux of water in R_o and $s(\xi)$ is defined as:

$$s = (1 - d_1^{-1} d_2)(\rho_{10} - \rho_{30}\nu) \tag{3.4}$$

where ρ_{10} and ρ_{30} are the constant bulk densities of B_1 and B_3 in R_o, respectively. Neglecting the gravitational effects and using Darcy's law, f_o is given as:

$$f_o = K_o \sigma_o \delta_o^{-1} \tag{3.5}$$

where K_o (positive number) is the hydraulic conductivity of R_o; σ_o and δ_o are defined as:

$$\sigma_o = P_n - P_o \quad , \quad P_o = P(0) \tag{3.6}$$

$$\delta_o = n_o - n \tag{3.7}$$

The boundary n_1 is a free boundary. The composition may be discontinuous at n_1 and the limiting value $\rho_3(\delta+)$ of ρ_3 as ξ approaches δ while ξ is in R_2 is given (Nakano 1994a,b) as:

$$\rho_3(\delta+) = \rho_{30} V_o (r + V_o)^{-1} \tag{3.8}$$

where r is the rate of heave given as:

$$r = d_2^{-1} f_o + \left(d_2^{-1} - d_1^{-1}\right)\rho_{10} V_o \tag{3.9}$$

The heat flux is discontinuous at n_1 and the jump condition is given as:

$$q(\delta+) = q(\delta-) + f_1(\delta-)\left[L + (c_1 - c_2)T_1\right] \tag{3.10}$$

where $q(\delta-)$ and $f_1(\delta-)$ are limiting values of q and f_1, respectively, as ξ approaches δ while ξ is in R_1 and T_1 is $T(\delta)$.

We will reduce (2.17) and (2.18) to a simpler form. Using (2.13), (2.14) and (2.15), we obtain:

$$cV = c_o V_o - (c_1 - c_2)\Lambda + c_1(f_o - f_1) \tag{3.11}$$

where

$$c_o = c_1\rho_{10} + c_3\rho_{30} \tag{3.12}$$

$$\Lambda(\xi) = \int_0^{\xi} \lambda_2 d\xi \ , \quad \xi \geq 0 \tag{3.13}$$

Using (3.11) and integrating (2.17), we obtain:

$$-k\,T' = k_o\alpha_o - k_o\beta T + L\Lambda \ , \quad 0 < \xi < \delta \tag{3.14}$$

where k_o and k are the thermal conductivities of R_o and R_1, respectively, α_o is the absolute value of the temperature gradient at $\xi = 0$, and $\beta(\xi)$ is defined as:

$$\beta = k_o^{-1}\left[c_1 f_o + c_o V_o - (c_1 - c_2)\Lambda\right] \tag{3.15}$$

Using (3.14) and neglecting sensitive heat terms, we will reduce (3.10) to:

$$k_1\alpha_1 - k_o\alpha_o = \left[f_1(\delta-) + \Lambda(\delta-)\right]L \tag{3.16}$$

where k_1 is the thermal conductivity of R_2 and α_1 is $-T'(\delta+)$.

Using the principle of mass and heat conservation, we have derived equations that must be satisfied by a quasi-steady solution of soil freezing. It is clear that these equations are not sufficient to solve the problem. We need a model of a frozen fringe that specifies $f_1(\xi)$ and $\Lambda(\xi)$.

4. Model Study

A model of a frozen fringe called M_1 was introduced by Nakano (1990) to explain empirical findings on the growth condition of an ice layer in freezing soils. The model has been modified as the empirical evaluation has progressed (Takeda and Nakano 1990, Nakano and Takeda 1991, Takeda and Nakano 1993, Nakano and Takeda 1994). The latest version assumes the validity of equations in R_1 given as:

$$\rho_3 = \rho_{30}\,, \quad \rho_1 = \nu(T)\rho_{30} \le \rho_{10}\,, \quad \rho_1(0+) = \rho_{10} \tag{4.1}$$

$$f \equiv f_1 = -K_1 P' - K_2 T' \tag{4.2}$$

$$K_2 / K_1 \to \gamma \quad \text{as} \quad f \to 0 \tag{4.3}$$

$$P(\delta-) = P_a = \sigma + P_n\,, \quad \sigma \ge 0 \tag{4.4}$$

$$P'(\delta -) \ge 0 \tag{4.5}$$

where γ is a constant (1.12 MPa/°C), $P(\xi)$ is the pressure of unfrozen water, P_a is the applied confining pressure (uniaxial stress), σ is the effective confining pressure, and $K_i(i = 1,2)$ is the transport property of a given soil that generally depends on the temperature and the composition of the soil. Since ρ_3 is a constant, we will assume that K_i is a function of T alone. The M_1 model is a generalization of somewhat simpler models (Ratkje et al. 1982, Derjaguin and Churaev 1978, Kuroda 1985, Horiguchi 1987) in which the ratio K_2/K_1 is equal to γ regardless of f. In addition to the above equations, we will also assume that $k = k_o$ and that k_1 is a given constant for the sake of simplicity.

When (4.1) holds true, v_3 vanishes and (3.3) is reduced to:

$$f(\xi) = f_o + \left(1 - d_1^{-1} d_2\right) (\rho_{10} - \rho_{30}\nu) V_o\,, \quad 0 < \xi < \delta \tag{4.6}$$

The $\Lambda(x)$ is given as:

$$\Lambda(\xi) = d_1^{-1} d_2 (\rho_{10} - \rho_{30}\nu) V_o \tag{4.7}$$

Neglecting small terms, we will reduce (3.15) to:

$$\beta = k_o^{-1}\left[c_1 f_o + \left(c_3\rho_{30} + c_2\rho_{10}\right)V_o\right] \tag{4.8}$$

According to M_1 the properties of a given soil are described by three empirically determined functions of T; K_1, K_2 and ν that are known to be continuous and increasing functions of T. The hydraulic field is specified by P_n, δ_o and P_a while the thermal field is specified by α_o and α_1. Our problem is to find constants $V_o \geq 0$, $\delta \geq 0$ and functions $f(\xi)$, $T(\xi) \leq 0$, $P(\xi)$ so that the following equations are satisfied:

From (4.6) we have:

$$f(\xi) = f_o + \left(1 - d_1^{-1}d_2\right)\left[\rho_{10} - \rho_{30}\nu\{T(\xi)\}\right]V_o \ , \quad 0 < \xi < \delta \tag{P.1}$$

From (4.2) we have:

$$f(\xi) = -K_1\{T(\xi)\}P'(\xi) - K_2\{T(\xi)\}T'(\xi) \ , \quad 0 < \xi < \delta \tag{P.2}$$

From (3.14) we have:

$$k_o T'(\xi) + k_o\alpha_o = k_o\beta T(\xi) - d_1^{-1}d_2\left[\rho_{10} - \rho_{30}\nu\{T(\xi)\}\right]LV_o \ , \quad 0 < \xi < \delta \tag{P.3}$$

where β is given by (4.8)
From (3.16) we have:

$$k_1\alpha_1 - k_o\alpha_o = Lf_o + \left[\rho_{10} - \rho_{30}\nu\{T(\delta)\}\right]LV_o \tag{P.4}$$

Boundary conditions are given as:

$$T(0) = 0 \tag{P.5}$$

$$P(\delta -) = P_a \tag{P.6}$$

Restraints are given as:

$$P'(\delta -) \geq 0 \tag{P.7}$$

$$K_2/K_1 \to \gamma \quad \text{as} \quad f \to 0 \tag{P.8}$$

Nakano (1994a,b) has obtained an approximate solution to the problem (P.1) through (P.8) that contains two parameters, α_o and α_1. The behavior of the solution is conveniently described in the diagram of temperature gradients as shown in Figure 2, where a quadrant ($\alpha_i \geq 0$) is divided into three regions, R_m, R_s and R_u by two boundaries, R_s^{**} and R_s^{*} .

(i) In R_m we have:

$$\alpha_o > (k_1 / k_o)\alpha_1 \tag{4.9}$$

In this region the heat input to R_1 exceeds the heat output. Hence, melting occurs.
(ii) On R_s^{**} we have:

$$\alpha_o = (k_1 / k_o)\alpha_1 \tag{4.10}$$

$$f = f_o = 0 \ , \quad V_o = 0 \tag{4.11}$$

$$\sigma = -\gamma T_1 \ , \quad T_1 = T(\delta) \tag{4.12}$$

On R_s^{**} the heat input to R_1 is equal to the heat output, the phase equilibrium of water holds true and an existing ice layer neither grows nor melts. The equation (4.12) is often called the generalized Clausius-Clapeyron equation and its validity has been shown empirically by Radd and Oertle (1973).
(iii) In R_s we have:

$$(k_1 / k_o)\alpha_1 > \alpha_o > k_1(k_o + Ly)^{-1}\alpha_1, \quad y = K_2(T_1) \tag{4.13}$$

$$f = f_o = -K_1(T_1)P'(\delta-) + y\alpha_o \ , \quad V_o = 0 \ , \quad P'(\delta-) > 0 \tag{4.14}$$

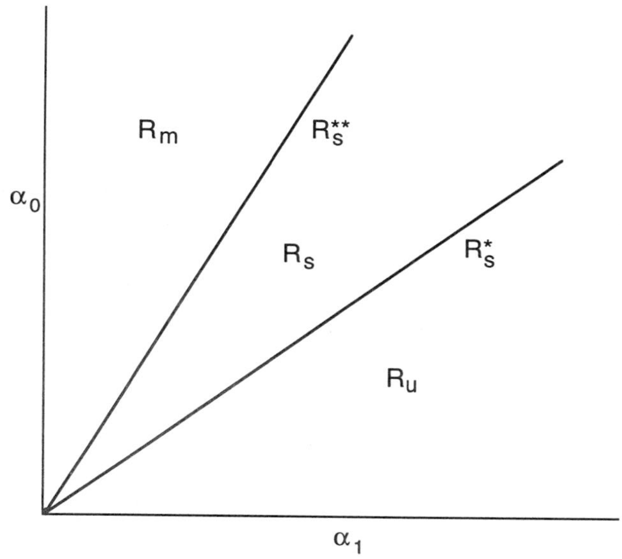

Figure 2. Diagram of temperature gradients α_1 and α_o.

In this region the growth of an ice layer (containing hardly any soil particles) occurs.
(iv) On R_s^* we have:

$$\alpha_o = k_1(k_o + Ly)^{-1}\alpha_1 \tag{4.15}$$

$$f = f_o = y\alpha_o \quad , \quad V_o = 0 \; , \quad P'(\delta-) = 0 \tag{4.16}$$

The boundary R_s^* is a curve stemming from the origin because y weakly depends on α_o (or α_1).
(v) In R_u we have:

$$k_1(k_o + Ly)^{-1}\alpha_1 > \alpha_o \tag{4.17}$$

$$f(\delta-) = f_o + \left(1 - d_1^{-1}d_2\right)(\rho_{10} - \rho_{30}\nu_1)V_o = \mathrm{by}\alpha_o \tag{4.18}$$

$$V_o > 0 \; , \quad P'(\delta -) = 0 \tag{4.19}$$

where

$$\nu_1 = \nu(T_1) \; , \quad b\alpha_o = \pi_1 - \pi_o\nu_1 \; , \quad \pi_o = d_1 d_2^{-1} k_o^{-1}\rho_{30} L V_o \; , \quad \pi_1 = \alpha_o + \pi_o \rho_{10}\rho_{30}^{-1}$$

In R_u the growth of an ice layer does not occur; instead frozen soil (a mixture of ice and soil particles) grows. A curve with a constant V_o in R_u is given as:

$$\alpha_o = (k_o + Lby)^{-1}\left[k_1\alpha_1 - d_1^{-1}d_2(\rho_{10} - \rho_{30}\nu_1)L V_o\right] \tag{4.20}$$

This curve is nearly parallel with R_s^* and the ice content of R_2 decreases with increasing V_o because of (3.8).

Before we discuss the two semi-empirical models, we will derive a few more equations needed in the discussion below. The rate of frost heave r is give as:

$$r = d_2^{-1}by\alpha_o + \left(d_2^{-1} - d_1^{-1}\right)\nu_1\rho_{30}V_o \tag{4.21}$$

When V_o is positive, the frost heave ratio h and the water intake ratio h_w are given as:

$$h \equiv r / V_o = d_2^{-1}by\alpha_o V_o^{-1} + \left(d_2^{-1} - d_1^{-1}\right)\nu_1\rho_{30} \tag{4.22}$$

$$h_w \equiv f_o / V_o = by\alpha_o V_o^{-1} - \left(1 - d_1^{-1}d_2\right)(\rho_{10} - \rho_{30}\nu_1) \tag{4.23}$$

Since y and ν_1 are functions of T_1, we need another equation to determine T_1. Integrating (4.2) from $\xi = 0$ to δ, we obtain:

$$\sigma + \delta_o K_o^{-1} f_o = -T_1 I(T_1) \tag{4.24}$$

$$I(T_1) = \gamma\phi_o - (f_o + s_1 V_o)(\pi_1 K_o)^{-1}[\phi_1 + w_o(\pi_o/\pi_1)\phi_2] + s_1 V_o(\pi_1 K_o)^{-1}\phi_2 \tag{4.25}$$

where

$$s_1 = (1 - d_1^{-1} d_2)\rho_{10}, \quad w_o = \rho_{10}/\rho_{30}$$

and ϕ_0, ϕ_1 and ϕ_2 are defined as:

$$\phi_o T_1 = \int_0^{T_1} (K_o/K_1)(K_2/K_{20})dT, \quad \phi_1 T_1 = \int_0^{T_1} (K_o/K_1)dT, \quad \phi_2 T_1 = \int_0^{T_1} (\nu/w_o)(K_o/K_1)dT$$

where K_{20} is $K_2(0+)$. We will define that $\phi_i(i = 0,1,2) = 1$ if $T_1 = 0$.

5. Signorini-Type Condition

We will introduce a Signorini-type condition given as:

$$P'(\delta-) \geq 0 \quad , \quad V_o P'(\delta-) = 0 \tag{P.9}$$

Now we study the problem stated as follows. Given non-negative constants $\alpha_o, \alpha_1, f_o, P_a$, find constants $V_o \geq 0$, $\delta \geq 0$ and functions $f(\xi) \geq 0$, $T(\xi) \leq 0$, $P(\xi)$ so that (P.1)–(P.6) and (P.9) are satisfied. In what follows we want to identify the conditions on α_o, α_1, f_o and P_a, guaranteeing the existence of a solution to the problem described above and to find criteria that discriminate between the cases of frost penetration ($V_o > 0$) and ice layer formation ($V_o = 0$).

From (P.4) we have a necessary condition for the solvability:

$$k_1\alpha_1 \geq k_o\alpha_o \tag{5.1}$$

Since the right-hand side of (P.4) is non-negative because of (4.1), the problem cannot be solved if (5.1) is violated. Now we will consider the limit case; $k_1\alpha_1 = k_o\alpha_o$.

Proposition 5.1. If $k_1\alpha_1 = k_o\alpha_o$, a solution to the problem exists if and only if $f_o = 0$.

Proof. From (P.4), since the two terms on the right-hand side are non-negative, we know that the only compatible hydraulic datum is $f_o = 0$. Moreover, $V_o = 0$. Hence, from (P.1), (P.3) and (P.5) we get:

$$f(\xi) = 0 \quad , \quad T(\xi) = -\alpha_o\xi \tag{5.2}$$

Because of (3.5) and (3.6), $f_o = 0$ implies that the water pressure $P(0)$ is equal to P_n. Integrating (P.2), we have:

$$\sigma = \alpha_o \int_0^{\delta} \frac{K_2(-\alpha_o \xi)}{K_1(-\alpha_o \xi)} d\xi \tag{5.3}$$

which has one unique solution $\delta \geq 0$ provided $\sigma \geq 0$.

Remark 5.1. Incidentally, we note that, still in the case $k_1\alpha_1 = k_o\alpha_o$, letting

$$\gamma(T) = \int_0^T \frac{K_2(s)}{K_1(s)} ds \quad ,$$

we reduce (5.3) to:

$$\sigma = -\gamma[T(\delta)] \tag{5.4}$$

whenever a linear approximation is reasonable for γ, then $\gamma(T) = \gamma T$ and we obtain the generalized Clausius-Clapeyron equation (4.12).

Next, we study the discrimination between ice formation and frost penetration. If ice formation has to occur, we want to find which conditions the data have to satisfy. Note that, in this case, δ has to be considered among the data. Of course, setting $V_o = 0$ in (P.1) and (P.3), we find:

$$f(\xi) = f_o \tag{5.5}$$

$$T(\xi) = \beta_o^{-1}\alpha_o[1 - \exp(\beta_o \xi)] \tag{5.6}$$

where $\beta_o = c_1 f_o / k_o$.

We define y as:

$$y = K_2[T(\delta)]$$

where $T(\delta)$ is calculated according to (5.6). In what follows we assume for simplicity that y is a constant; the general discussion is much more complicated.

Proposition 5.2. If the following holds true

$$k_o\alpha_o < k_1\alpha_1 < (k_o + Ly)\alpha_o \tag{5.7}$$

then the growth of an ice layer occurs provided that f_o and δ satisfy the constraints expressed by (5.8) and (5.9) below.

Proof. Recalling (P.9) and (P.2), we deduce from (5.5) and (5.6) the following condition:

$$f_o - y\alpha_o \exp(\beta_o \delta) \leq 0 \tag{5.8}$$

this means that a necessary condition is

$$f_o \leq y\alpha_o \tag{5.9}$$

Once (5.9) is satisfied, (5.8) becomes a constraint on δ.

Now we turn to (P.4) which now

$$k_1\alpha_1 - k_o\alpha_o = Lf_o \tag{5.10}$$

From (5.9) and (5.10) we find (5.7). It is straightforward to see that $T(\xi)$ given by (5.6) and $P(\xi)$ derived from (P.2) and (P.6) solve the problem completely.

Let us consider the case of frost penetration. According to (P.2) and (P.9), we have:

$$f(\delta-) = -y\,T'(\delta) \tag{5.11}$$

This is a relationship between $T(\delta)$ and $T'(\delta)$ because of (P.1). Incidentally, we note that in the special case $d_1 = d_2$, (5.11) corresponds to the specification of $T'(\delta)$.

Substituting (5.11) in (P.3), we find:

$$f(\delta-)y^{-1} = \alpha_o - \beta T(\delta) + (d_1k_o)^{-1}d_2\left[\rho_{10} - \rho_{30}\nu\{T(\delta)\}\right]LV_o \tag{5.12}$$

Hence, we need:

$$f_o > y\alpha_o \tag{5.13}$$

Indeed if (5.13) is not satisfied, (5.12) cannot hold even for $V_o = 0$. For every possible V_o, we can solve (P.1), (P.3) and use (5.11) to find δ.

Now we turn to (P.4), which can be used to determine V_o, provided that:

$$k_1\alpha_1 - k_o\alpha_o > Lf_o \tag{5.14}$$

i.e. that (recall (5.13)):

$$k_1\alpha_1 > (k_o + Ly)\alpha_o \tag{5.15}$$

At this point, (P.2) is used to find $P(\xi)$.

6. Semi-Empirical Models

Konrad and Morgenstern (1980, 1981) empirically found that the mass flux of water f_o in R_o at the formation of the so-called final ice lens is proportional to the average temperature gradient T_f' in R_1. This may be written as:

$$S_p = -f_o / T_f' \tag{6.1}$$

The positive proportionality factor S_p is termed the segregation potential. Konrad and Morgenstern (1982b) also found empirically that S_p is a decreasing function of σ_o and σ. Extending the concept of segregation potential, Konrad and Morgenstern (1982a) introduced a semi-empirical model (SP model) of soil freezing. It has been shown (Nakano 1992) that the segregation potential corresponds to y when a point (α_1, α_o) belongs to R_s^*, namely:

$$S_p = y = K_2(T_1) \text{ and } V_o = 0, \text{ when } (\alpha_1, \alpha_o) \text{ on } R_s^* \tag{6.2}$$

We will study the functional dependence of y on given hydraulic and thermal fields below by using (4.24).

It is known (Andersland and Anderson 1978) that the empirically determined function $\nu(T)$ takes a form given as:

$$\nu(T) = a_o |T|^{-a_1} \tag{6.3}$$

where a_o and a_1 are positive constants. Experimental methods were proposed to determine K_1 (Williams and Burt 1974, Horiguchi and Miller 1983) and K_2 (Perfect and Williams 1980). Horiguchi and Miller (1983) empirically found that K_1 of several frozen porous media also takes the same form as (6.3). Recently Nakano and Takeda (1994) empirically found that K_2 of Kanto loam can be described in the same form. Since ν, K_1 and K_2 are known to be bounded, we will use forms given as:

$$K_1 = \begin{cases} K_o & A \le T < 0 \\ K_o (A/T)^{b_1} & A > T \end{cases} \tag{6.4}$$

$$K_2 = \begin{cases} K_{20} & A \le T < 0 \\ K_{20} (A/T)^{b_2} & A > T \end{cases} \tag{6.5}$$

$$\nu = \begin{cases} w_o & A \le T < 0 \\ w_o (A/T)^{b_3} & A > T \end{cases} \tag{6.6}$$

where A is a small negative constant and b_i is a positive number. Because of (4.3) K_0 and K_{20} are related as:

$$K_{20} / K_0 = \gamma \tag{6.7}$$

The values of parameters in (6.4), (6.5) and (6.6) determined empirically (Nakano and Takeda 1994) for Kanto loam are: $K_o = 1.77 \times 10^3$ g/(cm • d • MPa), $K_{20} = 1.98 \times 10^3$ g/(cm • d • °C), $w_o = 0.740$, $b_1 = 0.520$, $b_2 = 1.04$, $b_3 = 0.110$ and $A = -1.5 \times 10^{-4}$°C where d denotes day.

Using (6.4)–(6.6)and neglecting small terms by referring to the data of Kanto loam given above, we will reduce (4.24) to:

$$C_1 y^{-\lambda_o} - J(\sigma) - E = C_o \delta_o f_o \tag{6.8}$$

$$f_o = \pi_1 \left[y\left(1 - C_3 Y^{\lambda}\right) - C_4\left(1 - Y^{\lambda}\right) \right] \tag{6.9}$$

$$J(\sigma) = \begin{cases} \lambda_1^{-1}(b_2 - b_1) + C_2\sigma & , \quad A < -\sigma/\gamma \\ \lambda_1^{-1}(C_2\sigma)^{\lambda_1} & , \quad A \geq -\sigma/\gamma \end{cases} \tag{6.10}$$

where $C_i (i = 0, \cdots, 4)$ is a positive number defined as:

$$C_o = -(AK_{20})^{-1} , \quad C_1 = \lambda_1^{-1} b_2 K_{20}^{\lambda_o} / (b_1 + 1) , \quad C_2 = -(A\gamma)^{-1}$$

$$C_3 = w_o \pi_o / \pi_1 , \quad C_4 = s_1 V_o / \pi_1$$

λ_o, λ_1 and λ are defined as

$$\lambda_o = (b_1 - b_2 + 1) / b_2 , \quad \lambda_1 = b_2 \lambda_o , \quad \lambda = b_3 / b_2$$

Y is defined as:

$$Y = y / K_{20}$$

An error term E, which is generally much less than the other two terms in the left side of (6.8) is given as:

$$E = C_5(C_3 y - C_4) y^{\lambda - \lambda_o - 1} - C_6 y^{2\lambda - \lambda_o} \tag{6.11}$$

where

$$C_5 = \lambda_2^{-1} b_3 K_{20}^{\lambda_o - \lambda} / (b_1 + 1) \,, \quad C_6 = \lambda_2^{-1} C_3 K_{20}^{\lambda_o - 2\lambda} \quad , \quad \lambda_2 = b_1 - b_3 + 1$$

A special case, where $\lambda_o = 0$, can be treated similarly but we will restrict our discussion to the case of positive λ_o.

Now we will find the behavior of S_p by using (6.8). First we will consider a special case of $\sigma = 0$. Because of (6.2), (6.8) is reduced to:

$$C_1 y^{-\lambda_o} - \lambda_1^{-1}(b_2 - b_1) = C_o \delta_o \alpha_o y \tag{6.12}$$

The first term of the left side of (6.12) is much greater than the second; y is given as:

$$y = (C_1 / C_0)^{1/(\lambda_0 + 1)} (\alpha_o \delta_o)^{-1/(\lambda_o + 1)} \tag{6.13}$$

Using (3.5) we may write (6.13) as:

$$y = \left[C_1 / (K_o C_o)\right]^{1/\lambda_o} (\sigma_o)^{-1/\lambda_o} \tag{6.14}$$

When $\sigma > 0$, the term $J(\sigma)$ is not negligible but it is easy to find that y is a decreasing function of σ. We have found that S_p can be described by either a decreasing function of α_o, δ_o and σ or that of σ_o and σ. In Figure 3 we plotted experimental data, y vs. α_o with $\delta_o = 2$ cm under $\sigma = 0$, 16.2, 48.7 and 195 kPa (Nakano and Takeda 1994) together with predicted y calculated by (6.8). It is easy to see that ln y and ln α_o are nearly linear when $\sigma = 0$. As σ increases the relationship between ln y and ln α_o becomes nonlinear as anticipated.

Takashi et al. (1978) conducted numerous freezing tests under quasi-steady conditions in order to determine empirical descriptions for the heave ratio h and the water intake ratio h_w. In their tests the unfrozen part of the soil was kept very close to 0°C, which implies that α_o was kept small. The empirical descriptions obtained are given as:

$$h = (m_1 / \sigma)\left[1 + (m_2 / V_o)^{1/2}\right] + m_o \tag{6.15}$$

$$h_w = d_2 (m_1 / \sigma)\left[1 + (m_2 / V_o)^{1/2}\right] - \left(1 - d_1^{-1} d_2\right) m_3 \tag{6.16}$$

where $m_i (i = 0, 1, \cdots, 3)$ are positive numbers that depend on a given soil. The sets of constants m_i for a few kinds of soils have been reported (Ohrai et al. 1991). In Figure 4 the values of h and h_w vs. V_o of Negishi silt (Ohrai et al. 1991) are presented with two different values of σ.

From Figure 4 it is easy to find that h (or h_w) is decreasing with increasing V_o and σ. It is important to note that h_w becomes negative for $V_o > 1.5$ cm/d when $\sigma = 0.5$ MPa. It is known that water may either be expelled or attracted to a freezing front n_1 when the front advances through the unfrozen part of soil. Coarse-grained sandy soils are known to expel water under most conditions while fine-grained soils expel water only at greater confining pressure σ (McRoberts and Morgenstern 1975). In the model M_1 there are two driving forces of water, gradients of temperature and pressure. The temperature gradient drives water toward a

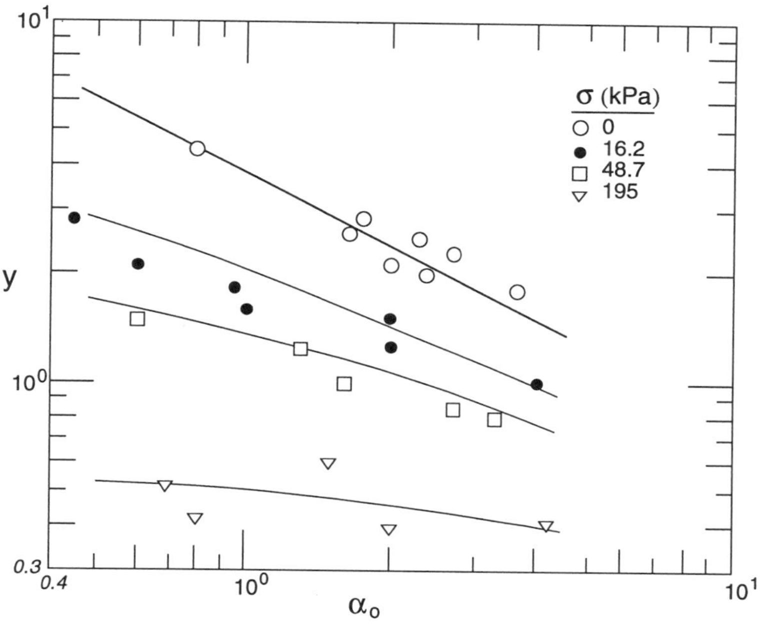

Figure 3. Experimental data of y[g(cm°C d)$^{-1}$]

freezing front, while the pressure gradient drives water to the opposite direction. As the pressure gradient increases with increasing σ, then f_o decreases and becomes even negative at the greater values of σ.

According to the model M_1, h and h_w are given by (4.22) and (4.23), respectively. Since the value of y is a positive root of (6.8) and y depends implicitly on α_o, δ_o, σ and V_o, it is difficult to obtain simpler forms such as (6.15) and (6.16). In order to find the behavior of (4.22) and (4.23), we calculated h and h_w of Kanto loam under a quasi-steady condition that is comparable to that of Takashi et al. (1978). Based on the size of soil samples, we will set $\delta_o = 2.0$ cm. Since α_o was kept small in their tests, we will assume that $\alpha_o = 10^{-2}$ °C/cm. The calculated values of h and h_w for Kanto loam with two different values of σ are presented in Figure 5. It is interesting to find that unlike Negishi silt the pressure of 0.5 MPa is not high enough to induce water expulsion. However, when $\sigma = 1.0$ MPa, h_w becomes negative. Efforts are being made to evaluate (4.22) and (4.23) experimentally.

We have discussed two semi-empirical models that are used for engineering practices. It is known that S_p depends on σ_o and σ. The empirical determination of such dependence is costly. On the other hand, it is also costly to determine empirically four positive numbers (6.27) and (6.28) in the Takashi model. An accurate mathematical model is needed that provides the dependence of S_p or m_i ($i = 0, 1, \cdots 3$) on pertinent variables specifying given thermal and hydraulic conditions in terms of well-defined functions (or parameters) describing the properties of a given soil.

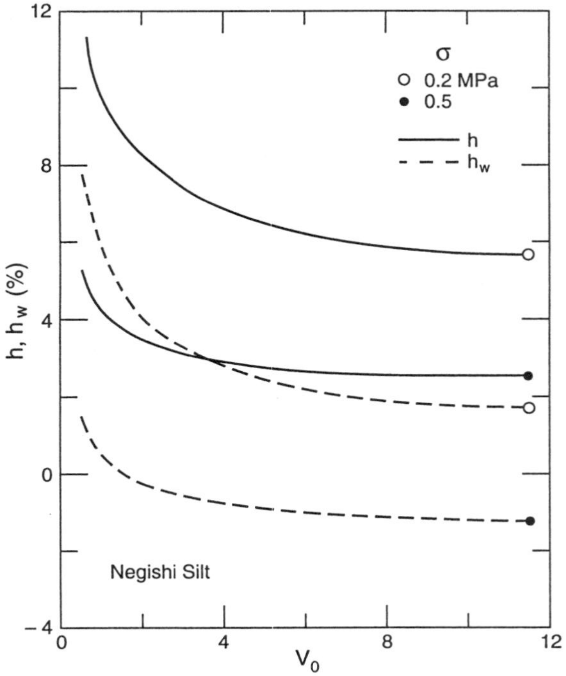

Figure 4. Empirically determined dependence of h and h_W on V_O for Negishi silt (Ohrai et al., 1991).

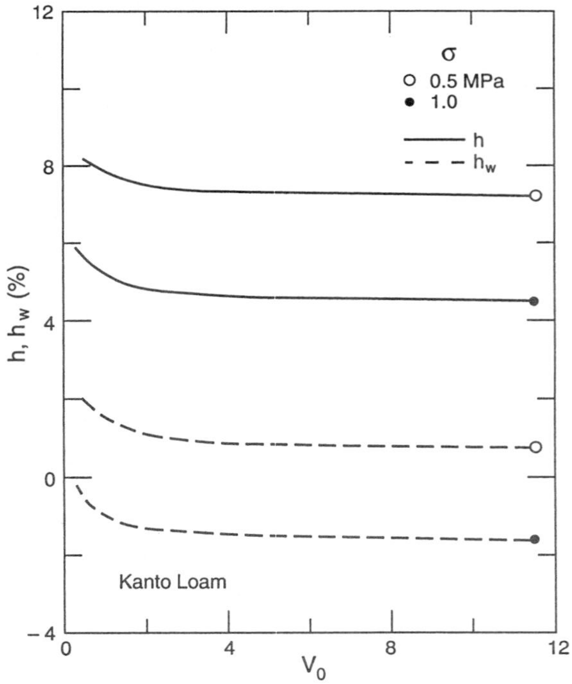

Figure 5. Predicted dependence of h and h_W on V_O for Kanto loam by (4.22) and (4.23).

Yoshisuke Nakano, U.S. Army Cold Regions Research and Engineering Laboratory, Hanover, NH 03755 USA

Mario Primicerio, Department of Mathematics "U. Dini," University of Firenze, Firenze, Italy

References

Andersland, O.B. and Anderson, D.M., Eds., 1978. Geotechnical Engineering for Cold Regions, McGraw-Hill, New York, pp. 74–79.

Beskow, G., 1935. Soil freezing and frost heaving with special attention to roads and railroads. Swed. Geol. Soc. Yearb., Ser. C, 26(3): 1–145.

Derjaguin, B.V. and Churaev, N.V. 1978. The theory of frost heaving. J. Colloid Interface Sci., 67: 391–396.

Ding, Z. and Talamucci, F., 1994. The existence of the solution for the problem of primary frost heave with lens formation. Dept. Math. Rept. 1994/5, University of Firenze, Firenze, Itally.

Drake, J.M. and Klafter, J., 1990. Dynamics of confined molecular systems. Physics Today, May:46–55.

Fasano, A. and Primicerio, M., 1984. Freezing of porous media: A review of mathematical models. Proc. German-Italian Symposium "Application of Mathematics in Technology." V. Boffi, H. Heunzert, Eds., Teubner, pp. 288–311.

Granick, S., 1991. Motions and relaxations of confined liquids. Science, 253: 1374–1379.

Horiguchi, K., 1987. An osmotic model for soil freezing. Cold Reg. Sci. and Tech., 14: 13–22.

Horiguchi, K. and Miller, R.D., 1983. Hydraulic conductivity functions of frozen materials. Proc. 4th Int. Conf. on Permafrost, Nat. Acad. Sci., Washington, D.C., 450–508.

Kay, B.D. and Perfect, E., 1988. Heat and mass transfer in freezing soils. Proc. 5th Int. Symp. Ground Freezing. Balkema, Rotterdam, Holland, 1: 3–21.

Konrad, J.M. and Morgenstern, N.R., 1980. A mechanistic theory of ice lens formation in fine-grained soils. Can. Geotech. J., 17: 473–486.

Konrad, J.M. and Morgenstern, N.R., 1981. The segregation potential of a freezing soil. Can. Geotech. J., 18: 482–491.

Konrad, J.M. and Morgenstern, N.R., 1982a. Prediction of frost heave in the laboratory during transient freezing. Can. Geotech. J., 19: 250–259.

Konrad, J.M. and Morgenstern, N.R., 1982b. Effects of applied pressure on freezing soils. Can. Geotech. J., 19: 494–505.

Kuroda, T., 1985. Theoretical study of frost heaving – Kinetic process at water layer between ice lens and soil particles. Proc. 4th Int. Symp. on Ground Freezing, Sapporo, Japan, A.A. Balkema, 1: 39–46.

McRoberts, E.C. and Morgenstern, N.R., 1975. Pore water expulsion during freezing. Can. Geotech. J., 12: 130–141.

Nakano, Y., 1990. Quasi-steady problems in freezing soils: I. Analysis on the steady growth of an ice layer. Cold Reg. Sci. Technol., 17: 207–226.

Nakano, Y., 1991. Transport of water through frozen soils. Proc. 5th Int. Symp. on Ground Freezing, Beijing, China, 1: 65–70.

Nakano, Y., 1992. Mathematical model on the steady growth of an ice layer in freezing soils. Physics and Chemistry of Ice. N. Maeno and T. Hondoh edited. Hokkaido Univ. Press. p. 364–369.

Nakano, Y., 1994a. Traveling wave solutions to the problem of quasi-steady freezing of soils. CRREL Rept. 94-3.

Nakano, Y., 1994b. Quasi-steady problems in freezing soils: IV. Traveling wave solutions. To appear in Cold Reg. Sci. Tech.

Nakano. Y. and Takeda, K., 1991. Quasi-steady problems in freezing soils: III. Analysis on experimental data. Cold Reg. Sci. Technol., 19: 225–243.

Nakano, Y. and Takeda, K., 1993. Evaluation of existing hypotheses used in the mathematical description of ice segregation in freezing soils. Free Boundary Problems in Fluid Flow with Applications. Pitman Research Notes in Mathematics Series 282: 225–242.

Nakano, Y. and Takeda, K., 1994. Growth condition of an ice layer in freezing soil under applied loads: II. Analysis. CRREL Rep. 94-1.

Ohrai, T. and Yamamoto, H., 1991. Frost heaving in artificial ground freezing. In Freezing and Melting Heat Transfer in Engineering. Cheng, K.C. and Seki, N. eds. Hemisphere Pub., New York, Chap. 17: 547–580.

Perfect, E. and Williams, P.J., 1980. Thermally induced water migration in saturated frozen soils. Cold Reg. Sci. Technol., 3: 101–109.

Radd, F.J. and Oertle, D.H., 1973. Experimental pressure studies of frost heave mechanism and the growth-fusion behavior of ice. Proc. 2nd Int. Conf. on Permafrost, Nat. Acad. Sci., Washington, D.C., 377–384.

Ratkje, S.K., Yamamoto, H., Takashi, T., Ohrai, T. and Okamoto, J., 1982. The hydraulic conductivity of soils during frost heave. Frost i Jord, 24: 22–26.

Ritter, M.B., Awschalom, D.D. and Shafer, M.W., 1988. Collective behavior of supercooled liquids in porous media. Physical Rev. Letters, 61(8): 966–969.

Taber, S., 1930. The mechanics of frost heaving. J. Geol., 38: 303–317.

Takashi, T., Yamamoto, H., Ohrai, T. and Masuda, M., 1978. Effect of penetration rate of freezing and confining stress on the frost heave ratio of soil. Proc. 3rd Int. Conf. on Permafrost, Nat. Res. Council Can., 1: 737–942.

Takeda, K. and Nakano, Y., 1990. Quasi-steady problems in freezing soils: II. Experiment on steady growth of an ice layer. Cold Reg. Sci. Technol., 18: 225–247.

Takeda, K. and Nakano, Y., 1993. Growth condition of an ice layer in freezing soil under applied loads: I. Experiment. CRREL Rep. 93–21.

Williams, P.J. and Burt, T.P., 1974. Measurement of hydraulic conductivity of frozen soils. Can. Geotech. J., 11: 647–650.

R H NOCHETTO AND C VERDI

Curvature driven interface motion

We survey convergence results for a fully discrete approximation of evolving interfaces whose normal velocity equals mean curvature plus a forcing function. The problem is first approximated via a singularly perturbed parabolic variational inequality with double obstacle ± 1, and then discretized with piecewise linear finite elements in space and backward differences in time. Convergence and interface error estimates past singularities are discussed. Estimates of transition layer width confirm the local structure observed in actual computations. Several simulations for tori illustrate the potential of this approach.

1. Fully Discrete Variational Formulation

We formulate the problem as a parabolic variational inequality in a cylindrical domain $Q = \Omega \times (0,T)$, where $\Omega \subset \mathbf{R}^n$ is bounded. Given $g = g(\mathbf{x},t)$, we consider the evolving oriented interface $\Sigma(t) \subset \Omega$ that emanating from Σ_0 propagates in the normal inward direction with velocity

$$V = \kappa + g,$$

with κ being the sum of the principal curvatures of $\Sigma(t)$ (positive if the enclosed set is locally mean convex). This evolution may exhibit singularities, topological changes, and even nonuniqueness (fattening).

Let the initial front $\Sigma_0 \subset \Omega$ be smooth and let d_0 be the signed distance function to Σ_0 (negative inside Σ_0). Let $0 < \varepsilon \ll 1$ be a relaxation parameter and γ be

$$\gamma(x) = -1 \text{ if } x < -\tfrac{\pi}{2}, \quad \gamma(x) = \sin x \text{ if } x \in [-\tfrac{\pi}{2}, \tfrac{\pi}{2}], \quad \gamma(x) = +1 \text{ if } x > \tfrac{\pi}{2}.$$

We indicate with L^2 and H^1 the usual energy spaces and introduce the convex set

$$\mathcal{K} = \{\varphi \in H^1(\Omega) : |\varphi| \le 1 \text{ in } \Omega, \ \varphi = 1 \text{ on } \partial\Omega\}.$$

We consider the following singularly perturbed parabolic variational inequality with double obstacle ± 1: *find* $\chi_\varepsilon \in L^2(0,\infty;\mathcal{K}) \cap H^1(0,\infty;L^2(\Omega))$ *such that* $\chi_\varepsilon(\cdot,0) = \gamma(\frac{d_0(\cdot)}{\varepsilon})$ *and, for a.e.* $t \in (0,T)$ *and all* $\varphi \in \mathcal{K}$,

$$\int_\Omega \Big(\varepsilon \partial_t \chi_\varepsilon(\varphi - \chi_\varepsilon) + \varepsilon \nabla \chi_\varepsilon \cdot \nabla(\varphi - \chi_\varepsilon) - \tfrac{1}{\varepsilon}\chi_\varepsilon(\varphi - \chi_\varepsilon) - \tfrac{\pi}{4} g(\varphi - \chi_\varepsilon)\Big) d\mathbf{x} \ge 0. \quad (1)$$

The zero level set $\Sigma_\varepsilon(t)$ of $\chi_\varepsilon(\cdot,t)$ can be viewed as an approximation of $\Sigma(t)$. Because of the double obstacle, χ_ε is forced to satisfy $|\chi_\varepsilon| \le 1$ irrespective of g, but more importantly $\chi_\varepsilon(\cdot,t)$ attains the values -1 or $+1$ outside a narrow transition layer $\mathcal{T}_\varepsilon(t)$ of size $\mathcal{O}(\varepsilon)$ in the vicinity of any regular point $\mathbf{x} \in \Sigma(t)$ [12].

The level set approach provides an alternative notion of evolution past singularities by representing $\Sigma(t)$ as the zero level set of a continuous function $w(\cdot,t)$, which satisfies

$$\partial_t w - |\nabla w| \operatorname{div}\left(\tfrac{\nabla w}{|\nabla w|}\right) - |\nabla w| g = 0 \quad \text{in } \mathbf{R}^n \times (0,T), \quad w(\cdot,0) = w_0(\cdot) \in C(\mathbf{R}^n), \tag{2}$$

in the viscosity sense, where $w_0 < 0$ inside Σ_0 [2,4,14]. The generalized evolution $\Sigma(t)$ and its inside $I(t)$ and outside $O(t)$ are defined as

$$\Sigma(t) = \{\mathbf{x} \in \mathbf{R}^n : w(\mathbf{x},t) = 0\},$$
$$I(t) = \{\mathbf{x} \in \mathbf{R}^n : w(\mathbf{x},t) < 0\}, \quad O(t) = \{\mathbf{x} \in \mathbf{R}^n : w(\mathbf{x},t) > 0\},$$

$\Sigma(t)$ is unique and coincides with the classical flow while it is smooth. Moreover $\Sigma_\varepsilon(t)$ converges to $\Sigma(t)$ as $\varepsilon \downarrow 0$ with a quadratic rate before the onset of singularities [8,9], and with linear order near regular points even past singularities [12].

We further discretize (1) in space with conforming finite elements, and in time with backward differences. Let $\mathcal{V}_h$ be the space of continuous piecewise linear functions over a quasi-uniform mesh $\mathcal{T}_h$ of size h, and set $\mathcal{K}_h := \mathcal{K} \cap \mathcal{V}_h$. Let $X^0 \in \mathcal{K}_h$ be the piecewise linear interpolant of $\gamma(\frac{d_0}{\varepsilon})$ and $\tau = \frac{T}{N} > 0$ be the uniform time-step. The fully discrete approximation of (1) reads: *for* $1 \le n \le N$ *seek* $X^n \in \mathcal{K}_h$ *such that for all* $\phi \in \mathcal{K}_h$

$$\int_\Omega \tfrac{\varepsilon}{\tau}(X^n - X^{n-1})(\phi - X^n) + \varepsilon \nabla X^n \cdot \nabla(\phi - X^n) - \tfrac{1}{\varepsilon} X^n(\phi - X^n) - \tfrac{\pi}{4} g^n(\phi - X^n) \ge 0, \tag{3}$$

where $g^n = g(\cdot, n\tau)$. Since the underlying operator $\varepsilon\partial_t v - \varepsilon\Delta v - \frac{1}{\varepsilon}v$ is linear, (3) is easy to implement and solve. Solvability, and also uniqueness, is guaranteed provided $\tau \le \varepsilon^2$. We denote by $X_{\varepsilon,h,\tau}$ the continuous-in-time piecewise linear interpolant of the discrete solution $\{X^n\}_{n=0}^N$. We also designate with $\Sigma_{\varepsilon,h,\tau}$ the zero level set of $X_{\varepsilon,h,\tau}$ and with $\mathcal{T}_{\varepsilon,h,\tau}$ the transition region, namely

$$\Sigma_{\varepsilon,h,\tau}(t) = \{\mathbf{x} \in \Omega : X_{\varepsilon,h,\tau}(\mathbf{x},t) = 0\}, \quad \mathcal{T}_{\varepsilon,h,\tau}(t) = \{\mathbf{x} \in \Omega : |X_{\varepsilon,h,\tau}(\mathbf{x},t)| < 1\}.$$

We show in §§2,3 that (3) retains the local structure of (1), and review convergence and interface error estimates of [13]. In §4 we further exploit such a local structure, and combine it with a variable relaxation parameter, to end up with the dynamic mesh algorithm of [11]. Several tori simulations conclude this paper.

2. Convergence Past Singularities

We suppose that $\mathcal{T}_h$ is *strongly acute*, which in $\mathbf{R}^2$ means that for any couple of adjacent triangles the sum of the opposite angles does not exceed $\pi - \theta$ for a constant $\theta > 0$. Consequently (3) satisfies the *discrete maximum principle*. Since no numerical integration is considered in (3), this method represents the best and simplest scenario in terms of convergence and error analysis.

The following result of [13] yields convergence of $\Sigma_{\varepsilon,h,\tau}(t)$ to $\Sigma(t)$ past singularities provided no fattening occurs, namely $\Sigma(t)$ does not develop interior.

PROPOSITION 1. *Let* $\mathbf{x} \in I(t)$ *(resp.* $\mathbf{x} \in O(t)$*) and let* $h = o(\varepsilon^{3/2}|\log\varepsilon|^{-4})$ *and* $\tau = o(\varepsilon^3|\log\varepsilon|^{-5})$. *Then there exists* $\varepsilon^* > 0$ *depending on* $(\mathbf{x}, t)$ *such that*

$$X_{\varepsilon,h,\tau}(\mathbf{x}, t) = -1 \quad (\textit{resp. } X_{\varepsilon,h,\tau}(\mathbf{x}, t) = 1) \quad \forall\, \varepsilon \le \varepsilon^*.$$

To prove Proposition 1 we construct fully discrete barriers as in [5] via a *parabolic projection*, which is shown to be quasi-optimal in L^∞ in both accuracy and regularity requirements for functions that only belong to $W_p^{2,1}(Q)$. This key result extends those in [3] for functions in $W_\infty^{2,1}(Q)$, and constitutes the sole restriction to 2D. The function obtained by projection has to be lifted a little bit to become a candidate for discrete supersolution. But then to compensate for the antimonotone nature of the term $-\frac{1}{\varepsilon}X^n$ (potential) in (3), we are forced to perturb the driving force g. This is already a major difference with respect to [5], that examines a monotone problem, and the question arises as to how viscosity solutions of (2) behave under perturbations of g. Exploiting their well known continuous dependence, and so choosing a perturbation of order $o(1)$, convergence of interfaces is guaranteed. We cannot expect, however, a rate of convergence without additional assumptions on $\Sigma(t)$. This issue is further explored next.

3. Interface and Layer Width Estimates

We examine the behavior of $X_{\varepsilon,h,\tau}$ in a vicinity of regular and certain singular points. Our study hinges upon the following three properties. The first one is *Lipschitz dependence* on the maximum norm of viscosity solutions w of (2) with respect to perturbations on g and w_0. The second property is *discrete nondegeneracy*, which asserts quadratic growth of $X_{\varepsilon,h,\tau}$ away from the obstacles ±1 as in [16]. The third crucial ingredient is *nondegeneracy* of w, which is explained below.

We say that $\mathbf{x} \in \Sigma(t)$ is a regular point if w is C^1 in a neighborhood of $(\mathbf{x}, t)$ and $\nabla w(\mathbf{x}, t) \neq 0$. Hence $w(\cdot, t)$ exhibits a linear growth away from x (nondegeneracy). If thick$(\mathcal{T}_{\varepsilon,h,\tau}(t); \mathbf{x}, \mathbf{n})$ denotes the thickness of $\mathcal{T}_{\varepsilon,h,\tau}(t)$ in the direction $\mathbf{n}$ across $\mathbf{x}$, we have the following result from [13].

PROPOSITION 2. *Let* $\mathbf{x} \in \Sigma(t)$ *be a regular point, and let* $h = \mathcal{O}(\varepsilon^2|\log\varepsilon|^{-4})$ *and* $\tau = \mathcal{O}(\varepsilon^4|\log\varepsilon|^{-5})$. *Then there exist constants* $\varepsilon^*, C > 0$ *such that for all* $\varepsilon \leq \varepsilon^*$

$$\operatorname{dist}\left(\mathbf{x}, \Sigma_{\varepsilon,h,\tau}(t)\right), \ \operatorname{thick}\left(\mathcal{T}_{\varepsilon,h,\tau}(t); \mathbf{x}, \tfrac{\nabla w(\mathbf{x},t)}{|\nabla w(\mathbf{x},t)|}\right) \leq \tfrac{C}{|\nabla w(\mathbf{x},t)|}\varepsilon.$$

This rate of convergence is rather unexpected in that it is valid even past singularities provided $\mathbf{x}$ is regular. The thickness estimate reveals that the local structure of the double obstacle formulation is inherited by the fully discrete scheme (3), thereby confirming a key computational property exploited in [11].

We finally elaborate upon certain singularities. Let w be C^1 in a vicinity of a singular point $\mathbf{x} \in \Sigma(t)$, namely $\nabla w(\mathbf{x}, t) = 0$, and possess the following asymptotic behavior [6]: *there exists a constant* β, *a set of orthonormal vectors* $\{\mathbf{e}_i\}_{i=1}^n$, *and nonvanishing constants* $\{\lambda_i\}_{i=1}^k$ *with* $1 \leq k \leq n$ *such that*

$$w(\mathbf{x}+\mathbf{y}, t+s) = \beta s + \sum_{i=1}^{k} \lambda_i y_i^2 + o(|s| + |\mathbf{y}|^2) \quad \text{for } \mathbf{y} = \sum_{i=1}^{k} y_i \mathbf{e}_i, \quad |\mathbf{y}|, |s| \ll 1. \tag{4}$$

This is certainly true if $w \in C^2$ in the vicinity of $(\mathbf{x}, t)$, in which case $\{\lambda_i\}_{i=1}^k$ are the nontrivial eigenvalues of the Hessian $D^2 w(\mathbf{x}, t)$. Substituting (4) into (2), it is not difficult to see that [6] $\lambda_i = \lambda \neq 0 \quad 1 \leq i \leq k$ and $\beta = 2(k-1)\lambda + g(\mathbf{x}, t)$, which combined with (4) yields the curvature blow-up rate of type I singularities [7]:

$$\kappa \approx |\mathbf{y}|^{-1} \approx \left|\tfrac{\lambda}{\beta s}\right|^{1/2} \quad \text{as } s \uparrow 0.$$

PROPOSITION 3. *Let* $\mathbf{x} \in \Sigma(t)$ *satisfy* (4) *with* $\beta, \lambda > 0$, *and let* $h = \mathcal{O}(\varepsilon^2|\log\varepsilon|^{-4})$ *and* $\tau = \mathcal{O}(\varepsilon^4|\log\varepsilon|^{-5})$. *Then there exist constants* $\varepsilon^*, C > 0$ *such that for all* $\varepsilon \leq \varepsilon^*$

$$\mathcal{X}_{\varepsilon,h,\tau}(\mathbf{x}+\mathbf{y}, t \pm s) = \pm 1 \text{ for } C\tfrac{\varepsilon}{\beta} \leq s \leq C\tfrac{\varepsilon^*}{\beta}, \ |\mathbf{y}|^2 < C\tfrac{\beta}{\lambda}s; \tag{5}$$

$$\text{there exists } |s| \leq C\tfrac{\varepsilon}{\beta} \text{ for which } \mathbf{x} \in \Sigma_{\varepsilon,h,\tau}(t+s); \tag{6}$$

$$\operatorname{thick}\left(\mathcal{T}_{\varepsilon,h,\tau}(t); \mathbf{x}, \mathbf{e}_i\right) \leq C\left(\tfrac{\varepsilon}{\lambda}\right)^{1/2} \text{ for all } 1 \leq i \leq k. \tag{7}$$

This result partially explains why it is feasible to approximate the time at which singularities occur, and in particular the extinction time. In view of (5), the approximate interface $\Sigma_{\varepsilon,h,\tau}(t)$ behaves much like the true one in that it either disappears locally in time, if $k = n$ (sphere), or at least it moves away from $\mathbf{x}$ when $2 \leq k < n$ (cylinder). Statement (7) quantifies a smearing effect observed in computations in the vicinity of singularities, which give rise to a much more diffuse transition layer. We then see that reducing the local size of ε via a density function $A \le a(\mathbf{x}) \le 1$, along with that of h and τ, could result in enhanced singularity resolution. This idea leads to the dynamic mesh algorithm of [11], briefly discussed in §4, but deserves further investigation. We refer to [10,15] for a study of the reaction-diffusion approach before the onset of singularities.

4. Simulations

Assuming axial symmetry we split $\mathbb{R}^n = \mathbb{R}^{n_1} \times \mathbb{R}^{n_2}$ and denote any point in $\mathbb{R}^n$ by $(\mathbf{x}_1, \mathbf{x}_2) \in \mathbb{R}^{n_1} \times \mathbb{R}^{n_2}$. We set $\mathbf{z} = (z_1, z_2)$ where $z_i = |\mathbf{x}_i|$ and $Z = \{\mathbf{z} \in \mathbb{R}^2 : (\mathbf{x}_1, \mathbf{x}_2) \in \Omega\}$. In practice, we use a variable relaxation parameter $\varepsilon a(\mathbf{z})$, and discretize the axially symmetric version of (3) in space over a locally refined mesh $\mathcal{T}_h$, with mesh density function $ha(\mathbf{z})$. Parameters h and ε are related so as to guarantee numerical resolution of the transition layer $\mathcal{T}_\varepsilon(t)$ as well as the optimal distribution of spatial degrees of freedom. Time stepping is now explicit, which guarantees that the discrete transition layer will move at most one layer of elements. Its coupling with mass lumping yields a trivial algebraic problem that reduces to a matrix-vector product followed by a componentwise truncation to meet the obstacle constraint; consequently there is no iteration involved. Since there is no need for an entire partition $\mathcal{T}_h$ of Ω into finite elements, but just for a partition of the transition layer, the *dynamic mesh algorithm* of [11] triangulates solely the transition region, and then updates the resulting mesh to follow the layer motion. Mesh updating is simple and thus practical. We refer to [11] for details on the algorithm. Theoretical results such as those in §§2,3 are yet to be derived.

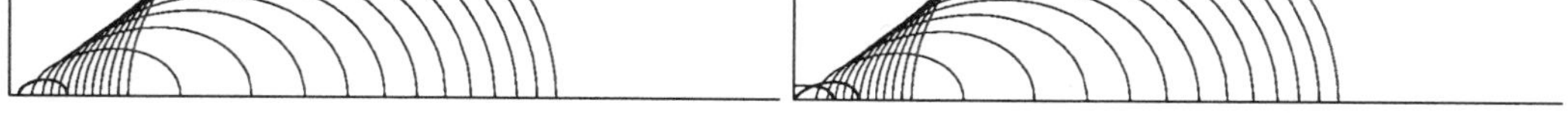

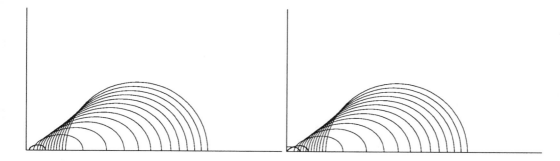

Figure 1. Torus $S^1 \times S^1$: interfaces at $t = 0.025i$ and t_*^λ, t_m^λ (bold lines)
(a) $\lambda = 0.64298575535$ ($t_m^\lambda = 0.296290$, $t_\dagger^\lambda = 0.298381$)
(b) $\lambda = 0.64298575536$ ($t_*^\lambda = 0.295622$, $t_m^\lambda = 0.297845$, $t_\dagger^\lambda = 0.299720$).

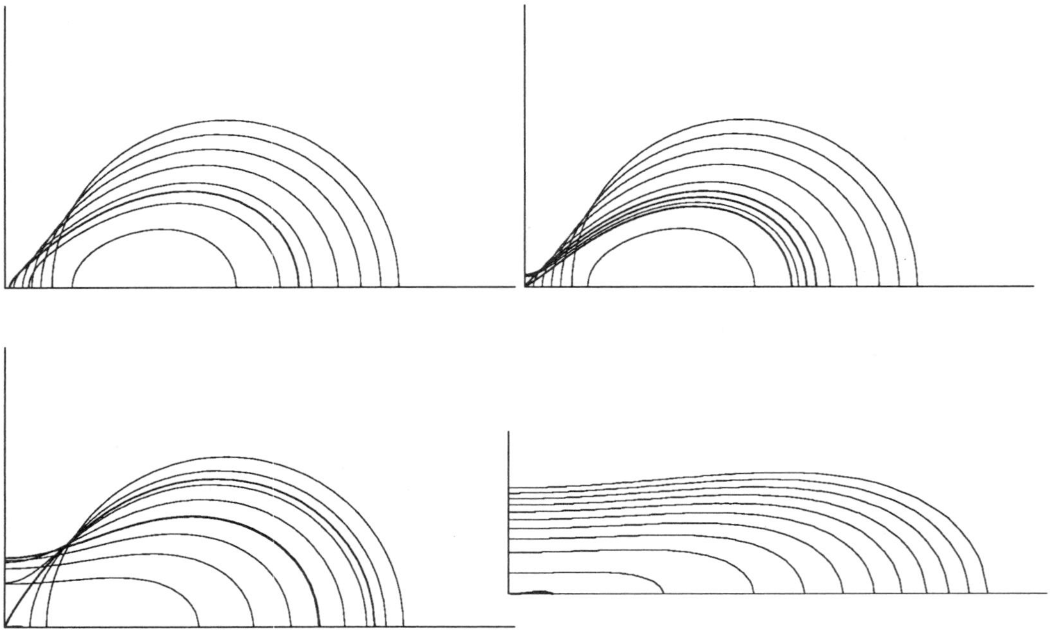

Figure 2. Torus $S^1 \times S^2$: interfaces at $t = 0.025i$ and t_*^λ, t_{**}^λ, t_m^λ (bold lines)
(a) $\lambda = 0.78556296$ ($t_m^\lambda = 0.111080$, $t_\dagger^\lambda = 0.170624$)
(b) $\lambda = 0.78556298$ ($t_*^\lambda = 0.111617$, $t_m^\lambda = 0.119092$, $t_{**}^\lambda = 0.129542$, $t_\dagger^\lambda = 0.170844$)
(c) $\lambda = 0.808$ ($t_*^\lambda = 0.040106$, $t_m^\lambda = 0.100950$, $t_{**}^\lambda = 0.198271$, $t_\dagger^\lambda = 0.198279$)
and zoom (interfaces at $t = 0.0025i \geq 0.175$ and t_{**}^λ).

We consider the evolution by mean curvature ($g = 0$) of two tori in 3D and 4D, with $Z = [0,3)^2$ and $\Sigma_0^\lambda = \{\mathbf{z} \in Z : w_0(\mathbf{z}) = \lambda\}$ being the initial interface, where $w_0(\mathbf{z}) = \left((z_1 - 1)^2 + z_2^2\right)^{1/2}$ and $0 < \lambda < 1$. The numerical experiments are performed with $a(\mathbf{z}) = \max(0.05, 0.07z_1 + 0.07z_2^2)$ and are intended to determine critical radii λ_* and corresponding extinction $t_\dagger$ and critical t_* times.

4.1. Torus $S^1 \times S^1 = \partial(S^1 \times D^2)$. Let $n_1 = 2$, $n_2 = 1$. There exists a critical radius λ_* such that for all $\lambda < \lambda_*$ the surface shrinks to $S^1 \subset \{z_2 = 0\}$ and then suddenly disappears at $t_\dagger^\lambda$, whereas for $\lambda > \lambda_*$ the surface focuses at the origin at some time t_*^λ, when a singularity occurs, then changes its topology to S^2 and eventually disappears at $t_\dagger^\lambda$. Uniqueness of the evolution has been proven together with the fact $t_\dagger^{\lambda_*} = t_*^{\lambda_*}$ [17]. Our numerical experiments confirm these results (see Figure 1) and provide the

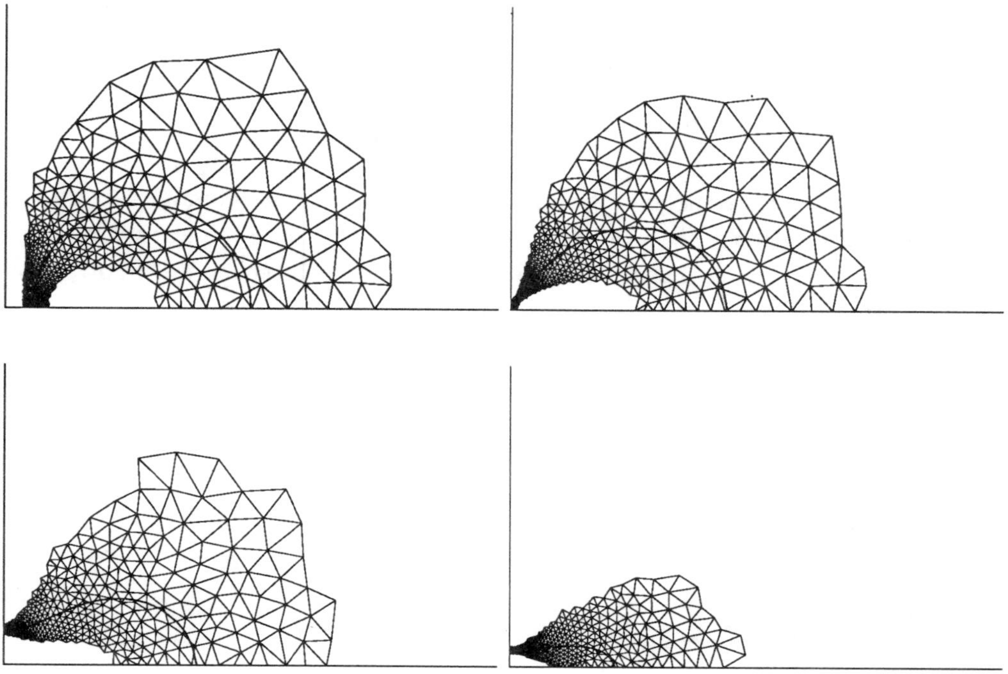

Figure 3. Torus $S^1 \times S^2$ ($\lambda = 0.797$, $h = 0.1, \varepsilon = 0.28$): mesh and interfaces at $t = 0, t_*^\lambda, 0.1, 0.175$.

following approximate λ_* for $h = 0.05, \varepsilon = 0.15$:

$$0.64298575535 < \lambda_* < 0.64298575536.$$

4.2. Torus $S^1 \times S^2 = \partial(S^1 \times D^3)$. Let $n_1 = n_2 = 2$. The numerical experiments show now the existence of two critical radii $0 < \lambda_* < \lambda_{**}$, along with evolutions quite different from those in §4.1. For $\lambda < \lambda_*$ the surface shrinks at $t_\dagger^\lambda$ to $S^1 \subset \{z_2 = 0\}$ and suddenly disappears at $t_\dagger^\lambda$. For $\lambda > \lambda_{**}$ a singularity develops at the origin at time t_*^λ, and then the surface becomes topologically equivalent to S^3 until it shrinks to the origin. For $\lambda_* < \lambda < \lambda_{**}$ the evolution is for a while similar to the case $\lambda > \lambda_{**}$, but a second singularity forms at the origin at a later time $t_{**}^\lambda > t_*^\lambda$ after which the surface becomes again a 4D-torus and shrinks at $t_\dagger^\lambda$ to $S^1 \subset \{z_2 = 0\}$ and disappears. A recent result by Angenent *et al.* [1] would suggest that the surface at time $t_*^{\lambda_*}$ is tangent to a 4D-cone at the origin with opening $\frac{\pi}{4}$ which develops fattening. Our simulations suggest that (see Figure 2):

$$\lim_{\lambda \downarrow \lambda_*} t_*^\lambda = \lim_{\lambda \downarrow \lambda_*} t_{**}^\lambda < \lim_{\lambda \downarrow \lambda_*} t_\dagger^\lambda \quad \text{and} \quad \lim_{\lambda \uparrow \lambda_{**}} t_{**}^\lambda = \lim_{\lambda \uparrow \lambda_{**}} t_\dagger^\lambda,$$

which would imply that the evolution is unique for both critical values λ_* and λ_{**}. Detecting fattening for λ_* is a delicate issue which thus deserves further numerical investigation; we refer to [11] for a further discussion. Moreover we found for $h = 0.05, \varepsilon = 0.15$:

$$0.78556296 < \lambda_* < 0.78556298, \quad 0.8090 < \lambda_{**} < 0.8092.$$

Of interest is also the sequence of dynamic meshes depicted in Figure 3.

Acknowledgement: We would like to thank Maurizio Paolini for his invaluable participation in the development of the dynamic mesh algorithm and the numerical simulations.

References

[1] S. ANGENENT, T. ILMANEN, AND J. VELASQUEZ, *Nonuniqueness of motion by mean curvature in dimensions four through seven,* in preparation.

[2] Y.G. CHEN, Y. GIGA, AND S. GOTO, *Uniqueness and existence of viscosity solutions of generalized mean curvature flow equation,* J. Differential Geom., **33** (1991), 749–786.

[3] K. ERIKSSON AND C. JOHNSON, *Adaptive finite element methods for parabolic problems II: optimal error estimates in $L_\infty L_2$ and $L_\infty L_\infty$,* SIAM J. Numer. Anal. (to appear).

[4] L.C. EVANS AND J. SPRUCK, *Motion of level sets by mean curvature. I,* J. Differential Geom., **33** (1991), 635–681.

[5] A. FETTER, *L^∞-error estimate for an approximation of a parabolic variational inequality,* Numer. Math., **50** (1987), 557–565.

[6] Y. GIGA AND S. GOTO, *Geometric evolution of phase-boundaries,* in *On the evolution of phase boundaries,* IMA VMA 43 (M.E. Gurtin and G.B. MacFadden, eds.), Springer-Verlag, New York, 1992, 51–65.

[7] G. HUISKEN, *Asymptotic behavior for singularities of the mean curvature flow,* J. Differential Geom., **31** (1990), 285–299.

[8] R.H. NOCHETTO, M. PAOLINI, AND C. VERDI, *Optimal interface error estimates for the mean curvature flow,* Ann. Scuola Norm. Sup. Pisa Cl. Sci. (4), **21** (1994), 193–212.

[9] R.H. NOCHETTO, M. PAOLINI, AND C. VERDI, *Sharp error analysis for curvature dependent evolving fronts,* Math. Models Methods Appl. Sci., **3** (1993), 771–723.

[10] R.H. NOCHETTO, M. PAOLINI, AND C. VERDI, *Double obstacle formulation with variable relaxation parameter for smooth geometric front evolutions: asymptotic interface error estimates,* Asymptotic Anal. (to appear).

[11] R.H. NOCHETTO, M. PAOLINI, AND C. VERDI, *A dynamic mesh method for curvature dependent evolving interfaces,* J. Comput. Phys. (submitted).

[12] R.H. NOCHETTO AND C. VERDI, *Convergence of double obstacle problems to the generalized geometric motion of fronts,* SIAM J. Math. Anal. (to appear).

[13] R.H. NOCHETTO AND C. VERDI, *Convergence past singularities for a fully discrete approximation of curvature driven interfaces,* SIAM J. Numer. Anal. (submitted).

[14] S. OSHER AND J.A. SETHIAN, *Fronts propagating with curvature dependent speed: algorithms based on Hamilton-Jacobi formulations, J. Comput. Phys.,* **79** (1988), 12-49.

[15] M. PAOLINI AND C. VERDI, *Asymptotic and numerical analyses of the mean curvature flow with a space-dependent relaxation parameter,* Asymptotic Anal., **5** (1992), 553–574.

[16] P. PIETRA AND C. VERDI, *On the convergence of the approximate free boundary for the parabolic obstacle problem,* Atti Accad. Naz. Lincei Rend. Cl. Sci. Fis. Mat. Natur. (8), **79** (1985), 159–171.

[17] H.-M. SONER AND P.E. SOUGANIDIS, *Singularities and uniqueness of cylindrically symmetric surfaces moving by mean curvature,* Comm. Partial Differential Equations, **18** (1993), 859–894.

Ricardo H. Nochetto
Department of Mathematics and Institute for Physical Science and Technology
University of Maryland
College Park, MD 20742, USA

Claudio Verdi
Dipartimento di Matematica
Università di Milano
Via Saldini 50
20133 Milano, Italy

C PFLAUM

Discretization of second order elliptic differential equations on sparse grids

Elliptic partial differential equations can be discretized with bilinear finite elements on a full grid with $O(N^2)$ grid points. Then, the finite element solution converges with order $O(N)$ in the energy norm and with order $O(N^{-2})$ in the L_2-norm. Using sparse grids the number of grid points and the dimension of the FE-space is reduced to $O(N \log N)$.
A FE-discretization is presented that solves elliptic partial differential equations with sparse grids in an efficient way. For this discretization, a convergence with order $O(N^{-1} \log N)$ in the energy norm and with order $O(N^{-1.5} \log N)$ in the L_2-norm is proved. Numerical results show a convergence with order $O(N^{-2} \log N)$ in the supremum norm. Thus, in comparison to full grids, by sparse grids the same accuracy is achieved but with much less grids points.

1. Introduction

Elliptic partial differential equations can be discretized with bilinear finite elements on a sparse grid. In case of general second order elliptic differential equations, the stiffness matrix of the finite element discretization can become very complicated. Then, it is not possible to store the stiffness matrix in a sparse grid data structure and to solve the discrete equation system in an efficient way. Therefore, it is necessary to approximate the bilinear form a corresponding to the elliptic equation. This gives stiffness matrices for which these problems don't arise.

The sparse grid interpolation with bilinear finite elements can be compared in the W_2^1-norm and the L_2-norm with the full grid interpolation with second order finite elements. In case of more smoothness this sparse grid interpolation can be compared in the L_2-norm even with third order finite elements on a full grid. These comparisons mean that nearly the same Sobolev spaces give nearly the same order of convergence with respect to the number of grid points. A sparse grid discretization of elliptic equations should have these properties too.

The bilinear form a is approximated by interpolating the variable coefficients by a piecewise constant sparse grid interpolant. Theorem 5 and 6 show that this discretization can be compared with a finite element discretization of second order on a full grid in the W_2^1- and the L_2-norm. Numerical experiments show that this sparse grid discretization can be compared with third order finite elements on a full grid in the L_2-norm, if the bilinear form a is strong elliptic. A multi-level-algorithm guarantees an efficient solution of the discrete equation system (see [7]).

For simplicity, the discretization in this paper is explained only for the regular sparse grids $\mathcal{D}_n$. However, it is possible to generalize the discretization for adaptive sparse grids. An algorithm has been implemented which solves general second order elliptic differential equations on adaptive sparse grids.

Throughout the paper, it is $h = 2^{-n}$, where $n \in \mathbb{N}$ and $\Omega =]0,1[^2$.

2. Sparse Grids and Sparse Grid Interpolation

For estimations of the sparse grid interpolation error, we need suitable Sobolev Spaces. Denote

$W_p^l(\Omega)$ and $\overset{\circ}{W}{}_p^l(\Omega)$ the standard Sobolev spaces (see [9]). For $l \in \mathbb{N}$ and $p \in [1,\infty]$ we define the mixed Sobolev spaces

$$W_p^{G,l}(\Omega) := \left\{ f \in L_p(\Omega) \,\middle|\, \frac{\partial^{i+j} f}{\partial x^i \partial y^j} \in L_p(\Omega), \text{ where } i,j \in \mathbb{N}_0,\ i+j \le l \text{ and } i,j < \frac{l}{2}+1 \right\}$$

with the norm

$$\|f\|_{W_p^{G,l}} := \left\| \left(\left\| \frac{\partial^{i+j} f}{\partial x^i \partial y^j} \right\|_{L_p} \right)_{i+j\le l,\ i,j<\frac{l}{2}+1} \right\|_{l_p},$$

where $\| - \|_{l_p}$ is the discrete l_p-norm. Furthermore, we define

$$\overset{\circ}{W}{}_p^{G,l}(\Omega) := \overline{C_0^\infty(\Omega)}^{W_p^{G,l}}.$$

Now we explain what a sparse grid is. Assume $i, j \in \mathbb{N}$. Let be

$$\begin{aligned} \Omega_{i,j} &:= \{(k \cdot h_x, l \cdot h_y) |\ h_x = 2^{-i},\ h_y = 2^{-j} \quad \text{and} \quad k = 0, \dots, 2^i,\ l = 0, \dots, 2^j\} \\ \overset{\circ}{\Omega}_{i,j} &:= \Omega_{i,j} \cap]0,1[^2 \end{aligned}$$

full grids of mesh size 2^{-i} and 2^{-j}. The regular sparse grids $\mathcal{D}_n$ and $\overset{\circ}{\mathcal{D}}_n$ are defined by

$$\mathcal{D}_n := \bigcup_{i+j\le n+1} \Omega_{i,j} \quad \text{and} \quad \overset{\circ}{\mathcal{D}}_n := \mathcal{D}_n \cap]0,1[^2,$$

where $n \in \mathbb{N}$ (see figure 1). A more detailed description of general abstract and adaptive sparse grids and their properties is given in [5] and [8].

Denote $V^{i,j} \subset \mathcal{C}(\bar{\Omega})$ the full grid finite element space on the grid $\Omega_{i,j}$ with bilinear finite elements. The regular finite element spaces for the regular sparse grids $\mathcal{D}_n$ are defined by

$$\begin{aligned} V_{\mathcal{D}_n} &:= \operatorname{span}_{\mathbb{R}}\{v | v \in V^{i,j} \quad \text{and} \quad i+j \le n+1\} \subset W_2^1(\Omega) \cap \mathcal{C}(\bar{\Omega}) \qquad \text{and} \\ \overset{\circ}{V}_{\mathcal{D}_n} &:= V_{\mathcal{D}_n} \cap \overset{\circ}{W}{}_2^1(\Omega) \subset \overset{\circ}{W}{}_2^1(\Omega) \cap \mathcal{C}(\bar{\Omega}). \end{aligned}$$

There are two ways for the definition of the sparse grid interpolation operator $\mathcal{I}_{\mathcal{D}_n}$. The first way is described in [10] with hierarchical basis functions. The second way is a little bit more simple, but it can be used only for the regular sparse grids $\mathcal{D}_n$. This way uses the combination formula. Let be Let be $I^{i,j} : \mathcal{C}(\bar{\Omega}) \mapsto V^{i,j}$ the full grid interpolation operator. The sparse grid interpolation operator is defined by

$$\mathcal{I}_{\mathcal{D}_n} = \sum_{i=1}^{n} I^{2^{-i},2^{i-n-1}} - \sum_{i=1}^{n-1} I^{2^{-i},2^{i-n}}.$$

The following lemma is proved in [2].

Lemma 1 *Assume $f \in \mathcal{C}(\bar{\Omega})$, then*

$$\mathcal{I}_{\mathcal{D}_n}(f)(z) = f(z) \qquad \forall z \in \mathcal{D}_n.$$

The sparse grid interpolation error is now.

Theorem 1 (Error of the Sparse Grid Interpolation with Piecewise Bilinear Functions) *There exists a constant $C > 0$ such that for $h = 2^{-n}$ the error in the supremum norm is*

$$\|f - \mathcal{I}_{\mathcal{D}_n}(f)\|_\infty \leq C\|f\|_{W_\infty^{G,4}} h^2 \log h^{-1} \qquad \text{for } f \in \mathcal{C}^4(\bar{\Omega}),$$

$$\|f - \mathcal{I}_{\mathcal{D}_n}(f)\|_\infty \leq C\|f\|_{W_\infty^{G,3}} h^{\frac{3}{2}} \qquad \text{for } f \in \mathcal{C}^3(\bar{\Omega}),$$

the error in the L_2-norm is

$$\|f - \mathcal{I}_{\mathcal{D}_n}(f)\|_2 \leq C\|f\|_{W_2^{G,4}} h^2 \log h^{-1} \qquad \text{for } f \in W_2^{G,4}(\Omega),$$

$$\|f - \mathcal{I}_{\mathcal{D}_n}(f)\|_2 \leq C\|f\|_{W_2^{G,3}} h^{\frac{3}{2}} \qquad \text{for } f \in W_2^{G,3}(\Omega),$$

the error in the W_2^1-norm is

$$\|f - \mathcal{I}_{\mathcal{D}_n}(f)\|_{W_2^1} \leq C\|f\|_{W_2^{G,4}} h \qquad \text{for } f \in W_2^{G,4}(\Omega), \quad \text{and}$$

$$\|f - \mathcal{I}_{\mathcal{D}_n}(f)\|_{W_2^1} \leq C\|f\|_{W_2^{G,3}} h \log h^{-1} \qquad \text{for } f \in W_2^{G,3}(\Omega).$$

The proof of the first and third inequality is given in [1] and [10]. The fifth inequality is given in a similar form in [1]. A proof of the last two inequalities is given below. We shall use another technique for the estimation of the hierarchical surplus as given in [1] and [10]. These techniques give a hint how to prove the other inequalities of Theorem 1.

PROOF Assume $h_x = 2^{-i}$ and $h_y = 2^{-j}$, where $i, j \in \mathbb{N}$. The full grid interpolant is

$$f_{h_x,h_y} := I^{h_x,h_y}(f).$$

The hierarchical surplus is defined by

$$v_{h_x,h_y} := f_{h_x,h_y} - f_{2h_x,h_y} - f_{h_x,2h_y} + f_{2h_x,2h_y}.$$

Denote $I_x^{h_x}$ and $I_y^{h_y}$ the interpolation operators, which interpolate only in the x- and y-direction, respectively. Observe $I^{h_x,h_y} = I_x^{h_x} \circ I_y^{h_x}$. By the one dimensional interpolation theory, we know that there is a constant $\tilde{C} > 0$, such that

$$\|f - I_x^{h_x}(f)\|_2 \leq \tilde{C}\left\|\frac{\partial f}{\partial x}\right\|_2 h_x \qquad \text{for } f \in \mathcal{C}^1(\bar{\Omega}),$$

$$\|f - I_x^{h_x}(f)\|_2 \leq \tilde{C}\left\|\frac{\partial^2 f}{\partial x^2}\right\|_2 h_x^2 \qquad \text{for } f \in \mathcal{C}^2(\bar{\Omega}) \quad \text{and}$$

$$\left\|\frac{\partial(f - I_x^{h_x}(f))}{\partial x}\right\|_2 \leq \tilde{C}\left\|\frac{\partial^2 f}{\partial x^2}\right\|_2 h_x \qquad \text{for } f \in \mathcal{C}^2(\bar{\Omega}).$$

Let us assume $f \in W_2^{G,4}(\Omega)$.

1.Step. (The hierarchical surplus is small.) We have to show that there is a constant $C > 0$ such that

$$\|v_{h_x,h_y}\|_{W_2^1} \leq C(h_x^2 h_y + h_x h_y^2)\|f\|_{W_2^{G,4}}. \tag{1}$$

Let us assume $f \in \mathcal{C}^4(\Omega)$. By the one dimensional interpolation formulas, we get

$$\left\|\frac{\partial v_{h_x,h_y}}{\partial x}\right\|_2 = \left\|\frac{\partial\left(\left(I_x^{2h_x} - I_x^{h_x}\right) \circ \left(I_y^{2h_y} - I_y^{h_y}\right)\right)(f)}{\partial x}\right\|_2 =$$

$$= \left\|\frac{\partial\left(I_x^{2h_x} - I_x^{h_x}\right)\left(\left(I_y^{2h_y} - I_y^{h_y}\right)(f)\right)}{\partial x}\right\|_2 \leq \tilde{C}h_x\left\|\frac{\partial^2\left(I_y^{2h_y} - I_y^{h_y}\right)(f)}{\partial x^2}\right\|_2 \leq$$

$$\leq \tilde{C}h_x\left\|\left(I_y^{2h_y} - I_y^{h_y}\right)\left(\frac{\partial^2 f}{\partial x^2}\right)\right\|_2 \leq \tilde{C}^2 h_x h_y^2\left\|\frac{\partial^4 f}{\partial x^2 \partial y^2}\right\|_2.$$

By symmetry, we conclude

$$|v_{h_x,h_y}|_{W_2^1} \le \tilde{C}^2\sqrt{(h_x h_y^2)^2 + (h_x^2 h_y)^2}\|f\|_{W_2^{G,4}} \le \tilde{C}^2(h_x h_y^2 + h_x^2 h_y)\|f\|_{W_2^{G,4}}.$$

Now we use the same trick for $\|v_{h_x,h_y}\|_2$. Therefore we get for a constant $C > 0$.

$$\|v_{h_x,h_y}\|_{W_2^1} \le C(h_x h_y^2 + h_x^2 h_y)\|f\|_{W_2^{G,4}}.$$

Observe that $\mathcal{C}^4(\bar{\Omega})$ is dense in $W_2^{G,4}(\Omega)$. By Sobolev's Lemma, $W_2^{G,4}(\Omega) \hookrightarrow \mathcal{C}(\bar{\Omega})$ is a continuous embedding. By a density argument, we conclude equation (1) for every $f \in W_2^{G,4}(\Omega)$.
2.Step. (The error of the full grid solution is small.)
By the interpolation theory on full grids, we know that there is a constant $C > 0$, such that

$$\|f - f_{h,h}\|_{W_2^1} \le hC\|f\|_{W_2^2}.$$

3.Step. (Zenger's sparse grid idea.)
A simple calculation (see [10] or [6]) shows

$$f_{h,h} - \mathcal{I}_{\mathcal{D}_n}(f) = \sum_{i=1}^{n}\sum_{j=n-i+1}^{n} v_{2^{-i},2^{-j}}.$$

By step 1, we get for a constant $\hat{C} > 0$

$$\begin{aligned}\|f_{h,h} - \mathcal{I}_{\mathcal{D}_n}(f)\|_{W_2^1} &= \left\|\sum_{i=1}^{n}\sum_{j=n-i+1}^{n} v_{2^{-i},2^{-j}}\right\|_{W_2^1} \le \\ &\le C\|f\|_{W_2^{G,4}}\sum_{i=1}^{n}\sum_{j=n-i+1}^{n}(2^{-(2i+j)} + 2^{-(2j+i)}) \le h\hat{C}\|f\|_{W_2^{G,4}}\end{aligned}$$

Step 2 and the triangle inequality complete the proof of the fifth inequality in Theorem 1.

Now, let us assume $f \in W_2^{G,3}(\Omega)$.
1.Step. (The hierarchical surplus is small.) We have to show that there is a constant $C > 0$ such that

$$\|v_{h_x,h_y}\|_{W_2^1} \le C h_x h_y \|f\|_{W_2^{G,3}}. \tag{2}$$

Let us assume $f \in \mathcal{C}^3(\Omega)$. By the one dimensional interpolation formulas, we get

$$\begin{aligned}\left\|\frac{\partial v_{h_x,h_y}}{\partial x}\right\|_2 &= \left\|\frac{\partial\left(I_x^{2h_x} - I_x^{h_x}\right)\left(\left(I_y^{2h_y} - I_y^{h_y}\right)(f)\right)}{\partial x}\right\|_2 \le \tilde{C}h_x\left\|\frac{\partial^2\left(I_y^{2h_y} - I_y^{h_y}\right)(f)}{\partial x^2}\right\|_2 \le \\ &\le \tilde{C}h_x\left\|\left(I_y^{2h_y} - I_y^{h_y}\right)\left(\frac{\partial^2 f}{\partial x^2}\right)\right\|_2 \le \tilde{C}^2 h_x h_y\left|\frac{\partial^3 f}{\partial x^2 \partial y}\right|_2.\end{aligned}$$

In the same way as above, we get (2).
2.Step. (The error of the full grid solution is small.)
By the interpolation theory on full grids, we know that there is a constant $C > 0$, such that

$$\|f - f_{h,h}\|_{W_2^1} \le hC\|f\|_{W_2^2}.$$

3.Step. (Zenger's sparse grid idea.)
In the same way as above, we get

$$\begin{aligned}\|f_{h,h} - \mathcal{I}_{\mathcal{D}_n}(f)\|_{W_2^1} &= \left\| \sum_{i=1}^{n} \sum_{j=n-i+1}^{n} v_{2^{-i},2^{-j}} \right\|_{W_2^1} \leq \\ &\leq C\|f\|_{W_2^{G,3}} \sum_{i=1}^{n} \sum_{j=n-i+1}^{n} 2^{-(i+j)} \leq hn\hat{C}\|f\|_{W_2^{G,3}}\end{aligned}$$

Step 2 and the triangle inequality complete the proof of the sixth inequality in Theorem 1.
□

Now, we want to define the sparse grid interpolation with piecewise constant functions. Denote $V_c^{i,j}$ the finite element space of piecewise constant finite elements on the full grid $\Omega_{i,j}$. The finite element space of piecewise constant functions on the regular sparse grid $\mathcal{D}_n$ is defined by

$$V_{\mathcal{D}_n}^c := \operatorname{span}_{\mathbf{R}}\{v | v \in V_c^{i,j} \quad \text{and} \quad i + j \leq n + 1\}.$$

Furthermore, denote V_c^i the finite element space of piecewise constant finite elements on the one dimensional full grid of mesh size 2^{-i}. The full grid interpolation operator in one dimension is defined by

$$\begin{aligned}I_c^i : \mathcal{C}([0,1]) &\mapsto V_c^i \\ I_c^i(f)\left(\left(k + \frac{1}{2}\right)h\right) &:= \frac{f(kh) + f((k+1)h)}{2} \quad \text{for } i = 0, \ldots, 2^i - 1 \text{ and } h = 2^{-i}.\end{aligned}$$

We can use this operator also for functions $f \in \mathcal{C}(\bar{\Omega})$

$$I_{c,x}^i(f)(x', y') := I_c^i(f(\cdot, y'))(x') \text{ and } I_{c,y}^i(f)(x', y') := I_c^i(f(x', \cdot))(y').$$

Now, we can define a full grid interpolation operator in two dimensions

$$\begin{aligned}I_c^{n,m} : \mathcal{C}(\bar{\Omega}) &\mapsto V_c^{n,m} \\ I_c^{n,m} &:= I_{c,x}^n \circ I_{c,y}^m.\end{aligned}$$

For simplicity, we define the sparse grid interpolation operator by the combination formula

$$\mathcal{I}_{\mathcal{D}_n}^c := \sum_{i=1}^{n} I_c^{2^{-i},2^{i-n-1}} - \sum_{i=1}^{n-1} I_c^{2^{-i},2^{i-n}}.$$

As in the proof of Theorem 1, we get

Theorem 2 (Error of the Sparse Grid Interpolation with Piecewise Constant Functions) *There exists a constant $C > 0$ such that the error in the supremum norm is for $h = 2^{-n}$*

$$\|f - \mathcal{I}_{\mathcal{D}_n}^c(f)\|_\infty \leq C|f|_{W_\infty^{G,2}} h \log h^{-1} \qquad \text{for } f \in \mathcal{C}^2(\bar{\Omega}).$$

The Bramble-Hilbert-Lemma gives the following well-known theorem:

Theorem 3 (Integration of the Full Grid Interpolant) *There exists a constant $C > 0$ such that for $h = 2^{-n}$*

$$\begin{aligned}\left|\int_0^1 (f - I_c^n(f))w \, dx\right| &\leq h^2 C\|f\|_{W_\infty^2} \|w\|_{W_1^1} \quad \text{for} \quad f \in W_\infty^2(]0,1[), \quad w \in W_1^1(]0,1[), \\ \left|\int_\Omega (f - I_c^{n,n}(f))w \, d(x,y)\right| &\leq h^2 C\|f\|_{W_\infty^2} \|w\|_{W_1^1} \quad \text{for} \quad f \in W_\infty^2(\Omega), \quad w \in W_1^1(\Omega).\end{aligned}$$

The analogous sparse grid result is the following theorem.

Theorem 4 (Integration of the Sparse Grid Interpolant) *There exists a constant $C > 0$ such that for $h = 2^{-n}$*

$$\left|\int_\Omega (f - \mathcal{I}^c_{\mathcal{D}_n}(f))w \; d(x,y)\right| \leq h^{\frac{3}{2}} C\|f\|_{W^{G,3}_\infty}\|w\|_{W^1_1} \; for \; f \in C^3(\bar{\Omega}) \;, \quad w \in W^1_1(\Omega).$$

PROOF Let $w \in W^1_1(\Omega)$ be fixed. Let's denote

$$u_{h_x,h_y} := \int_\Omega I^{i,j}_c(f)w \; d(x,y),$$

where $h_x = 2^{-i}$ and $h_y = 2^{-j}$. The two dimensional hierarchical surplus of the integration is defined by

$$v_{h_x,h_y} := u_{h_x,h_y} - u_{2h_x,h_y} - u_{h_x,2h_y} + u_{2h_x,2h_y}.$$

1.Step. (The hierarchical surplus is small.)
We have to show that there is a constant $C > 0$ such that

$$|v_{h_x,h_y}| \leq C \min(h_x^2 h_y, h_x h_y^2)\|w\|_{W^1_1}\|f\|_{W^{G,3}_\infty}.$$

By symmetry, it suffices to prove

$$|v_{h_x,h_y}| \leq C h_x^2 h_y \|w\|_{W^1_1}\|f\|_{W^{G,3}_\infty}.$$

By Theorem 3 and the definition of $I^{i,j}_c$, we get

$$\begin{aligned}
|v_{h_x,h_y}| &= \\
&= \left|\int_\Omega \left(I^{i-1,j-1}_c - I^{i,j-1}_c - I^{i-1,j}_c + I^{i,j}_c\right)(f)w \; d(x,y)\right| = \\
&= \left|\int_\Omega \left(I^{i-1}_{c,x} - I^i_{c,x}\right) \circ \left(I^{j-1}_{c,y} - I^j_{c,y}\right)(f)w \; d(x,y)\right| \leq \\
&\leq h_x^2 C\left(\|w\|_{L_1} + \left\|\frac{\partial w}{\partial x}\right\|_{L_1}\right) \cdot \\
&\quad \cdot \max\left(\|(I^{j-1}_{c,y} - I^j_{c,y})(f)\|_\infty, \left\|\frac{\partial}{\partial x}(I^{j-1}_{c,y} - I^j_{c,y})(f)\right\|_\infty, \left\|\frac{\partial^2}{\partial x^2}(I^{j-1}_{c,y} - I^j_{c,y})(f)\right\|_\infty\right) \leq \\
&\leq h_x^2 C\|w\|_{W^1_1} \cdot \\
&\quad \cdot \max\left(\|(I^{j-1}_{c,y} - I^j_{c,y})(f)\|_\infty, \left\|(I^{j-1}_{c,y} - I^j_{c,y})\left(\frac{\partial f}{\partial x}\right)\right\|_\infty, \left\|(I^{j-1}_{c,y} - I^j_{c,y})\left(\frac{\partial^2 f}{\partial x^2}\right)\right\|_\infty\right).
\end{aligned}$$

It is well-known that $\|g - I^j_c(g)\|_\infty \leq C' h_y \left\|\frac{\partial g}{\partial y}\right\|_\infty$ for a constant C' and every $g \in C^1([0,1])$. Therefore, there is a constant C'' such that

$$|v_{h_x,h_y}| \leq h_x^2 h_y C''\|w\|_{W^1_1}\|f\|_{W^{G,3}_\infty}.$$

2.Step. (The error of the full grid solution is small.)
Let's denote $u := \int_\Omega fw \; d(x,y)$. By Theorem 3, we know

$$|u - u_{h,h}| \leq h^{\frac{3}{2}} C\|f\|_{W^{G,3}_\infty}\|w\|_{W^1_1}.$$

3.Step. (Zenger's sparse grid idea.)
Let's denote

$$u_h^K := \int_\Omega \mathcal{I}^c_{\mathcal{D}_n}(f) w \; d(x,y) = \sum_{i=1}^{n} u_{2^{-i},2^{i-n-1}} - \sum_{i=1}^{n-1} u_{2^{-i},2^{i-n}}.$$

A simple calculation (see [10] or [6]) shows

$$u_{h,h} - u_h^K = \sum_{i=1}^{n} \sum_{j=n-i+1}^{n} v_{2^{-i},2^{-j}}.$$

By step 1, we get for a constant $\tilde{C} > 0$

$$\begin{aligned}
|u_{h,h} - u_h^K| &= \left| \sum_{i=1}^{n} \sum_{j=n-i+1}^{n} v_{2^{-i},2^{-j}} \right| \leq \\
&\leq C\|w\|_{W_1^1} \|f\|_{W_\infty^{G,3}} \sum_{i=1}^{n} \sum_{j=n-i+1}^{n} \min(2^{-(2i+j)}, 2^{-(2j+i)}) \leq \\
&\leq h^{\frac{3}{2}} \tilde{C} \|w\|_{W_1^1} \|f\|_{W_\infty^{G,3}}
\end{aligned}$$

Step 2 and the triangle inequality complete the proof. □

3. Discretization of Elliptic Equations

We use the same notation as in [4]. Let $f \in (L_2(\Omega))'$ and

$$\begin{aligned}
a : \mathring{W}_2^1(\Omega) \times \mathring{W}_2^1(\Omega) &\mapsto \mathbb{R} \\
(u, v) &\mapsto \int_\Omega \sum_{|\alpha|,|\beta| \leq 1} a_{\alpha,\beta} (D^\alpha u)(D^\beta v) \; d(x,y),
\end{aligned}$$

where α, β are multiindices and $A = (a_{\alpha,\beta})_{|\alpha|,|\beta|\leq 1} \in \left(\mathcal{C}(\bar{\Omega})\right)^{3\times 3}$. Let us assume that a is continuous and $\mathring{W}_2^1(\Omega)$-elliptic. We are looking for a solution $u \in \mathring{W}_2^1(\Omega)$ of the equation

$$a(u,v) = f(v) \qquad \text{for all } \; v \in \mathring{W}_2^1(\Omega). \tag{3}$$

The problem is now that we can not replace $\mathring{W}_2^1(\Omega)$ by the finite element space $\mathring{V}_{\mathcal{D}_n}$ and use the same bilinear form a. If we did so, we would get a manifold of stiffness matrices of dimension more than $O(2^n n)$ for this class of elliptic equations. Then, we would not be able to store the stiffness matrix in a sparse grid data structure. Therefore, we replace the bilinear form a by the bilinear form

$$\begin{aligned}
a_h : \mathring{W}_2^1(\Omega) \times \mathring{W}_2^1(\Omega) &\mapsto \mathbb{R} \\
(u, v) &\mapsto \int_\Omega \sum_{|\alpha|,|\beta| \leq 1} \mathcal{I}^c_{\mathcal{D}_n}(a_{\alpha,\beta}) (D^\alpha u)(D^\beta v) \; d(x,y).
\end{aligned}$$

A discretization of equation (3) is now:

Find a $u_h \in \mathring{V}_{\mathcal{D}_n}$ such that

$$a_h(u_h, v_h) = f(v_h) \quad \text{for all } \; v_h \in \mathring{V}_{\mathcal{D}_n}. \tag{4}$$

We need some preparations for convergence theorems. We define for a matrix $B = (b_{\alpha,\beta})_{|\alpha|,|\beta|\leq 1} \in (L^\infty(\Omega))^{3\times 3}$

$$\|B\|_\infty := \max_{|\alpha|,|\beta|\leq 1}(\|b_{\alpha,\beta}\|_\infty)$$

and for $B \in \left(W^{G,n}_\infty(\Omega)\right)^{3\times 3}$

$$\|B\|_{W^{G,n}_\infty} := \max_{|\alpha|,|\beta|\leq 1}(\|b_{\alpha,\beta}\|_{W^{G,n}_\infty}).$$

The proof of the following lemma is left to the reader.

Lemma 2 *Assume $B \in (L^\infty(\Omega))^{3\times 3}$. Then we get*

$$\left|\int_\Omega \sum_{|\alpha|,|\beta|\leq 1} b_{\alpha,\beta}(D^\alpha w)(D^\beta v)\, d(x,y)\right| \leq 9\|w\|_{W^1_2}\|v\|_{W^1_2}\|B\|_\infty \quad \textit{for all} \quad w,v \in \mathring{W}^1_2(\Omega).$$

By Lemma 2 and Theorem 2, it follows

Lemma 3 *Assume $A \in \left(\mathcal{C}^2(\bar{\Omega})\right)^{3\times 3}$. Then, there is a constant $C > 0$ such that*

$$|a(w,v) - a_h(w,v)| \leq h \log h^{-1} C\|A\|_{W^{G,2}_\infty}\|w\|_{W^1_2}\|v\|_{W^1_2} \quad \textit{for all} \quad w,v \in \mathring{W}^1_2(\Omega).$$

The next lemma shows that the bilinear form a_h is continuous and $\mathring{W}^1_2(\Omega)$-elliptic with constants independent of h for sufficient small h.

Lemma 4 *Assume $A \in \left(\mathcal{C}^2(\bar{\Omega})\right)^{3\times 3}$. There are constants $C_1, C_2 > 0$ independent of h and a mesh size $h_0 > 0$, such that*

$$\begin{aligned} a_h(w,v) &\leq C_1\|w\|_{W^1_2}\|v\|_{W^1_2} \quad \textit{for all} \quad w,v \in \mathring{W}^1_2(\Omega) \quad \textit{and} \\ \|v\|^2_{W^1_2} &\leq C_2\, a_h(v,v) \quad \textit{for all} \quad v \in \mathring{W}^1_2(\Omega) \quad \textit{and} \quad h < h_0. \end{aligned}$$

PROOF By the assumption that a is continuous and $\mathring{W}^1_2(\Omega)$-elliptic, we know the existence of constants $\tilde{C}_1, \tilde{C}_2 > 0$ such that $|a(w,v)| \leq \tilde{C}_1\|w\|_{W^1_2}\|v\|_{W^1_2}$ and $\|v\|^2_{W^1_2} \leq \tilde{C}_2 a(v,v)$. By Lemma 3, we conclude

$$\begin{aligned} |a_h(w,v)| &\leq |a(w,v)| + |a_h(w,v) - a(w,v)| \leq \\ &\leq \left(\tilde{C}_1 + Ch\log h^{-1}\|A\|_{W^{G,2}_\infty}\right)\|w\|_{W^1_2}\|v\|_{W^1_2} \leq \\ &\leq \left(\tilde{C}_1 + C\|A\|_{W^{G,2}_\infty}\right)\|w\|_{W^1_2}\|v\|_{W^1_2} \quad \text{and} \\ a_h(v,v) &\geq a(v,v) - |a(v,v) - a_h(v,v)| \geq \left(\frac{1}{\tilde{C}_2} - Ch\log h^{-1}\|A\|_{W^{G,2}_\infty}\right)\|v\|^2_{W^1_2}. \end{aligned}$$

For sufficient small h we can achieve

$$\left(\frac{1}{\tilde{C}_2} - Ch\log h^{-1}\|A\|_{W^{G,2}_\infty}\right) > \frac{1}{2\tilde{C}_2}.$$

□

Lemma 4 provides that there is a unique solution u_h of equation (4) for sufficient small h. Theorem 5 shows a good convergence in the energy norm.

Theorem 5 (Convergence in the Energy Norm) *Assume* $A \in \left(\mathcal{C}^2(\bar{\Omega})\right)^{3\times 3}$ *and* $u \in W_2^{G,3}(\Omega)$. *Then,* u_h *converges to* u *in the* W_2^1*-norm for sufficient small* h *with order*

$$\|u - u_h\|_{W_2^1} = O(h \log h^{-1}).$$

PROOF Let's assume $h < h_0$ (see Lemma 4). By Strang's First Lemma (see p.195 [3]), we get a constant $C' > 0$ such that

$$\|u - u_h\|_{W_2^1} \le C' \inf_{v_h \in \mathring{V}_{\mathcal{D}_n}} \left(\|u - v_h\|_{W_2^1} + \sup_{z_h \in \mathring{V}_{\mathcal{D}_n}} \frac{|a(v_h, z_h) - a_h(v_h, z_h)|}{\|z_h\|_{W_2^1}} \right).$$

Let $v_h = \mathcal{I}_{\mathcal{D}_n}(u)$. Then, by Lemma 3 and Theorem 1, we conclude that there is a constant $C'' > 0$ such that

$$\|u - u_h\|_{W_2^1} \le C'' h \log h^{-1} \left(\|u\|_{W_2^{G,3}} + \|\mathcal{I}_{\mathcal{D}_n}(u)\|_{W_2^1} \right).$$

Theorem 1 shows that there is a constant $C''' > 0$ such that

$$\|\mathcal{I}_{\mathcal{D}_n}(u)\|_{W_2^1} \le \|u\|_{W_2^1} + \|\mathcal{I}_{\mathcal{D}_n}(u) - u\|_{W_2^1} \le C''' \|u\|_{W_2^{G,3}}.$$

The last two inequalities complete the proof. □

Let us assume that there is a constant $C > 0$ such that

$$|w|^2_{W_2^1} \le C a(w, w) = C \int_\Omega \sum_{|\alpha|,|\beta| \le 1} a_{\alpha,\beta} (D^\alpha w)^2 \, d(x, y) \quad \forall \quad w \in W_2^1(\Omega).$$

For a good convergence in the L_2-norm we need the regularity of the adjoint problem.

Lemma 5 (Regularity of the Adjoint Problem) *Assume* $A \in \left(\mathcal{C}^1(\bar{\Omega})\right)^{3\times 3}$ *and* $g \in L_2(\Omega)$. *Then, the solution* $w \in \mathring{W}_2^1(\Omega)$ *of the equation*

$$a(v, w) = \int_\Omega g v \, d(x, y) \quad \forall v \in \mathring{W}_2^1(\Omega)$$

is contained in $W_2^2(\Omega)$. *There is a constant* $C > 0$ *independent of* $g \in L_2(\Omega)$ *such that*

$$\|w\|_{W_2^2(\Omega)} \le C \|g\|_{L_2(\Omega)}.$$

We only want to give a short hint for a proof of this regularity lemma.
HINT FOR A PROOF Extend the functions w, v, f and the bilinear form a symmetrically to functions on the torus like in [6]. Generalize Lemma 20.1 in [9] to functions on the torus. Now go one similar to the proof of Theorem 7.2 in [6]. The regularity of the solution on the torus gives the regularity of the solution w on the unit square Ω. □

With the trick of Aubin and Nitsche, we get the following theorem.

Theorem 6 (Convergence in the L_2-Norm) *Assume* $A \in \left(\mathcal{C}^3(\bar{\Omega})\right)^{3\times 3}$ *and* $u \in W_2^{G,3}(\Omega)$. *Then,* u_h *converges to* u *in the* L_2*-norm for sufficient small* h *with order*

$$\|u - u_h\|_{L_2} = O(h^{\frac{3}{2}} \log h^{-1}).$$

Proof Let's assume $h < h_0$ (see Lemma 4). Let $w \in \mathring{W}_2^1(\Omega)$ be the solution of the equation

$$a(v, w) = \int_\Omega (u - u_h)v \; d(x, y) \quad \forall v \in \mathring{W}_2^1(\Omega).$$

Assume $v = u - u_h$ and let $v_h \in \mathring{V}_{\mathcal{D}_n}$. This gives

$$\begin{aligned} \|u - u_h\|_{L_2}^2 &= a(u - u_h, w) = a(u - u_h, w - v_h) + a(u - u_h, v_h) = \\ &= a(u - u_h, w - v_h) + f(v_h) - a(u_h, v_h) = \\ &= a(u - u_h, w - v_h) + a_h(u_h, v_h) - a(u_h, v_h) = \\ &= a(u - u_h, w - v_h) + \\ &\quad + a_h(u_h - u, v_h) - a(u_h - u, v_h) + \\ &\quad + a_h(u, v_h - w) - a(u, v_h - w) + \\ &\quad + a_h(u, w) - a(u, w) \end{aligned}$$

Observe that $\Omega_{[\frac{n}{2}],[\frac{n}{2}]} \subset \mathcal{D}_n$. Therefore, we can choose $v_h = I^{[\frac{n}{2}],[\frac{n}{2}]}(w)$ the full grid interpolant. By the interpolation theory of full grids, there is a constant $C' > 0$ such that

$$\|w - v_h\|_{W_2^1} \le \sqrt{h}C'\|w\|_{W_2^2} \quad \text{and} \quad \|v_h\|_{W_2^1} \le C'\|w\|_{W_2^2}.$$

The general Cauchy-Schwarz inequality gives

$$\sum_{|\alpha|,|\beta| \le 1} \|(D^\alpha u)(D^\beta v)\|_{W_1^1} \le 27\|u\|_{W_2^2}\|w\|_{W_2^2}.$$

Thus, by Theorem 4, we get

$$\begin{aligned} |a_h(u, w) - a(u, w)| &\le h^{\frac{3}{2}}C\|A\|_{W_\infty^{G,3}} \sum_{|\alpha|,|\beta| \le 1} \|(D^\alpha u)(D^\beta v)\|_{W_1^1} \le \\ &\le h^{\frac{3}{2}}27C\|A\|_{W_\infty^{G,3}}\|u\|_{W_2^2}\|w\|_{W_2^2}. \end{aligned}$$

By Lemma 3 and Theorem 5, we conclude that there is a constant $C'' > 0$ such that

$$\|u - u_h\|_{L_2}^2 \le h^{\frac{3}{2}} \log h^{-1} C''\|w\|_{W_2^2}.$$

By Lemma 5, it follows for a constant $C''' > 0$

$$\|u - u_h\|_{L_2}^2 \le h^{\frac{3}{2}} \log h^{-1} C'''\|u - u_h\|_{L_2}.$$

This shows $\|u - u_h\|_{L_2} \le h^{\frac{3}{2}} \log h^{-1} C'''$.

□

4. Numerical Results

Let $\Psi = \{(x, y) \in]0, 1[^2 | 0 < x < 1 \text{ and } \frac{1}{2} \cdot (1 + \sin(\pi \cdot x)) > y > x \cdot \frac{1}{4}\}$. The function

$$u = \exp(x) \cdot \exp((1.0 - y)/(-4.0)) \cdot \cos(\sqrt{15.0} \cdot (1.0 - y)/4.0) \in W_2^1(\Psi)$$

is the solution of the equation

$$a(u, v) = \int_\Psi (\nabla u)^T \begin{pmatrix} 4 & 1 \\ 1 & 4 \end{pmatrix} \nabla v \; d(x, y) = 0 \quad \text{for all } v \in \mathring{W}_2^1(\Psi) \tag{5}$$

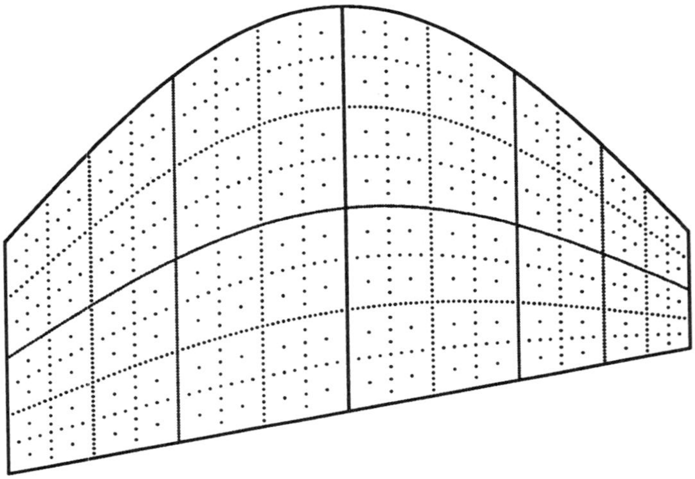

Figure 1: Sparse Grid on the Domain Ψ

n	3	4	5	6	7	8	9	10
$\|\|-\|\|_{\infty,\mathcal{D}_n}$	2.93-2	1.1e-2	3.7e-3	1.1e-3	3.5e-4	1.0e-4	3.0e-5	8.4e-6
$\|\|-\|\|_{2,\mathcal{D}_n}$	6.9e-3	2.2e-3	6.8e-4	2.0e-4	5.8e-5	1.6e-5	4.6e-6	1.3e-6
$\frac{\|\|-\|\|_{\infty,\mathcal{D}_{n-1}}}{\|\|-\|\|_{\infty,\mathcal{D}_n}}$	2.4	2.5	3.1	3.3	3.3	3.4	3.5	3.5
$\frac{\|\|-\|\|_{2,\mathcal{D}_{n-1}}}{\|\|-\|\|_{2,\mathcal{D}_n}}$	2.4	3.1	3.3	3.4	3.5	3.5	3.6	3.6

Table 1: Convergence of the discrete solution

with Dirichlet boundary conditions. We transform the domain $\Omega =]0,1[^2$ onto Ψ by the function Φ, which is linear in the y-direction.

Now we can define the bilinear form

$$\hat{a}(\hat{u},\hat{v}) := a(\hat{u}\circ\Phi^{-1},\hat{v}\circ\Phi^{-1})$$

and solve the equation $\hat{u} \in W_2^1(\Omega)$

$$\hat{a}(\hat{u},\hat{v}) = 0 \quad \text{for all} \quad \hat{v} \in \mathring{W}_2^1(\Omega) \quad \text{and suitable Dirichlet boundary conditions and} \tag{6}$$

with the discretization in section 3. This gives the discrete solution $\hat{u}_h$ of equation (6). By

$$u_h := \hat{u}_h \circ \Phi^{-1},$$

we get approximate solutions of equation (5). Figure 1 shows a transformed sparse grid on the domain Ψ with 2817 grid points.

We use the following discrete norms

$$\|w\|_{\infty,\mathcal{D}_n} := \max_{z\in\mathcal{D}_n} |w(z)| \quad \text{and}$$

$$\|w\|_{2,\mathcal{D}_n} := \sqrt{\frac{\sum_{z\in\mathcal{D}_n}|w(z)|^2}{|\mathcal{D}_n|}}.$$

Table 1 leads to the conjecture that u_h converges to u with the order

$$\|u_h - u\|_\infty = O(h^2 \log h^{-1}) \quad \text{and}$$
$$\|u_h - u\|_2 = O(h^2 \log h^{-1}).$$

Acknowledgement. I would like to thank Professor C. Zenger for many helpful discussions.

References

[1] H.-J. Bungartz. An adaptive poisson solver using hierarchical bases and sparse grids. In P. de Groen and R. Beauwens, editors, *in Proceedings of the IMACS International Symposium on Iterative Methods in Linear Algebra, Brüssel, April, 1991*. Elsevier, Amsterdam, 1992.

[2] M. Griebel, M. Schneider, and C. Zenger. A combination technique for the solution of sparse grid problems. In P. de Groen and R. Beauwens, editors, *in Proceedings of the IMACS International Symposium on Iterative Methods in Linear Algebra, Brüssel, April, 1991*. Elsevier, Amsterdam, 1992.

[3] C. Großmann and H.-G. Roos. *Numerik partieller Differentialgleichungen*. Teubner, Stuttgart, 1992.

[4] W. Hackbusch. *Theorie und Numerik elliptischer Differentialgleichungen*. Teubner, Stuttgart, 1986.

[5] C. Pflaum. Anwendung von Mehrgitterverfahren auf dünnen Gittern. Technische Universität München, 1992. Diplomarbeit.

[6] C. Pflaum. Convergence of the combination technique for the finite element solution of poisson's equation. SFB-Report 342/14/93 A, Technische Universität München, 1993.

[7] C. Pflaum. A multi-level-algorithm for the finite-element-solution of general second order elliptic differential equations on adaptive sparse grids. SFB-Report 342/12/94 A, Technische Universität München, 1994.

[8] C. Pflaum and U. Rüde. Gauss adaptive relaxation for the multilevel solution of partial differential equations on sparse grids. SFB-Report 342/13/93 A, Technische Universität München, 1993.

[9] J. Wloka. *Partielle Differentialgleichungen*. Teubner, Stuttgart, 1982.

[10] C. Zenger. Sparse grids. In W. Hackbusch, editor, *Parallel Algorithms for Partial Differential Equations: Proceedings of the Sixth GAMM-Seminar, Kiel, January 1990*, volume 31 of *Notes on Numerical Fluid Mechanics*. Vieweg, Braunschweig, 1991.

Christoph Pflaum
Institut für Informatik, Technische Universität München
D-80290 München, Germany
e-mail: pflaum@informatik.tu-muenchen.de

R RANNACHER

Hydrodynamic stability and a-posteriori error control in the solution of the Navier–Stokes equations

Summary. The fundamental question in computational fluid dynamics is that for the computability of real flows with a reliable bound of the error. Recently, in joint work with C. Johnson, [5], [7], and [8], a new approach towards quantitative error control has been developed for the solution of the incompressible Navier-Stokes equations. Combining the concepts of so-called "strong dual stability" and "Galerkin orthogonality" sharp a-priori as well as a-posteriori error estimates can be obtained with computable error constants. The dependence of these error constants on the Reynolds number is linked to the stability of the flow considered and appears crucial for its computability.

1. Formulation of the Problem

We consider the fundamental question of computational fluid dynamics ("CFD"), namely that for the numerical computability of real flows and for an accompanying control of the error. By "flows", of course, we mean "exact" solutions of the underlying differential equations describing conservation of mass, momentum and energy. At first glance, these questions seem superfluous, in view of the appearent success of CFD in modeling many practically relevant flows and the abundant number of research papers dealing with the various aspects of numerical methods and their implementation. In fact, reviewing the existing literature, particularly on the finite element method, one may gain the impression that most questions are more or less settled. Various discretization schemes have been designed together with seemingly rigorous convergence analysis including even optimal-order error estimates. The problem caused by the incompressibility constraint is well understood today, theoretically as well as algorithmically. Dominant convection is handled by upwinding or streamline diffusion techniques. For the nonstationary case, robust and accurate time-stepping schemes have been developed and even the problem of turbulence is tackled by employing certain concepts from dynamical system theory. The range of computer simulations and the efficiency of the solution processes has enormously been improved by the use of new-generation supercomputers and the implementation of fast multi-level methods. But this enormous increase in computational power and numerical technology has also led to a new attitude concerning the requirements on the reliability in numerical flow simulations.

Consider, for example, the Navier-Stokes equations which model the flow of a viscous incompressible Newtonian fluid decribed by the velocity $\mathbf{u} = \mathbf{u}(x,t): \Omega\times I \to R^d$ and the pressure $p = p(x,t): \Omega\times I \to \mathbf{R}$:

$$\partial_t \mathbf{u} - \nu\Delta\mathbf{u} + \mathbf{u}\cdot\nabla\mathbf{u} + \nabla p = f, \quad \nabla\cdot\mathbf{u} = 0, \quad \text{in } \Omega\times I, \tag{1.1}$$

$$\mathbf{u}_{|t=0} = \mathbf{u}^0, \quad \mathbf{u}_{|\Gamma_0} = 0, \quad \mathbf{u}_{|\Gamma_{in}} = \mathbf{u}_{in}, \quad (\nu\partial_n\mathbf{u} - p\mathbf{u})_{|\Gamma_{out}} = 0. \tag{1.2}$$

Here, Ω is a bounded domain in $\mathbf{R}^d$ ($d = 2,3$) with sufficiently regular boundary and $I = [0,T]$ a time interval. The characteristic length L and velocity U are assumed to be normalized, such that the Reynolds number is $Re = 1/\nu$. Accordingly, the relevant time scale is $T \approx Re$ and the minimum spatial scale to be resolved is $h_{min} \approx \sqrt{\nu}$ (see [4], [7]). We restrict our discussion to the regime of "laminar" flows occuring for $Re \approx 10^1\text{-}10^5$.

The questions we want to address in this paper are as follows:

Question 1: *Is there any theoretical support that certain practically interesting flows (i.e., solutions of the Navier-Stokes equations) can actually be computed numerically?*

If the answer is "no", everybody should be worried about this deficiency of the theory!

Question 2: *Is it possible to have quantitative control on the error in flow computations?*

This appears to be of utmost importance in order to lift CFD from the still dominating principle of "try and error" in flow computations to the level of being a reliable tool with rigorous mathematical basis.

2. Critical Review of Available Error Analysis

Unfortunately, the available theory in CFD provides only little help for answering the questions raised above. In the literature one finds various approaches to the numerical solution of the Navier-Stokes equations (see, e.g., [14], and the literature cited therein).

- Spatial discretization by Finite-Difference, Finite-Volume, or Finite-Element methods with grid-size denoted by h.
- Time discretization by implicit Backward-Euler, Crank-Nicolson or Fractional-Step schemes with time-step size denoted by k.
- Nonlinearity treated explicitly or by predictor-corrector techniques.
- Convection stabilized by upwinding or by streamline diffusion.

The typical results from the literature to support the functionality of these methods are asymptotic error estimates of the form

$$\max_I \|U^n - \mathbf{u}(t_n,\cdot)\| \le C(\mathbf{u},T,\nu)\{h^2+k^2\}, \tag{2.1}$$

where $\|\cdot\|$ is an appropriate spatial norm, e.g., the L^2-norm, and the constant $C = C(\mathbf{u},T,\nu)$ depends on the regularity of the solution $\mathbf{u}$, on the length of the time interval T, and on the viscosity ν (and, of course, on the characteristics of the discretization method). This implies that the error bound reduces by a factor $1/4$ if the mesh sizes are halved. But, although mathematically rigorous, such a result appears meaningless for practical purposes unless the dependence of the error constant C on the Reynolds-number $Re \approx 1/\nu$ can be made explicit. In normal situations of strongly parabolic problems, e.g., the heat equation, one expects $C \approx 1\text{-}10^4$ which may be acceptable from a numerical point of view. However, the Navier-Stokes problem is not a "normal problem" for interesting case, i.e., for larger Reynolds numbers. The dependence of the theoretical error estimates on the Reynolds number is due to various sources:

- *Structure of the differential operator:* In traditional error analysis, the nonlinearity is treated as a perturbation, i.e., the related error terms are absorbed into the diffusive part $-\nu\Delta\mathbf{u}$ (in the stationary case by using Young's inequality) resulting in $C \approx K/\sqrt{\nu}$, or into the acceleration term $\mathbf{u}_t$ (in the stationary case by using Gronwall's inequality) resulting in $C \approx \exp(KT/\sqrt{\nu})$, where $K \approx \max_{\Omega\times I}|\nabla u| \approx 1/\sqrt{\nu}$. Using a least-squares stabilization in the scheme, one can formally remove the explicit ν-dependence in the exponent getting $C \approx e^{KT}$ (see [5] and [13]).

- *Regularity of the solution:* In the presence of rigid boundaries (no-slip boundary condition), the regularity of the solution is essentially determined by the width of the boundary layer which is $\delta \approx \sqrt{\nu}$. Hence, $C \approx \max_{\Omega\times I}|\nabla\mathbf{u}| \approx 1/\nu$. This dependence on ν could in principle be handled by appropriately refining the mesh in the boundary layer.

- *Stability of the solution:* The stability of the solution is normally introduced into the analysis by the linearization of the problem, e.g., in the stationary case through the coercivity constant $\gamma(\mathbf{u})$ of the Fréchet derivative of the nonlinear operator taken at $\mathbf{u}$, in the form $C \approx \gamma(\mathbf{u})$. The dependence of $\gamma(\mathbf{u})$ on Re can range between $\gamma(\mathbf{u}) \approx Re$ and, in the worst case, $\gamma(\mathbf{u}) \approx e^{KRe}$ (see the examples in [1], [7], [14]). In the nonstationary case the stability constant has to be expected to behave like $\gamma(\mathbf{u}) \approx e^{KReT}$ in general reflecting the possible onset of turbulence, i.e., chaotic behavior. Since the stability of the flow to be computed is inherent to the given physical problem and cannot be improved by numerical tricks, the dependence of the error constant C on this property is crucial. It distinguishes "laminar" flows which are amenable to numerical computation from "turbulent" ones which in principle cannot be computed on their finest scales. For $Re \approx 10^1$-10^5, a dependence like $C \approx Re^2$ or even $C \approx e^{cRe}$ is not acceptable ($e^{20} > 10^8$, $e^{100} > 10^{43}$, and $e^{1000} \approx \infty$!).

Hence, to justify numerical computation for laminar flow, it is necessary to identify appropriate stability concepts for *a-priori* error estimates (to guarantee convergence) and *a-posteriori* error estimates (for adaptive accuracy control). Quantitatively most stability concepts are equivalent (see [3; Part 2]) but this strongly depends on the Reynolds number. Choice of the wrong norm may lead to unfavorable dependence of the error constants on Re, e.g., like $C \approx Re^2$. We have to specify cases where the relevant error constants are of "optimal" size, i.e., $C \approx Re$. We emphasize that only a sharp *a-posteriori* estimation of the error with realistic constants leads to a reliable and efficient adaptive control of the discretization. However, until now, such quantitative error estimates have not been available even for such very basic situations as Couette (constant sheer) flow and Poisseuille (constant pipe) flow. The aim of this paper is to describe a new approach towards realistic error estimation in certain numerical schemes for flow computations and its application to very simple but proto-typical cases.

3. Discretization of the Navier-Stokes Problem

For illustrational purposes we consider a rather simple low-order discretization of the Navier-Stokes problem which inherits the essential properties required for a rigorous a-posteriori error analysis. This is the lowest-order "discontinuous" Galerkin method (so-called "DG(0)-method") in time combined with a conforming, exactly "divergence-free", finite element method (called "FE-method") in space (for details see [2] and [3]). Although, the idealized properties of this discretization ansatz are not realized to full extent in practice, it may serve as a proto-typical situation for our approach to a-posteriori error control. Most of the arguments which will be used below can be modified to apply also to realistic discretization schemes for the Navier-Stokes equations. For further simplifying the presentation, we suppose that the flow is driven only by a body force $\mathbf{f}$, i.e., particularly that $\Gamma_0 = \partial\Omega$.

Below, we will use a somewhat condensed notation in order to abstract as much as possible from the insignificant particularities of the discretization scheme considered. We begin with a variational formulation of the Navier-Stokes problem on the space-time domain $\Omega \times I$ which is the basis for the discontinuous Galerkin discretization in time. On the function spaces

$$\mathbf{V} \equiv \{\mathbf{v} \in \mathbf{H}_0^1(\Omega): \nabla \cdot \mathbf{v} = 0\}, \quad \mathbf{V}(I) \equiv H^1(I;\mathbf{V}),$$

the following linear forms are defined,

$$(\mathbf{u},\mathbf{v}) \equiv (\mathbf{u},\mathbf{v})_{L^2}\,, \quad \|v\| \equiv (v,v)^{1/2}\,,$$

$$a(\mathbf{u},\mathbf{v}) \equiv \nu(\nabla\mathbf{u},\nabla\mathbf{v})\,, \quad b(\mathbf{u},\mathbf{v},\mathbf{w}) \equiv (\mathbf{u}\cdot\nabla\mathbf{v},\mathbf{w})\,,$$

$$(\mathbf{u},\mathbf{v})_I \equiv \int_I (\mathbf{u},\mathbf{v})\,dt\,, \quad a_I(\mathbf{u},\mathbf{v}) \equiv \int_I a(\mathbf{u},\mathbf{v})\,dt\,, \quad b_I(\mathbf{u},\mathbf{v},\mathbf{w}) \equiv \int_I b(\mathbf{u},\mathbf{v},\mathbf{w})\,dt\,,$$

$$A(\mathbf{u};\mathbf{v},\mathbf{w}) \;\equiv\; (\mathbf{v}_t,\mathbf{w})_I + a_I(\mathbf{v},\mathbf{w}) + b_I(\mathbf{u},\mathbf{v},\mathbf{w}) + (\mathbf{v}(0),\mathbf{w}(0))\,,$$

$$F(\mathbf{w}) \;=\; (\mathbf{f},\mathbf{w})_I + (\mathbf{u}^0,\mathbf{w}(0))\,.$$

Using this notation, the weak formulation of the Navier-Stokes problem reads as follows,

$$\text{(P)} \qquad \mathbf{u}\in\mathbf{V}(I): \quad A(\mathbf{u};\mathbf{u},\phi) = F(\phi) \quad \forall\,\phi\in\mathbf{V}(I)\,. \tag{3.1}$$

Next, we define the FE-DG(0)-method. Let $\mathbf{V}_h \subset \mathbf{V}$ be a finite element subspace of trial functions which are piecewise polynomial with respect to some decomposition of Ω into "elements" K, i.e., triangles or rectangles (in 2-d) and tetrahedra or octahedra (in 3-d). For the usual regularity requirements on such decompositions see, e.g., [2]. The quality of the decompositions is characterized by a mesh size function $h(x)$, such that $h(x) \equiv h_K$ for each K. This spatial discretization is supposed to be at least of second order, i.e., the polynomial ansatz contains all linears. Further, the time interval I is split into subintervals $I_n = (t_n, t_{n+1})$ of length $k_n = t_{n+1} - t_n$, where the discrete times satisfy $0 = t_0 < t_1 < \ldots < t_N < t_{n+1} = T$. Further, we introduce a step-size function $k(t)$ by setting $k(t) \equiv k_n$ on I_n.

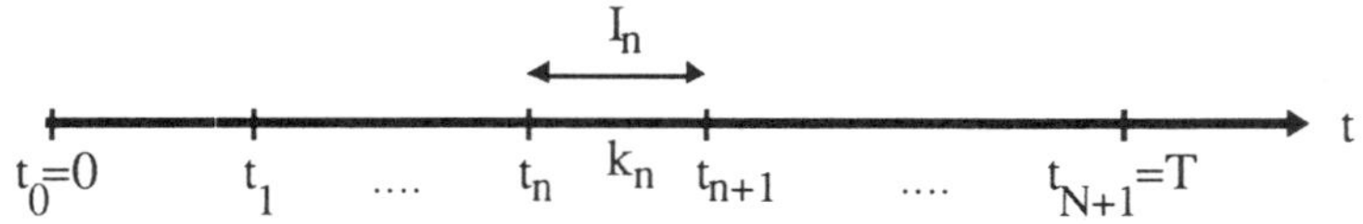

Accordingly, the discrete function spaces are introduced by

$$\mathbf{V}_{h,k} \;\equiv\; \{v\in L^2(I;\mathbf{V}_h):\; v \text{ constant in time on each } I_n\}\,,$$

and the discrete forms by

$$A_k(\mathbf{u};\mathbf{v},\mathbf{w}) \;\equiv\; \sum_{n=0}^{N} \{(\mathbf{v}_t,\mathbf{w})_n + a_n(\mathbf{v},\mathbf{w}) + b_n(\mathbf{u},\mathbf{v},\mathbf{w})\} + \sum_{n=0}^{N} ([\mathbf{v}^n],\mathbf{w}_+^n) + (\mathbf{v}_+^0,\mathbf{w}_+^0)\,,$$

where $v_+^n \equiv \lim_{s\to 0_+} v(t_n+s)$, $v_-^n \equiv \lim_{s\to 0_+} v(t_n-s)$, $[v^n] \equiv v_+^n - v_-^n$. With this notation, the fully discrete Navier-Stokes problem reads as follows:

$$(P_k) \qquad U\in \mathbf{V}_{h,k}(I): \quad A_k(U;U,\phi) = F(\phi) \quad \forall\, \phi\in \mathbf{V}_{h,k}(I)\,. \tag{3.2}$$

We remark that the DG(0)-scheme is almost equivalent to the well-known backward Euler time-stepping scheme if one identifies $\mathbf{U}^n \equiv \mathbf{U}_-^n$,

$$k_n a(\mathbf{U}^{n+1},\phi) + k_n b(\mathbf{U}^{n+1},\mathbf{U}^{n+1},\phi) + (\mathbf{U}^{n+1}-\mathbf{U}^n,\phi) = \int_{t_n}^{t_{n+1}} (\mathbf{f},\phi)\, dt \quad \forall\, \phi\in \mathbf{V}_h\,.$$

The scheme (P_k) contains the essential features of a "real" discretization of the Navier-Stokes problem. For more practically relevant discretizations and details on the efficient solution of the resulting algebraic problems, we refer, e.g., to [2], [3], [12], and [15].

4. A-Posteriori Error Analysis

We begin by stating the typical result of an a-posteriori error analysis of the discretization scheme described above:

$$\max_I\|\mathbf{U}-\mathbf{u}\| \le cL_kC_s\Big\{\max_I\|kR_{time}(\mathbf{U})\| + \max_I\|h^2R_{space}(\mathbf{U})\|\Big\}\,, \tag{4.1}$$

where $R_{time}(\mathbf{U})$ and $R_{space}(\mathbf{U})$ are the (computable) residuals of $\mathbf{U}$ in time and in space, respectively. In detail we have

$$\|kR_{time}(\mathbf{U})_{|I_n}\| \le \|kd_t\mathbf{U}^n\| + \max_{I_n}\|k\mathbf{f}\|\,,$$

$$\|h^2R_{space}(\mathbf{U})_{|I_n}\| \le \Big\{\sum_K \Big(h_K^3\|[\nu\partial_n\mathbf{U}-p\mathbf{n}]\|_{\partial K}^2 + h_K^4\|\mathbf{f}-\mathbf{U}_t+\nu\Delta\mathbf{U}-\mathbf{U}\cdot\nabla\mathbf{U}-\nabla P\|_K^2\Big)\Big\}^{1/2}\,,$$

where $d_t\mathbf{U}^n \equiv k_n^{-1}(\mathbf{U}_+^n-\mathbf{U}_-^n)$, $[\cdot]_{|\partial K}$ denotes the maximal jump across the element boundary ∂K , and P is the discrete pressure (determined by a pressure Poisson equation) associated to the velocity $\mathbf{U}$. The constant C_s measures the stability of a certain linearized "dual" problem corresponding to (P) which will be specified below and is independent of $\mathbf{U}, \nu$, h, and k , while c only depends on the local interpolation properties of the underlying finite element spaces. Further, for technical reasons there also occurs a constant of the from $L_k = \max_{0\le n\le N}(1+|\log(k_n)|)^{1/2}$.

On the basis of the estimate (4.1), the error control mechanism would then decrease the local mesh sizes h_K and k_n, until a prescribed tolerance level TOL is reached:

$$cL_kC_s\Big\{\max_I\|kR_{time}(U)\| + \max_I\|h^2R_{space}(U)\|\Big\} \approx \text{TOL}.$$

This requires a reasonable upper bound for the stability constant C_s. Clearly, this error control strategy is feasible only in cases with (at worst) $C_s \approx Re$. A dependence like $C_s \approx Re^2$ or worse would destroy the reliability in a numerical computation since the prescribed error tolerance would stay beyond reach with practical mesh sizes. The error growth proportionally to Re seems to be generic for viscous flow problems and cannot be removed by modifying the discretization.

We will now explain the theoretical error analysis of the error in the scheme (P_k) step by step starting from simple model cases (for a general description of this approach see [6]). First, we illustrate the principle ideas at the solution of a linear algebraic system $Ax = b$ with a regular coefficient matrix A. Let $\tilde{x}$ be an approximate solution. We want to estimate the error $e \equiv x-\tilde{x}$ in terms of the (computable) "residual" $r = b-A\tilde{x}$. To this end, we use the (fictitous) solution y of the "dual problem" $A^Ty = e$, to obtain

$$\|e\|^2 = (e,A^Ty) = (Ae,y) = (b-A\tilde{x},y) \leq \|r\|\,\|y\| \tag{4.2}$$

and, consequently $\|e\| \leq C_s\|r\|$ with the "(dual) stability constant" $C_s = \|(A^T)^{-1}\|$. In the case of an infinite dimensional problem, this simple argument is complicated by the freedom of chosing various norms and related stability constants C_s.

As the next level of complexity, we consider the discretization of the (stationary) Poisson equation

$$-\nu\Delta u = f \quad \text{in } \Omega, \quad u_{|\partial\Omega} = 0, \tag{4.3}$$

with the variational formulation

$$u\in V: \quad a(u,\phi) = (f,\phi) \quad \forall\, \phi\in V, \tag{4.4}$$

where $V = H_0^1(\Omega)$. Using finite element subspaces $V_h \subset H_0^1(\Omega)$, as descibed above, the corresponding discrete problems read

$$u_h\in V_h: \quad a(u_h,\phi) = (f,\phi) \quad \forall\, \phi\in V_h. \tag{4.5}$$

Combining (4.4) and (4.5), we obtain the typical "Galerkin orthogonality property"

$$e \equiv u-u_h, \quad a(e,\phi) = 0 \quad \forall\, \phi\in V_h. \tag{4.6}$$

In the context of a Galerkin method the natural quantity entering an a-posteriori error estimate is the "residual" $R(u_h) \in V^* \equiv H^{-1}(\Omega)$ of the approximate solution defined by

$$<R(u_h),\phi> \equiv (f,\varphi) - a(u_h,\varphi) \qquad \varphi \in V . \tag{4.7}$$

However, the evaluation of $R(u_h)$ in the natural norm of V^* is not very illustrative as it does not provide any information on the size of the error depending on the local mesh-size. We would rather prefer to measure it in an h-dependent L^2-norm. To this end, we employ a so-called "duality argument". Let $v \in V$ be the solution of the auxiliary dual problem

$$a'(v,\phi) = (e,\phi) \qquad \forall\ \phi \in V , \tag{4.8}$$

such that $||e||^2 = a'(v,e) = a(e,v) = <R(u_h),v>$. Using the Galerkin orthogonality, $<R(u_h),\cdot> = 0$ on V_h , we conclude that

$$||e||^2 = <R(u_h),v-\phi> = \sum_K \left\{(f+\Delta u_h,v-\phi)_K - (\partial_n u_h,v-\phi)_{\partial K}\right\}$$

$$\leq \sum_K \left\{||f+\Delta u_h||_K||v-\phi||_K + ||[\partial_n u_h]||_{\partial K}||v-\phi||_{\partial K}\right\}$$

$$\leq \sum_K \left\{||f+\Delta u_h||_K c_i h_K^2||\nabla^2 v||_K + ||[\partial_n u_h]||_{\partial K} c_i h_K^{3/2}||\nabla^2 v||_{\partial K}\right\}$$

$$\leq c_i\left\{\sum_K \left\{h_K^4||f+\Delta u_h||_K^2 + h_K^3||[\partial_n u_h]||_{\partial K}^2 c_i h_K^{3/2}\right\}\right\}^{1/2} ||\nabla^2 v|| ,$$

with certain "interpolation" constants c_i . Finally, using the well-known a-priori estimate $v||\nabla^2 v|| \leq C_s||e||$, we arive at

$$||e|| \leq c_i C_s\left\{\sum_K \left\{h_K^4||f+\Delta u_h||_K^2 + h_K^3||[\partial_n u_h]||_{\partial K}^2 c_i h_K^{3/2}\right\}\right\}^{1/2} . \tag{4.9}$$

We note that $C_s = 1$, if Ω is convex. The interpolation constants can be estimated to be of size $c_i \approx 1$ on reasonable meshes. By an analogous argument one can derive a-posteriori error estimates in various other norms, e.g, in the H^1-norm or in the L^∞-norm, for which the relevant stability constant C_s may take different values.

For the Navier-Stokes problem the analysis proceeds along the same line of argument including additionally a linearization step. We give a sketch of the proof concentrating on the time discretization (for further details see [7] and [8]). The linearized duality argument in this case reads

$$\mathbf{v}\in \mathbf{V}(I): \quad L'(\mathbf{u},\mathbf{U};\mathbf{v},\phi) = (\phi(T),\mathbf{e}(T)) \quad \forall\ \phi\in \mathbf{V}(I)\,, \tag{4.10}$$

where

$$L'(\mathbf{u},\mathbf{U};\mathbf{v},\phi) \equiv (-\mathbf{v}_t,\phi)_I + a_I(\mathbf{v},\phi) + b_I(\mathbf{u},\phi,\mathbf{v}) + b_I(\phi,\mathbf{U},\mathbf{v}) - (\phi(T),\mathbf{v}(T))\,,$$

is the (dual) Frechét derivative of $A(\mathbf{u};\cdot,\cdot)$ "between" $\mathbf{u}$ and $\mathbf{U}$. This leads us to the error representation

$$\|\mathbf{e}(\cdot,T)\|^2 = L'(\mathbf{u},\mathbf{U};\mathbf{v},\mathbf{e}) = L(\mathbf{u},\mathbf{U};\mathbf{e},\mathbf{v}) = A(\mathbf{u};\mathbf{u},\mathbf{v}-\bar{\mathbf{v}}) - A_k(\mathbf{U};\mathbf{U},\mathbf{v}-\bar{\mathbf{v}})$$

$$= F(\mathbf{v}-\bar{\mathbf{v}}) - A_k(\mathbf{U};\mathbf{U},\mathbf{v}-\bar{\mathbf{v}}) = \sum_{n=0}^{N} (\mathbf{f},\mathbf{v}-\bar{\mathbf{v}})_n - \sum_{n=0}^{N} ([\mathbf{U}^n],\mathbf{v}_+^n-\bar{\mathbf{v}}_+^n)$$

$$\le c_i L_k^{1/2}\left\{\|\mathbf{v}\|_N + L_k^{-1/2}\int_0^{T-k_N} \|\mathbf{v}_t\|\,dt\right\}\left\{\max_I\|[\mathbf{U}^n]\| + k_n \max_{I_n}\|\mathbf{f}\|\right\},$$

where $\bar{\mathbf{v}}$ denotes the mean value of $\mathbf{v}$ over each subinterval I_n , and $L_k \equiv 1+|\log(k_N)|$. For the combined discretization in space and time, we need the dual a-priori estimate

$$\max_I\|\mathbf{v}\| + L_k^{-1/2}\int_0^{T-k_N} \left\{\|\mathbf{v}_t\| + \nu\|\Delta\mathbf{v}\|\right\} dt \le C_s\|\mathbf{e}(\cdot,T)\|\,. \tag{4.10}$$

This then leads to an a-posteriori error estimate of the form (4.1).

For the (numerical) evaluation of the crucial stability constant C_s , one would have to identify $L(\mathbf{u},\mathbf{u}_h;\cdot,\cdot) \approx L(\mathbf{u}_h,\mathbf{u}_h;\cdot,\cdot)$ in a "backward" sensitivity analysis accompanying the "forward" computation of $\mathbf{U}$. First attempts in this direction for Poiseuille flow show that the stability constant may be of rather moderate size, $C_s \approx Re/50$ (see [1]).

5. Stability Constants

The traditional stability theory for flow problems is of *qualitative* nature, as being based on eigenvalue criteria through a linearized stability argument. The thereby obtained results are in fairly good agreement with experiments concerning the critical Reynolds numbers for the first bifurcation for, e.g., the Benard and the Taylor-Couette problems. But they do not fit with experiments for the other fundamental cases of parallel Poiseuille and Couette flow.

- Poiseuille flow (between two parallel plates) is predicted to turn turbulent at $Re \approx 5772$ through 2d Tollmien-Schlichting waves, while experiments show instability with essential 3d aspects for $Re \approx$ 1,000-10,000.
- Couette flow (parallel shear flow) is supposed to be even stable for all $Re > 0$, but experiments show instability for $Re \approx$ 300-1,500.

This failure of the theory was usually blamed on the deficiencies of lineraized stability theory being valid only for small perturbations. But recent results show that this is not necessarily true (L.N. Trefethen, et al., [13]). Linearized stability theory is correct but wrongly interpreted. In *nonsymmetric* eigenvalue problems, not only critical eigenvalues matter but also the size of the amplification factor for initial perturbations related to a small positive eigenvalue. For example, in Poisseuille an amplification factor of about 10^3 occurs for $Re \geq 549$.

A model example of quantitative "instability" is given by the following simple ODE-system, with a small parameter $\varepsilon > 0$, $\dot{v}_1 + \varepsilon v_1 + v_2 = 0$, $\dot{v}_2 + \varepsilon v_2 = 0$, which has the solution $v_1(t) = e^{-t\varepsilon} v_1^0 - te^{-t\varepsilon} v_2^0$, $v_2(t) = e^{-t\varepsilon} v_2^0$. The 2-fold eigenvalue $\lambda = \varepsilon$ of the system matrix is positive, but small. This implies that a small initial perturbation $|\mathbf{v}^0| \approx \varepsilon$ may grow over the time interval $[0,\varepsilon^{-1}]$ to size $|\mathbf{v}(\varepsilon^{-1})| \approx 1$. This effect also occurs in nonlinear problems. But in this case, the final decay to zero, for $\varepsilon^{-1} \leq t \to \infty$, is practically irrelevant when the "nonlinear" instability has already taken over.

A flow problem which resembles the structure of the above model system is x_1-independent parallel pipe flow, i.e., flow in an infinitely long straight pipe $\Omega = \mathbf{R} \times \omega$ in the x_1-direction, where ω is the cross-section in the (x_2,x_3)-plane. The solution of this problem is $\mathbf{u} = (u_1,0,0)$, $u_1 = u_1(x_2,x_3,t)$, $p = p(x_1,t)$, satisfying

$$u_{1,t} - \nu\Delta u_1 + p_{,1} = f_1 , \quad \text{in } \omega\times I , \quad u_1 = 0 \quad \text{on } \partial\omega\times I ,$$

where $\mathbf{f} = (f_1,0,0)$, $f_1 = f_1(x_2,x_3,t)$. In this case, the linearized "backward" Navier-Stokes equations (used in the linearization process) reads

$$\begin{aligned}
w_{1,t} - \nu\Delta w_1 + u_1 w_{1,1} + u_{1,2} w_2 + u_{1,3} w_3 + q_{,1} &= 0 \quad \text{in } \omega\times I , \\
w_{2,t} - \nu\Delta w_2 + u_1 w_{2,1} + \qquad\qquad\qquad + q_{,2} &= 0 \quad \text{in } \omega\times I , \\
w_{3,t} - \nu\Delta w_3 + u_1 w_{3,1} + \qquad\qquad\qquad + q_{,3} &= 0 \quad \text{in } \omega\times I , \\
w_{2,3} + w_{3,3} = 0 \quad \text{in } \omega\times I , \quad \mathbf{w}_{|\partial\omega} = 0 , \quad \mathbf{w}_{|t=0} &= \mathbf{w}^0 .
\end{aligned}$$

However, this is still too difficult for deriving an explicit analytical a-priori bound. As a further simplification, we assume that the pressure does not depend on x_1 , i.e., $p_{,1} \equiv 0$, and that also the perturbations are x_1-independent, i.e., $\mathbf{w}_{,1} = 0$ and $q_{,1} = 0$. Examples are flows in a long vertical pipe filled with fluid under gravity or in a long rotating pipe with variable rotation speed. The stability of this simple flow, can be analysed by elementary energy arguments yielding a constant of the form $C_s \leq 2\nu^{-1}\max_{\omega\times I}|\nabla u_1|$ (see [7] and [8]). The active mechanism is again a decoupling of the w_2- and w_3-equations in the perturbation equation from the w_1-equation which allows an independent estimation of the (w_2,w_3)-component of the perturbation. Slightly more complex situations possibly modeling the transition to turbulence are analysed in [9].

For the described simple situation the explicit estimate for the stability constant C_s gives us the a-posteriori error estimate

$$\max_I\|\mathbf{U}-\mathbf{u}\| \leq cM_{h,k}L_k\max\{T,\nu^{-1}\}\Big\{\max_I\|kR_{time}(\mathbf{U})\| + \max_I\|h^2R_{space}(\mathbf{U})\|\Big\},$$

where $M_{h,k} \equiv \max_{\omega\times I}|\nabla U_1|$. To our knowledge, this is the first quantitative error estimate for a numerical scheme for solving the Navier-Strokes equations.

6. Conclusion

A new approach to a-posteriori error estimation has been presented for the simulation of real flows by solving the Navier-Stokes equations. This approach relies on the specific properties of a Galerkin method and on a precise analysis of the stability of the flow to be computed, i.e., on the quantitative determination of certain stability constants. On the basis of analytical bounds for such constants rigorous error estimates can be given at least in certain simple model situations. This is a first step towards a reliable and efficient error control in flow simulations. For more realistic problems the required stability constants have to be determined through numerical experiments. The next steps to go are to extend the a-posteriori error analysis to more practical schemes including nonconforming finite elements, streamline diffusion damping, and numerical integration, and then to convert the described concept of a-posteriori error estimation into a practical tool for adaptive error control. This program is presently under work.

References

1. Eriksson, N., On the stability of pipe flow, Master of Sciences Thesis, Math. Dept., Chalmers University of Technology, Gothenburg, 1993.
2. Girault, V., Raviart, P.-A., Finite Element Methods for the Navier-Stokes Equations, Springer, Heidelberg, 1986.
3. Heywood, J.G., Rannacher, R., Finite element approximation of the nonstationary Navier-Stokes problem. Part 1, SIAM J.Numer.Anal. 19, 275-311 (1982), Part 2, ibidem, 23, 750-777 (1986), Part 3, ibidem, 25, 489-512 (1988), Part 4, ibidem, 27, 353-384 (1990).
4. Henshaw, W.D., Kreiss, H.O., Reyna, L.G., On the smallest scale for the incompressible Navier-Stokes equations, ICASE Report No. 88-8, 1988.
5. Johnson, C., The streamline diffusion finite element method for compressible and incompressible fluid flow, in "Finite Element Method in Fluids VII", Huntsville, 1989.
6. Johnson, C., A new paradigm for adaptive finite element methods, Proc. MAFELAP Conf., Brunel Univ. 93, John Wiley, to appear.
7. Johnson, C., Rannacher, R., Boman, M., Numerics and hydrodynamic stability: Towards error control in CFD, Preprint No. 93-12(SFB 359), IWR Universität Heidelberg, SIAM J.Numer.Anal., to appear.
8. Johnson, C., Rannacher R., On error control in CFD, in Proc. Conf. "Numerical Methods for the Navier-Stokes Equations", Heidelberg, Oct. 25-28, 1993 (F.-K. Hebeker, R. Rannacher, G. Wittum, eds.), pp. 133-144, Vieweg, 1994.
9. C. Johnson, R. Rannacher, M. Boman, On transition to turbulence and error control in CFD, Preprint, Universität Heidelberg, August 1994.
10. Rannacher, R., *On the numerical solution of the incompressible Navier-Stokes equations*; in Proc. GAMM Conf. 1992, Z.Angew.Math.Mech. 73, 203-216 (1993).
11. Rannacher, R., On the numerical solution of the incompressible Navier-Stokes equations, Z.Angew.Math.Mech. 73, 203-216 (1993).
12. Schreiber, P., Turek, S., An efficient finite element solver for the nonstationary incompressible Navier-Stokes equations in two and three dimensions, in Proc. Conf. "Numerical Methods for the Navier-Stokes Equations", Heidelberg, Oct. 25-28, 1993 (F.-K. Hebeker, R. Rannacher, G. Wittum, eds.), pp. 133-144, Vieweg, 1994.
13. Trefethen, L.N., Trefethen, A.,E., Reddy, S.C., Driscoll, T.A., A new direction in hydrodynamic stability: Beyond eigenvalues, Technical Report CTC92TR115 12/92, Cornell Theory Center, Cornell University, 1992.

14. Tobiska, L., Verfürth, R., Analysis of a streamline diffusion finite element method for the Stokes and Navier-Stokes equations, to appear in SIAM J.Numer.Anal.

15. Turek, S., Tools for simulating nonstationary incompressible flow via discretely divergence-free finite element models, Int. J. Numer. Meth. Fluids, 18, 71-105 (1994).

Institut für Angewandte Mathematik
Universität Heidelberg
INF 293, D-69120 Heidelberg, Germany
e-mail: rannacher@gaia.iwr.uni-heidelberg.de

J REULING

Partial regularity for incompressible materials with approximation methods

Abstract

We minimize the stored energy functional $F(u) := \int_\Omega |\nabla u|^p\,dx$ where $\Omega \subset \mathbb{R}^n$ and $p \geq n \geq 2$ with respect to the constraint $\det \nabla u = 1$ a.e. on Ω. We approximate F by a sequence of regularized functionals F_δ whose minimizers converge strongly to an F–minimizing function, and prove partial regularity results for the F_δ–minimizers.

AMS Classification: 73 C 50, 49 N 60
Key Words: partial regularity, nonlinear elasticity

We study a special class of polyconvex variational integrals related to nonlinear elasticity. The physical behaviour of socalled hyperelastic materials can be characterized with variational methods by minimizing a stored energy functional

$$F(v) = \int_\Omega f(\nabla v)dx.$$

Here $f : \mathbb{R}^{n\times n} \to \mathbb{R}$ is a polyconvex function, which has to satisfy some additional properties, for example nondegeneracy i.e.

$$f(\nabla u) \longrightarrow \infty \quad \text{as} \quad \det \nabla u \longrightarrow 0. \tag{1}$$

Condition (1) expresses the fact that an infinite amount of energy is needed to shrink a finite volume to zero. For more information about the physical and mathematical background in the setting of elasticity see for instance [C] or [D].

In a recent paper [FR] we obtained some partial regularity results where the stored energy density has the form

$$f(\xi) = |\xi|^p + h(\det \xi)$$

for p a real number, $p \geq n$. The real valued function $h : (0,\infty) \to [0,\infty)$ has the property $\lim\limits_{t\to 0} h(t) = \infty$ according to condition (1). For technical reasons h is assumed to be of class C^2 and strictly convex. We now weaken these restrictions on h. Our aim is to study incompressible materials, this means that we have to minimize the stored energy functional under the constraint $\det \nabla u = 1$ a.e. on Ω. We do this in modifying the energy functional to

$$F(u) = \int_\Omega |\nabla u|^p + h(\det \nabla u)\, dx,$$

where h has the special form

$$h(t) = \begin{cases} 0, & t = 1 \\ \infty, & t \neq 1 \end{cases}.$$

For this reason we have to allow the function h to take the value $+\infty$.

Our method is to replace F by a sequence of more regular functionals

$$F_\delta(u) = \int_\Omega |\nabla u|^p + h_\delta(\det \nabla u)\, dx \tag{2}$$

and to consider minimizers u_δ of F_δ in $\mathcal{C}_0$. We approximate h monotonicly from below by convex $C^{1,1}$ functions h_δ. In our special case this can be done by quadratic polynomials

$$h_\delta(t) = \frac{1}{\delta^3}(t-1)^2.$$

In a more general setting we assume h to be a convex map $h : \mathbb{R} \longrightarrow [0, \infty]$ with Lipschitz first derivative on $I := \{x \in \mathbb{R} \mid h(x) < \infty\}$. In this case we construct approximations h_δ with the following properties:

$$\begin{aligned} & h_\delta : \mathbb{R} \longrightarrow [0,\infty) \text{ is convex}, \\ & h'_\delta \in \mathrm{Lip}(\mathbb{R}), \\ \forall t \in \mathbb{R} \quad & h_\delta(t) \longrightarrow h(t) \text{ as } \delta \to 0 \text{ and} \\ \forall t \in \mathbb{R} \quad & h_\delta(t) \geq h_\varepsilon(t) \text{ if } \delta < \varepsilon. \end{aligned} \tag{3}$$

The first condition ensures that F_δ is a polyconvex functional and therefore the existence of u_δ is well known (see [D]). We now give an explicit description of these h_δ depending on the shape of I. Since h is convex we know that I is a connected subset of $\mathbb{R}$. We distinguish three cases:

(a) If $I = \{a\}$ consists of a single point, we proceed as above and define for $0 < \delta < \delta_0 = 1$: $J_\delta := (a - \frac{\delta}{2}, a + \frac{\delta}{2})$ and

$$h_\delta(t) = \frac{1}{\delta^3}(t-a)^2.$$

(b) If $I = (a, b)$ is a bounded interval, it is clear from convexity that $\lim_{t \to a^-} h(t) = \lim_{t \to b^+} h(t) = \infty$. We choose δ_0 small enough such that $J_\delta := (a+\delta, b-\delta) \subset (a, b)$ for $0 < \delta < \delta_0$ and define h_δ as the first Taylor approximation:

$$h_\delta(t) = \begin{cases} h(a+\delta) + h'(a+\delta)(t - (a+\delta)) & : \ t \leq a+\delta \\ h(t) & : \ t \in J_\delta \\ h(b-\delta) + h'(b-\delta)(t - (b-\delta)) & : \ t \geq b-\delta. \end{cases}$$

(c) If $I = (a, \infty)$, we define for $0 < \delta < \delta_0$ small enough such that $a + \delta_0 < \frac{1}{\delta_0}$: $J_\delta := (a + \delta, \frac{1}{\delta})$ and the Taylor approximation

$$h_\delta(t) = \begin{cases} h(a+\delta) + h'(a+\delta)(t - (a+\delta)) & : \ t \leq a+\delta \\ h(t) & : \ t \in J_\delta \\ h(a+\frac{1}{\delta}) + h'(a+\frac{1}{\delta})(t - (a+\frac{1}{\delta})) & : \ t \geq a+\frac{1}{\delta}. \end{cases}$$

We have nothing to define for $I = \mathbb{R}$ or $I = \emptyset$. If $I = (-\infty, b)$ we can proceed as in (c). We remark that from a physical point of view $I \subset \mathbb{R}^+$ is guaranteed.

It is easy to check that conditions (3) are fullfilled.

As a first step we prove strong convergence of the sequence of F_δ-minimizers. The proof goes back to an idea of Fuchs and Seregin [FS].

Theorem 1 (Approximation) *Let u_δ be minimizers of F_δ in $\mathcal{C}_0$ where h_δ are as defined above, assume that there exists $\hat{u}$ in $\mathcal{C}_0$ with $F(\hat{u}) < \infty$. Then there is a subsequence of $\{u_\delta\}$ in $H^{1,p}(\Omega, \mathbb{R}^n)$ converging strongly to a minimizer u of F in $\mathcal{C}_0$.*

Proof: Choose u_δ a F_δ-minimizer, then we have the estimate:

$$F_\delta(u_\delta) \leq F_\delta(\hat{u}) \longrightarrow F(\hat{u}) < \infty \text{ as } \delta \to 0.$$

Therefore $||u_\delta||_{1,p}$ is bounded independently of δ and $\{u_\delta\}_{0<\delta<\delta_0}$ converges weakly in $H^{1,p}(\Omega, \mathbb{R}^n)$ to some u:

$$u_\delta \rightharpoonup u \text{ in } H^{1,p}(\Omega, \mathbb{R}^n).$$

Now we show that u minimizes F in $\mathcal{C}_0$:
Let $0 < \varepsilon < \delta_0$, then

$$F_\varepsilon(u) \leq \liminf_{\delta \to 0} F_\delta(u_\delta) \leq \limsup_{\delta \to 0} F_\delta(u_\delta) \leq \limsup_{\delta \to 0} F_\delta(u) = F(u),$$

and thus

$$F(u) = \lim_{\delta \to 0} F_\delta(u_\delta) = \lim_{\delta \to 0} F_\delta(u). \tag{4}$$

Let v be an arbitrary function in $\mathcal{C}_0$ then

$$F_\delta(u_\delta) \leq F(v) \longrightarrow F(v) \text{ as } \delta \to 0,$$

and by (4) we have $F(u) \leq F(v)$. The next step is to show strong convergence $u_\delta \to u$ in $H^{1,p}(\Omega, \mathbb{R}^n)$: Since $L^p(\Omega, \mathbb{R}^n)$ is a uniformly convex space we have to show

$$\int_\Omega |\nabla u_\delta|^p dx \longrightarrow \int_\Omega |\nabla u|^p dx \text{ as } \delta \to 0. \tag{5}$$

We have

$$\int_\Omega |\nabla u_\delta|^p = F_\delta(u_\delta) - \int_\Omega h_\delta(\det \nabla u_\delta) - F(u) + \int_\Omega |\nabla u|^p + \int_\Omega h(\det \nabla u),$$

and thus

$$\limsup_{\delta\to 0} \int_\Omega |\nabla u_\delta|^p = \tag{6}$$

$$\limsup_{\delta\to 0} (F_\delta(u_\delta) - F(u)) + \limsup_{\delta\to 0} \left\{ \int_\Omega h(\det \nabla u) - \int_\Omega h_\delta(\det \nabla u_\delta) \right\} + \int_\Omega |\nabla u|^p.$$

Now we consider the polyconvex functional $\int_\Omega h_\varepsilon(\det \nabla u)$. Weak lower semicontinuity gives

$$\int_\Omega h_\varepsilon(\det \nabla u) \leq \liminf_{\delta\to 0} \int_\Omega h_\varepsilon(\det \nabla u_\delta) \leq \liminf_{\delta\to 0} \int_\Omega h_\delta(\det \nabla u_\delta).$$

Now observe

$$\int_\Omega h(\det \nabla u) = \lim_{\varepsilon\to 0} \int_\Omega h_\varepsilon(\det \nabla u) \leq \lim_{\varepsilon\to 0} \liminf_{\delta\to 0} \int_\Omega h_\delta(\det \nabla u_\delta).$$

Thus we have

$$\limsup_{\delta\to 0} \left\{ \int_\Omega h(\det \nabla u) - \int_\Omega h_\delta(\det \nabla u_\delta) \right\} \leq 0. \tag{7}$$

Inserting (7) and (4) in (6) we obtain

$$\limsup_{\delta\to 0} \int_\Omega |\nabla u_\delta|^p \leq \int_\Omega |\nabla u|^p,$$

which implies (5) and completes the proof of Theorem 1.

The next step is to establish partial regularity properties of the approximating sequence.

Theorem 2 (Partial Regularity) *There exists a subset $\Omega_\delta \subset \Omega$ such that the minimizers u_δ of F_δ are in the class $C^1(\Omega_\delta, \mathbb{R}^n)$, and further $\mathcal{L}^n(\Omega \setminus \Omega_\delta) \longrightarrow 0$ as $\delta \to 0$.*

Proof: Suppose that we are given $A_0 \in \mathbb{R}^{n\times n}$ such that

$$a_0 := \det A_0 \in J_\delta,$$

where J_δ is defined, as in section 1, depending on the form of I. Then we can calculate $\sigma = \sigma(A_0, \delta)$ such that

$$\det A \in J_\delta$$

holds for all $A \in \mathbb{R}^{n\times n}$, $|A - A_0| \leq \sigma$. Now we can quote an energy decay estimate, which is proved in [FR, Main Lemma] by a blow–up argument and using a compactness method introduced by Evans and Gariepy [EG].

Lemma 1 *There is a constant $c_* = c_*\big(A_0, p, \|h''_\delta\|_{C^2((a_0-\varepsilon, a_0+\varepsilon))}\big)$ with the following property: For each $t \in (0,1)$ there exists $\varepsilon = \varepsilon(A_0, t, \delta)$ such that, for every ball $B_R(x_0) \subset \Omega$, the conditions*

$$|(\nabla u_\delta)_{x_0,R} - A_0| \leq \sigma,$$

$$E\big(u_\delta, B_R(x_0)\big) := \fint_{B_R(x_0)} |\nabla u_\delta - (\nabla u_\delta)_{x_0,R}|^2 + |\nabla u_\delta - (\nabla u_\delta)_{x_0,R}|^p \, dx < \varepsilon^2$$

imply

$$E\big(u_\delta, B_{tR}(x_0)\big) \leq c_* t^2 E\big(u_\delta, B_R(x_0)\big).$$

From this result the regularity statement of Theorem 2 follows in a routine manner: Let $\Omega_\delta := \{x \in \Omega : \det \nabla u_\delta(x) \in J_\delta, x \text{ is a Lebesgue–point of } \nabla u_\delta,\ E(u_\delta, B_r(x)) \longrightarrow 0 \text{ as } r \to 0\}$. Iteration of the Main Lemma gives $u_\delta \in C^1(\Omega_\delta, \mathbb{R}^n)$, see [G] for details.

It remains to show $|\Omega \setminus \Omega_\delta| \longrightarrow 0$ as $\delta \to 0$. The definition of h_δ and minimality of u_δ implies

$$\mathcal{L}^n(\Omega \setminus \Omega_\delta) \min_{t\in\partial J_\delta} h(t) \leq \int_{\Omega\setminus\Omega_\delta} h(\det \nabla u_\delta)\, dx \leq \int_\Omega |\nabla u_0|^p + h(\det \nabla u_0)\, dx \leq F(\hat{u}) < \infty.$$

In the case $I = (a,b)$ or $I = (a,\infty)$ the property $\min_{t\in\partial J_\delta} h(t) \longrightarrow \infty$ follows by convexity of h. If $I = \{a\}$ we have

$$h_\delta\left(a - \frac{\delta}{2}\right) = h_\delta\left(a + \frac{\delta}{2}\right) = \frac{1}{\delta^3}\left(\frac{\delta}{2}\right)^2 = \frac{1}{4\delta} \longrightarrow 0 \text{ as } \delta \to 0.$$

Therefore we obtain

$$|\Omega \setminus \Omega_\delta| \leq \frac{F(\hat{u})}{\min_{t\in\partial J_\delta} h(t)} \longrightarrow 0 \text{ as } \delta \to 0,$$

and the proof of Theorem 2 is complete.

We now return to our original problem of incompressible materials. We want to minimize $F(u) = \int_\Omega |\nabla u|^p dx$ with respect to the constraint $\det \nabla u = 1$ a.e. For this reasonig we look for minimizers in $\mathcal{C}_1 := \{v \in H^{1,p} \,|\, v = u_0 \text{ on } \partial\Omega, \det \nabla v = 1 \text{ a.e. in } \Omega\}$.

Theorem 3 (Incompressible Materials) *If $\mathcal{C}_1$ is nonempty then there exists a sequence $\{u_\delta\}$ which converges strongly in $H^{1,p}$ to a minimizer u of F in $\mathcal{C}_1$. Moreover u_δ is of class C^1 on a subset Ω_δ and $\mathcal{L}^n(\Omega \setminus \Omega_\delta) \longrightarrow 0$ as $\delta \to 0$.*

The above problem is equivalent to $\hat{F}(u) := \int_\Omega |\nabla u|^p + h(\det \nabla u)dx \to \min$ in $\mathcal{C}_0$, where

$$h : \mathbb{R} \to [0, \infty], \quad h(t) = \begin{cases} 0 & t = 1 \\ \infty & t \neq 1 \end{cases}.$$

We are in case (a) with $I = \{1\}$. Applying Theorem 1 we get an approximating sequence $\{u_\delta\}$ converging strongly in $H^{1,p}(\Omega, \mathbb{R}^n)$ to u. Letting $0 < \varepsilon < 1$, since $\hat{F}_\varepsilon \leq \hat{F}_\delta$ as $\delta < \varepsilon$ we get from weak lower semicontinuity

$$\hat{F}_\varepsilon(u) \leq \liminf_{\delta \to 0} \hat{F}_\varepsilon(u_\delta) \leq \liminf_{\delta \to 0} \hat{F}_\delta(u_\delta) \leq \hat{F}(\hat{u}) < \infty.$$

Thus

$$\sup_{0<\varepsilon<1} \int_\Omega h_\varepsilon(\det \nabla u) < \infty \quad \text{and} \quad \int_\Omega h(\det \nabla u) < \infty$$

i.e.

$$\det \nabla u = 1 \text{ a.e. in } \Omega.$$

Partial regularity of u_δ follows by Theorem 2.

Remark 1: $\mathcal{C}_1 \neq \emptyset$ is ensured, if we impose smoothness on the boundary data, for example if u_0 is a C^1– diffeomorphism, then $\mathcal{C}_1$ is nonempty.

Remark 2: Theorem 3 is also an existence result, which we obtained without using the elaborate arguments due to Ball [B] or Mller [M].

As another application of our theorems we can consider compressible materials, where $\det \nabla u$ is restricted to values in $(0, 1]$. This means that in every subset $\Omega' \subset \Omega$ we have volume–contraction of the elastic material. Let $\tilde{h} : (0, 1) \to \mathbb{R}$, $\tilde{h}$ convex, describe this compression as a function of the determinant of ∇u, then we have to choose h as

$$h(t) = \begin{cases} \tilde{h}(t) & \text{if } 0 < t < 1 \\ \infty & \text{otherwise.} \end{cases}$$

References

[B] Ball, J.M.: Convexity Conditions and Existence Theorems in Nonlinear Elasticity. Arch.Rat.Mech.Anal. **63** (1977) 337–403.

[C] Ciarlet, Philippe: Mathematical Elasticity Vol. 1, North-Holland 1989.

[D] Dacorogna, B.: Direct Methods in the Calculus of Variations, Springer Verlag 1989.

[EG] Evans, L.C., Gariepy, R.F.: Blow up, compactness and partial regularity in the calculus of variations. Ind.Univ. Math.J. Vol. **36**, No. 2 (1987) 361–371.

[FR] Fuchs, M., Reuling, J.: Partial Regularity for certain Classes of Polyconvex Functionals related to Nonlinear Elasticity, Preprint No. 338 SFB 256 Bonn 1994.

[FS] Fuchs, M., Seregin, G.: Partial Regularity of the Deformation Gradient for some Model Problems in Nonlinear twodimensional Elasticity. Preprint No. 314 SFB 256 Bonn 1993.

[G] Giaquinta, M.: Multiple Integrals in the Calculus of Variations and Nonlinear Elliptic Systems, Princeton U.P., Princeton 1983.

[M] Müller, S.: Higher Integrability of Determinants and Weak Convergence in L^1, J.Reine Angew.Math. **412** (1990), 20–34 .

Jürgen Reuling
Fachbereich 9 Mathematik
Universität des Saarlandes
Postfach 151150
D - 66041 Saarbrücken

T ROUBÍČEK

A note about relaxation of vectorial variational problems

1. Introduction; the original variational problem and its relaxation.

The contribution deals with a vectorial variational problem

$$\text{(VP)} \qquad \int_\Omega \varphi(x,y(x),\nabla y(x))\mathrm{d}x \;\rightarrow\; \inf, \qquad y \in W^{1,p}(\Omega;\mathbb{R}^m)\ ,$$

where $\Omega \subset \mathbb{R}^n$ is a Lipschitz bounded domain and $\varphi : \Omega \times \mathbb{R}^m \times \mathbb{R}^{n\times m} \rightarrow \mathbb{R}$ is a potential density, supposed to be coercive Carathéodory function with a p-growth, $1 < p < +\infty$. The pecularity of the problem is that $\varphi(x,r,\cdot) : \mathbb{R}^{n\times m} \rightarrow \mathbb{R}$ is not supposed to be quasiconvex so that (VP) need not possess any solution and its extension (=relaxation) must be done.

We will basically use a continuous extension like the Young-measure setting, cf. [2, 3, 4]. We employ a suitable convex locally compact hull of $L^p(\Omega;\mathbb{R}^{n\times m})$, constructed as follows: Let us define the linear space of integrands $\mathrm{Car}^p(\Omega;\mathbb{R}^{n\times m}) = \{h:\Omega\times\mathbb{R}^{n\times m} \rightarrow \mathbb{R}$ Carathéodory; $|h(x,A)| \leq a(x) + b|A|^p,\ a \in L^1(\Omega), b \in \mathbb{R}\}$, equiped with the natural norm $\|h\| = \inf\{\|a\|_{L^1(\Omega)} + b;\ |h(x,A)| \leq a(x) + b|A|^p\}$. Furthermore, let us take a suitable (typically separable) subspace H of $\mathrm{Car}^p(\Omega;\mathbb{R}^{n\times m})$, and define the (norm,weak*)-continuous imbedding $i_H : L^p(\Omega;\mathbb{R}^{n\times m}) \rightarrow H^* : u \mapsto (h \mapsto \int_\Omega h(x,u(x))\mathrm{d}x)$. Supposing, as we may, that H contains a coercive integrand (i.e. $h = |A|^p \in H$), we may define

$$Y^p_H(\Omega;\mathbb{R}^{n\times m}) \;=\; \text{w*-cl}\ i_H(L^p(\Omega;\mathbb{R}^{n\times m}))\ ,$$

which is a convex, closed, locally compact, σ-compact hull of $L^p(\Omega;\mathbb{R}^{n\times m})$ if H^* is considered in its weak* topology. It is natural to address the elements of $Y^p_H(\Omega;\mathbb{R}^{n\times m})$ as "generalized Young functionals" because, for $H = L^1(\Omega;C_0(\mathbb{R}^{n\times m}))$ with C_0 denoting continuous functions vanishing at infinity, this set contains basically the funtionals introduced by L.C.Young in [20], which can be in this case identified with special elements of $L^\infty_{\mathrm{w}}(\Omega;M(\mathbb{R}^{n\times m})) \cong L^1(\Omega;C_0(\mathbb{R}^{n\times m}))^*$, called the Young measures; cf. [19] or also [1, 7, 17]. Nevertheless, the choice $H = L^1(\Omega;C_0(\mathbb{R}^{n\times m}))$ is not much suitable because such H cannot not contain coercive integrands. A more suitable example is rather the measures developed by DiPerna and Majda [8].

Now we can made readily a continuous extension of (VP). For $\eta \in H^*$ and $h \in H$ let us define $h \bullet \eta \in M(\bar\Omega) \cong C(\bar\Omega)^*$ by $\langle h \bullet \eta, g\rangle = \langle \eta, g\cdot h\rangle$ to be valid for all $g \in C(\bar\Omega)$. Furthemore, let us define the set of "gradient generalized Young functionals" by

$$G^p_H(\Omega;\mathbb{R}^{n\times m}) \;=\; \{\eta \in Y^p_H(\Omega;\mathbb{R}^{n\times m});\ \exists y_\xi \in W^{1,p}(\Omega;\mathbb{R}^m) :\ i_H(\nabla y_\xi) \rightarrow \eta \text{ weakly*}\}\ .$$

We will suppose that φ is coercive in the sense

$$\varphi(x,r,s) \geq a(x) + b|s|^p \tag{1}$$

with some $a \in L^1(\Omega)$ and b positive, and that φ satisfies

$$\forall y \in L^q(\Omega;\mathbb{R}^m): \qquad \varphi \circ y \in H\ , \tag{2}$$

$$\exists a_1 \in L^1(\Omega)\ \exists b_1, c_1 \in \mathbb{R}^+: \quad |\varphi(x,r,s)| \ \leq\ a_1(x) + b_1|r|^q + c_1|s|^p\ , \tag{3}$$

$$\left.\begin{array}{l}\exists a_2 \in L^{q/(q-1)}(\Omega)\ \exists b_2, c_2 \in \mathbb{R}^+: \\ |\varphi(x,r_1,s)-\varphi(x,r_2,s)| \leq (a_2(x) + b_2|r_1|^{q-1} + b_2|r_2|^{q-1} + c_2|s|^{p(q-1)/q})|r_1-r_2|\end{array}\right\} \tag{4}$$

with some $1 \leq q < np/(n-p)$ or $1 \leq q$ arbitrary provided $p \geq n$ (so that $W^{1,p}(\Omega;\mathbb{R}^m)$ is compactly imbedded into $L^q(\Omega;\mathbb{R}^m)$). Note that the assumptions (2)–(4) quarantees that the mapping $(y,\eta) \mapsto \langle \eta, \varphi \circ y\rangle$ is (weak×weak*)-continuous. The relaxed problem will look like:

$$(\mathsf{RP}) \qquad \left\{\begin{array}{ll}\text{minimize} & \langle \eta, \varphi \circ y\rangle \\ \text{subject to} & (1 \otimes \mathrm{id}) \bullet \eta = \nabla y\ , \\ & y \in W^{1,p}(\Omega;\mathbb{R}^m)\ , \quad \eta \in G^p_H(\Omega;\mathbb{R}^{n\times m}),\end{array}\right.$$

where we suppose $\varphi \circ y \in H$, with $[\varphi \circ y](x,A) = \varphi(x,y(x),A)$, and id: $\mathbb{R}^{n\times m} \to \mathbb{R}^{n\times m}$ being the identity. It can be shown (see [14]) that (RP) is actually a proper relaxation of (VP) in the sense that it has always a solution, inf(VP) = min(RP), every solution to (RP) is attainable by a minimizing net for (VP) and, conversely, every minimizing net (esp. sequence) for (VP) has a cluster point and each such a cluster point solves (RP).

The aim of this short note is to discuss various possibilities of a numerical approximation of the relaxed problem (RP) which has been also tested by computer experiments, though the results will not be presented here.

2. A finite-element approximation.

A first step in a numerical approximation of (RP) which we can certainly do quite easily consists in a finite-element approximation of (RP). Let us take a triangulation $\mathcal{T}_d$ of Ω, $d > 0$ being a mesh size, and put $V_d = \{y \in W^{1,p}(\Omega;\mathbb{R}^m);\ y$ piecewise affine on $\mathcal{T}_d\}$ and $Y_d \subset Y^p_H(\Omega;\mathbb{R}^{n\times m})$ defined by $Y_d = A^*_d(Y^p_H(\Omega;\mathbb{R}^{n\times m}))$ where $P_d : \mathrm{Car}^p(\Omega;\mathbb{R}^{n\times m}) \to \mathrm{Car}^p(\Omega;\mathbb{R}^{n\times m}) : h \mapsto h_d$ with $h_d(x,A) = \int_\triangle h(\tilde{x},A)\mathrm{d}\tilde{x}/\mathrm{meas}(\triangle)$ for $x \in \triangle \in \mathcal{T}_d$; cf. also [13] for details. It allows us to define the approximate relaxed problem:

$$(\mathsf{RP}_d) \qquad \left\{\begin{array}{ll}\text{minimize} & \langle \eta, \varphi \circ y\rangle \\ \text{subject to} & (1 \otimes \mathrm{id}) \bullet \eta = \nabla y\ , \\ & y \in V_d\ , \quad \eta \in G^p_H(\Omega;\mathbb{R}^{n\times m}) \cap Y_d\ ,\end{array}\right.$$

Proposition 1. *Let φ satisfy (1)–(4). Then* (RP_d) *converges to* (RP) *in the sense:*

$$\lim_{d\to 0}\min(\mathsf{RP}_d) = \min(\mathsf{RP})\ , \qquad \operatorname*{Limsup}_{d\to 0}\ \mathrm{Argmin}(\mathsf{RP}_d) \subset \mathrm{Argmin}(\mathsf{RP})\ ,$$

where "Limsup" (i.e. the upper Kuratowski limit) denotes the set of all cluster points of selected nets, and "Argmin" stands for the set of solutions to the problem indicated.

Sketch of the proof. Since apparently the set of admissible pairs for (RP_d) is smaller than for (RP), we have $\min(\mathsf{RP}_d) \geq \min(\mathsf{RP})$.

Now we want to prove that every (y,η) admissible for (RP) can be approximated by suitable admissible pairs for (RP_d) when $d \to 0$. We can easily see that there is a net $\{y_\iota\}_{\iota\in I} \in W^{1,p}(\Omega;\mathbb{R}^m)$ such that $y_\iota \to y$ weakly in $W^{1,p}(\Omega;\mathbb{R}^m)$ and $i_H(\nabla y_\iota) \to \eta$ weakly* in H^*. Moreover, mollifing (if necessary) suitably this net, we can even suppose that $y_\iota \in C^\infty(\bar\Omega)$. Let $\Pi_d y_\iota \in V_d$ be the linear interpolant of y_ι on the triangulation $\mathcal{T}_d$. For ι fixed and $d \to 0$, we have $\Pi_d y_\iota \to y_\iota$ strongly in $W^{1,p}(\Omega;\mathbb{R}^m)$ because of the regularity of y_ι. Therefore $i_H(\nabla\Pi_d y_\iota) \to i_H(\nabla y_\iota)$ weakly* in H^*. At the same time, the pair $(\Pi_d y_\iota, i_H(\nabla\Pi_d y_\iota))$ is admissible for (RP_d), and therefore

$$\langle i_H(\nabla\Pi_d y_\iota), \varphi\circ\Pi_d y_\iota\rangle \ \geq\ \min(\mathsf{RP}_d)\ \geq\ \min(\mathsf{RP})\ .$$

Supposing that (y,η) is a solution of (RP), we get by the continuity argument (because (2)–(4) makes the mapping $(y,\eta)\mapsto\langle\eta,\varphi\circ y\rangle$ weakly$\times$weakly* continuous; cf. [17, Lemma 4.3.6]) that

$$\lim_{\iota\in I}\lim_{d\to 0}\ \langle i_H(\nabla\Pi_d y_\iota), \varphi\circ\Pi_d y_\iota\rangle \ =\ \langle y, \eta\circ y\rangle \ =\ \min(\mathsf{RP})\ ,$$

so that $\min(\mathsf{RP}_d) \to \min(\mathsf{RP})$ for $d \to 0$.

The rest of the assertion follows immediately by the standard compactness arguments, taking into account the coercivity of the problem. □

3. A further approximation of (RP_d).

The essential problem is that an explicit description of $G^p_H(\Omega;\mathbb{R}^{n\times m})$ is generally not known, which implies that the admissible domain of (RP_d), i.e.

$$\mathcal{D}(\mathsf{RP}_d) \ =\ \{(y,\eta)\in V_d\times(G^p_H(\Omega;\mathbb{R}^{n\times m})\cap Y_d);\ (1\otimes\mathrm{id})\bullet\eta = \nabla y\}\ ,$$

is not effectively defined. A certain way to handle this problem is to confine ourselves to approximations of $\mathcal{D}(\mathsf{RP}_d)$. In principle, one can think either of an inner or of an outer approximation of it.

An inner approximations of $\mathcal{D}(\mathsf{RP}_d)$ has been recently proposed by Nicolaides and Walkington [11]. Namely, for $k\in\mathbb{N}$ they defined

$$\begin{aligned}\mathcal{D}_k(\mathsf{RP}_d) &= \{(y,\eta) \in V_d \times Y_d;\ (1 \otimes \mathrm{id}) \bullet \eta = \nabla y,\\ &\langle \eta, h\rangle = \sum_{i=1}^{2^k} \int_\Omega \lambda_i(x) h(x, u_i(x))\mathrm{d}x, \qquad \lambda_i,\ u_i \text{ elementwise constant},\\ &\lambda_i = \prod_{j=1}^{k} \lambda_{[i/j]+1,j}, \qquad u_i = A_{ik}, \qquad \lambda_{2i,j}A_{2i,j} + \lambda_{2i-1,j}A_{2i-1,j} = A_{i,j-1},\\ &\lambda_{2i,j} + \lambda_{2i-1,j} = 1, \qquad \lambda_{2i,j}, \lambda_{2i-1,j} \geq 0, \quad \mathrm{Rank}(A_{2i,j} - A_{2i-1,j}) \leq 1,\\ &i = 1, ..., 2^{j-1}, \quad j = 1, ..., k, \qquad A_{i,1} = \nabla y\}\ .\end{aligned}$$

In other words, the approximation $\mathcal{D}_k(\mathsf{RP}_d) \subset \mathcal{D}(\mathsf{RP}_d)$ consists of "element-wise constant" generalized Young functionals composed, on each element, from 2^k pairwise rank-1 connected matrices, which can be certainly reached by gradients. The essential theoretical disadvantage of this approximation is that, in general, $\overline{\bigcup_{k \in N} \mathcal{D}_k(\mathsf{RP}_d)} \neq \mathcal{D}(\mathsf{RP}_d)$ because otherwise the minimum of the functional $(y, \eta) \mapsto \langle \eta, \varphi \circ y\rangle$ over $\mathcal{D}_k(\mathsf{RP}_d)$ would have to approach min(RP) for $k \to \infty$ but, as shown by Dacorogna [7, Sect.5.1.1.2], it converges from above only to Rank-1 envelope of (VP). Therefore, if $\varphi(x, r, \cdot)$ has a quasiconvex envelope which is not rank-1 convex, then the approximation proposed by Nicolaides and Walkington cannot converge.

Nevertheless, we can also use an outer approximation of $\mathcal{D}(\mathsf{RP}_d)$. Let us put $\Xi = \{\xi = (v_1, ..., v_k);\ k \in \mathbb{N},\ v_j : \mathbb{R}^{n \times m} \to \mathbb{R} \text{ quasiconvex},\ |v_j(s)| \leq o(|s|^p)\}$, ordered by the inclusion. This makes Ξ a directed set so that we can use it to index generalized sequences (=nets). For any $\xi \in \Xi$ we put

$$\mathcal{D}^\xi(\mathsf{RP}_d) = \{(y, \eta) \in V_d \times Y_d;\ \forall v \in \xi :\ (1 \otimes v) \bullet \eta \geq v(\nabla y)\}\ .$$

Always, $\mathcal{D}^\xi(\mathsf{RP}_d) \supset \mathcal{D}(\mathsf{RP}_d)$ and, as a consequence of recent results by Kinderlehrer and Pedregal [10], also $\mathcal{D}(\mathsf{RP}_d) \supset \bigcap_{\xi \in \Xi} \mathcal{D}^\xi(\mathsf{RP}_d) \cap \{\eta \in Y_H^p(\Omega; \mathbb{R}^{n \times m}) \text{ is } p\text{-nonconcentrating}\}$, where "$p$-nonconcentrating" means that $\eta \in Y_H^p(\Omega; \mathbb{R}^{n \times m})$ is attainable by a sequence $\{i_H(u_k)\}_{k \in \mathbb{N}}$ such that the set $\{|u_k|\}_{k \in \mathbb{N}}$ is relatively weakly compact in $L^1(\Omega)$. This result suggests the following approximate problem:

(RP_d^ξ) $\qquad$ Minimize $\langle \eta, \varphi \circ y\rangle$ s.t. $(y, \eta) \in \mathcal{D}^\xi(\mathsf{RP}_d)$.

Proposition 2. *Let φ satisfy (1)–(4), H be separable and there is $G \otimes V$ dense in H with $G \subset L^\infty(\Omega)$ and $V \subset C(\mathbb{R}^{n \times m})$, and $\forall v \in V\ \exists v_l \in V$ with a growth strictly less than p such that $v_l \to v$ uniformly on bounded subsets of $\mathbb{R}^{n \times m}$. Then*

$$\lim_{\xi \in \Xi} \min(\mathsf{RP}_d^\xi) = \min(\mathsf{RP}_d)\ , \qquad \operatorname*{Limsup}_{\xi \in \Xi}\ \mathrm{Argmin}(\mathsf{RP}_d^\xi) \subset \mathrm{Argmin}(\mathsf{RP}_d)\ .$$

In other words, if (y_ξ, η_ξ) solves (RP_d^ξ), then the net $\{(y_\xi, \eta_\xi)\}_{\xi \in \Xi}$ has a (weak×weak)-cluster point (y, η) in $W^{1,p}(\Omega; \mathbb{R}^m) \times H^*$ and each such a cluster point solves (RP_d).*

Sketch of the proof. As $\min(\mathsf{RP}_d^\xi)$ is certainly bounded from above (e.g. by $\langle i_H(0), \varphi\circ 0\rangle < +\infty$) and the coercivity (1) is assumed, the net in question is bounded and therefore it has a (weak×weak*)-cluster point (y,η). We want to show that (y,η) must solve the auxiliary problem

$$(\mathsf{AP}_d) \qquad \begin{cases} \text{minimize} & \langle \eta, \varphi\circ y\rangle \\ \text{subject to} & (1\otimes v)\bullet\eta \geq v(\nabla y) \quad \forall v \text{ quasiconvex with a growth } < p, \\ & y\in V_d\,, \quad \eta\in Y_d\,. \end{cases}$$

As certainly $\min(\mathsf{RP}_d^\xi) \leq \min(\mathsf{AP}_d)$ and the mapping $\xi \mapsto \min(\mathsf{RP}_d^\xi)$ is nondecreasing, we have guaranteed $\lim_{\xi\in\Xi}\min(\mathsf{RP}_d^\xi) \leq \min(\mathsf{AP}_d)$. Supposing $\lim_{\xi\in\Xi}\min(\mathsf{RP}_d^\xi) < \min(\mathsf{AP}_d)$, by the coercivity of the problem we could choose a finer net than $\{(y_\xi,\eta_\xi)\}_{\xi\in\Xi}$ converging to some (y_0,η_0) satisfying all the constraints involved in (AP) but such that $\langle\eta_0,\varphi\circ y_0\rangle < \min(\mathsf{AP}_d)$, which is a contradiction. Therefore, $\lim_{\xi\in\Xi}\min(\mathsf{RP}_d^\xi) = \min(\mathsf{AP}_d)$ and then also (y,η) must solve (AP_d).

Then η must be p-nonconcentrating in the sense that there is a net $\{u_\alpha\}$ bounded in $L^p(\Omega;\mathbb{R}^{n\times m})$ such that $i_H(u_\alpha)\to\eta$ and the set $\{|u_\alpha|^p\}$ is relatively weakly compact in $L^1(\Omega)$. Indeed, if it would not be the case, the p-nonconcentrating modification of η would reach a strictly lower cost than η and all the constraints of (AP_d) would by satisfied as well, which is a contradiction; we refer to [16] for details.

As $p>1$ is supposed, the inequality constraints of (AP_d) include, in particular, also the constraint $(1\otimes\mathrm{id})\bullet\eta = \nabla y$ involved in (RP_d). Therefore, to prove that (y,η) solves also (RP_d), it suffices to show that $\min(\mathsf{AP}_d)\leq\min(\mathsf{RP}_d)$ (which, however, follows immediately from $\mathcal{D}(\mathsf{AP}_d)\supset\mathcal{D}(\mathsf{RP}_d)$) and that $\eta\in G_H^p(\Omega;\mathbb{R}^{n\times m})$.

First, we can localize our considerations on a current element so that it suffices to investigate only homogeneous Young functionals. As in [13], one can show that η cannot be separated from the set $M_y = \{\eta\in G_H^p(\Omega;\mathbb{R}^{n\times m});\ (1\otimes\mathrm{id})\bullet\eta = \nabla y\}$ by any test function with the growth strictly less than p. However, taking a general $1\otimes v$ and v^l with growth strictly less than p and such that $v^l\to v$ uniformly on bounded sets in $\mathbb{R}^{n\times m}$, then one can show that $\langle\eta, 1\otimes v^l\rangle \to \langle\eta, 1\otimes v\rangle$ for any $\eta\in Y_H^p(\Omega;\mathbb{R}^{n\times m})$ p-nonconcentrating; cf. [13, Example 3.1]. This shows that η in question cannot be separated from the closed convex set M_y by any test function of the form $1\otimes v$, and therefore also by any $\sum_{\text{finite}} g_i\otimes v_i$, so that it must belong to $G_H^p(\Omega;\mathbb{R}^{n\times m})$ which is closed. □

Though having a convergence guaranteed, the fatal disadvantage of the previous method is that the index set Ξ is very large and not effectively defined, so that it has a theoretical significance only.

Anyhow, both approximate methods mentioned above certainly give a general two-side estimate:

$$\min_{(y,\eta)\in\mathcal{D}^\xi(\mathsf{RP}_d)}\langle\eta,\varphi\circ y\rangle \;\leq\; \min(\mathsf{RP}_d) \;\leq\; \min_{(y,\eta)\in\mathcal{D}_k(\mathsf{RP}_d)}\langle\eta,\varphi\circ y\rangle\,. \tag{5}$$

Nevertheless, for practically reasonable indices $\xi\in\Xi$ and $k\in\mathbb{N}$, this estimate might be still very rough. Therefore, it is reasonable to inquire special situations where possibly the equalities for ξ or k large enough can appear.

As an example let us mention the case $\xi = \{\pm\mathrm{adj}_l;\ l = 1, ..., \min(n,m)\}$. Then the first equality in (5) takes place provided $\varphi(x,r,\cdot) : \mathbb{R}^{n\times m} \to \mathbb{R}$ has a polyconvex quasiconvexification. In this case, one can even derive (cf. [15]) optimality conditions for (RP_d^ξ), which takes the form

$$\mathcal{H}_{y,\lambda} \bullet \eta \ = \ \max_{A\in\mathbb{R}^{n\times m}} \mathcal{H}_{y,\lambda}(x,A)$$

with the "discrete Hamiltonian" $\mathcal{H}_{y,\lambda}(x,A) = -P_d\varphi(x,y(x),A) + \sum_{l=1}^{\min(n,m)} \lambda_l(x)\cdot\mathrm{adj}_l A$ and with $\lambda_l \in L^{p/l}(\Omega;\mathbb{R}^{\sigma(l)})$, $\sigma(l) = \binom{m}{l}\binom{n}{l}$, satisfying

$$\mathrm{div}\left(\sum_{l=1}^{\min(n,m)} \lambda_l \cdot \frac{\partial \mathrm{adj}_l}{\partial A}(\nabla y)\right) \ = \ (\frac{\partial\varphi}{\partial y} \circ y) \bullet \eta \ .$$

From these conditions we can deduce that there always exists a minimizer in the form of a convex combinations of at most $\prod_{l=1}^{\min(n,m)} \sigma(l) + 2$ atoms on each element, which eventually allows an effective computer implementation of (RP_d^ξ).

Remark 1. The requirement on $\varphi(x,r,\cdot)$ to have a polyconvex quasiconvexification is certainly not realistic in a general case so that one is forced to try to take larger indices ξ. Each such a choice gives some problem whose minimum is in between the polyconvexified problem and inf(VP). This is philosophically similar to the approach by Firoozye [9] who also proposed some envelope with such property.

Remark 2. A general feature of the resulted approximate problems is that they admit a partial decomposition, having always the form of a co-operative Stackelberg game. Namely, the leader controls the displacement y, seeking the minimum of the total energy, while the followers (one on each element) seek the minimum of the deformation energy on a current element for ∇y set up by the leader. Thus each follower is to solve repeatedly convex problems parametrized by ∇y.

Remark 3. If H is small enough it may happen that, for every solution (y,η) to (RP_d^ξ), η belongs to $G_H^p(\Omega;\mathbb{R}^{n\times m})$. It immediately implies that (y,η) solves also (RP_d) provided ξ contains at least linear functions adj_1. For example, if $\varphi(x,r,s) = v_0(s)$ with v_0 having a polyconvex quasiconvexification, then this feature takes place if $H = G \otimes V$ with V being contained in the linear hull of all minors and v_0; cf. [13].

References

[1] BALL, J.M.: A version of the fundamental theorem for Young measures. In: *PDEs and Continuum Models of Phase Transition.* (Eds. M.Rascle, D.Serre, M.Slemrod.) Lecture Notes in Physics **344**, Springer, Berlin, 1989, pp.207–215.

[2] BALL, J.M., JAMES, R.D.: Fine phase mixtures as minimizers of energy. *Archive Rat. Mech. Anal.* **100** (1988), 13–52.

[3] Ball, J.M., James, R.D.: Proposed experimental tests of a theory of fine microstructure and the two-well problem. *Phil. Trans. Royal Soc. London* A **338** (1992), 389–450.

[4] Chipot, M., Kinderlehrer, D.: Equilibrium configurations of crystals. *Arch. Rational Mech. Anal.* **103** (1988), 237–277.

[5] Collins, C., Luskin, M.: Numerical modelling of the microstructure of crystals with symmetry-related variants. In: *US-Japan Workshop on Smart/Inteligent Materials and Systems* (Eds. I.Ahmad et al.), Technomic Publ. Comp., Lancaster, 1990, pp.309–318.

[6] Collins, C., Luskin, M., Riordan J.: Computational results for a two-dimensional model of crystalline microstructure. In: IMA Vol. in Math. and Applications *Microstructure and Phase Transitions* (Eds. R.James, D.Kinderlehrer, M.Luskin), Springer, New York, to appear.

[7] Dacorogna, B.: *Direct Methods in the Calculus of Variations*, Springer, Berlin, 1989.

[8] DiPerna, R.J., Majda, A.J.: Oscillations and concentrations in weak solutions of the incompressible fluid equations. *Comm. Math. Physics* **108** (1987), pp.667–689.

[9] Firoozye, N.B.: Optimal use of the translation method and relaxations of variational problems. *Comm. Pure Appl. Math.* **44** (1991), 643–678.

[10] Kinderlehrer, D., Pedregal, P.: Remarks about gradient Young measures generated by sequences in Sobolev spaces. *Carnegie-Mellon Research Report* No. 92-NA-007, March 1992.

[11] Nicolaides, R.A., Walkington, N.J.: Computation of microstructure utilizing Young measure representations. Res. Report 92-NA-001, Center for Nonlinear Analysis, Carnegie Mellon University, 1992.

[12] Roubíček, T.: Convex compactifications and special extensions of optimization problems. *Nonlinear Analysis, Th., Methods, Appl.* **16** (1991), 1117–1126.

[13] Roubíček, T.: Effective characterization of generalized Young measures generated by gradients. *Boll. Unione Mat. Italiana* (accepted)

[14] Roubíček, T.: Relaxation of vectorial variational problems. Rapport de Recherche UMPA-93-no.88, ENS Lyon, 1993 (submitted)

[15] Roubíček, T.: Numerical approximation of relaxed variational problems. (submitted)

[16] Roubíček, T.: Nonconcentrating generalized Young functionals. (submitted)

[17] Roubíček, T.: *Relaxation in Optimization Theory and Variational Calculus.* W. de Gruyter, Berlin (in preparation)

[18] Roubíček, T., Hoffmann, K.-H.: Theory of convex local compactifications with applications to Lebesgue spaces. *Nonlinear Analysis, Th., Methods, Appl.* (accepted)

[19] Young, L.C.: Generalized curves and the existence of an attained absolute minimum in the calculus of variations. *Comptes Rendus de la Société des Sciences et des Lettres de Varsovie*, Classe III **30** (1937), 212–234.

[20] Young, L.C.: Generalized surfaces in the calculus of variations. *Ann. Math.* **43** (1942), part I: 84–103, part II: 530–544.

Author's address:
Institute of Information Theory and Automation,
Academy of Sciences of the Czech Republic,
Pod vodárenskou věží 4,
CZ-182 08 Praha 8, Czech Republic.

S SAINTLOS, M EL HAFI, A RIGAL AND J MAUSS

Numerical simulation of boundary layer problem with separation

In the context of the laminar steady two - dimensional flow of a newtonian fluid, the asymptotic modelling of the Navier - Stokes equations leads to a boundary layer problem with separation. Indeed, using methods of asymptotic analysis applied in singular perturbation problems and a knowledge of the physical situation, the analysis of the boundary structure in the vicinity of a hump is able to put in evidence multistructures for various flows. Moreover, we establish triple deck and double structures for the laminar steady two - dimensional flow of an incompressible Newtonian fluid for large values of the Reynolds number or of the Rayleigh number. So, the Navier Stokes equations degenerate, so that in the vicinity of the perturbation (which we take as a bump), we obtain a boundary layer problem for the viscous sublayer.

One of the interest of this boundary layer problem is that it is encountered it in many Fluid Mechanics problems. For example, in supersonic flows [1], in hypersonic flows [2], in non - confined flows of an incompressible fluid (accidented plate in Aerodynamics [3,4,5], in natural convection [6,7], in mixed convection [2]), in confined flow of an incompressible fluid (axisymmetric pipes, symmetric and non - symmetric channels [8,9]).

It is so interesting from the mathematical point of view : the equation of motion obtained in the viscous sublayer are " mixed parabolic " when boundary layer separation occurs, induced by the hump of the wall : the presence of a "forward" and "backward" flow with free boundary between the two flows comes from the elliptic nature of the global problem (Navier Stokes equations).

The purpose of this paper is to propose a numerical method to solve the boundary layer problem with separation.

1. MATHEMATICAL MODEL

The perturbation considered in this work is a geometrical indentation. After adimensionnalisation, near the wall $Y^* = F(X^*)$, the boundary problem is formulated with the following transformation :

$$X^* = X \qquad\qquad U^* = U$$
$$Y^* = Y + F(X) \qquad\qquad V^* = V + F'(X)$$

where U^*, V^* are the longitudinal and transverse velocities in the "physical" plane X^*, Y^*. The main interest of this transformation concerns the flow domain which becomes a half plane (because the shape of the wall becomes uniform) : this transformation does not modify the boundary layer equations ; the only modifications concern the boundary and matching conditions.

In the transformed flow field, the mathematical problem, which we may be considered as a universal problem (we will see later in what sense) for boundary layer theory, can be written as:

$$k\left[U\frac{\partial U}{\partial X} + V\frac{\partial U}{\partial Y}\right] = -\frac{dP}{dX} + \frac{\partial^2 U}{\partial Y^2} \qquad (1)$$

$$\frac{\partial U}{\partial X} + \frac{\partial V}{\partial Y} = 0 \qquad (2)$$

with the boundary conditions :

$$U = V = 0 \text{ on } Y = 0$$

and the matching conditions :

$$\lim_{Y \to \infty} U = \lambda\,(Y + F(X) + A(X)) \qquad (3)$$

$$\lim_{X \to -\infty} U = \lambda\, Y \qquad \text{as} \qquad \lim_{X \to -\infty} F(X) = 0$$

and a coupling relationship, depending on the physical problem, between A(X) the displacement thickness of the boundary layer and P(X) the pressure, where k is a normalization factor, connected with the large parameter (Reynolds number or Rayleigh number), which allows this last to be varied, and λ characterizes the basis flow.

The last condition shows that, far upstream of the bump, the flow motion is not modified. P is the pressure and k a normalization factor, connected with the Reynolds number, which allows the Reynolds number to be varied.

For example, we give any coupling relationships, each coupling relationship is connected to a special accident at the wall.

- In confined flows :
  - In a channel

$$A(X) = (G(X) - F(X))/2$$

where F(X) and G(X) are the equations respectively of the lower and upper walls ;

$$A(X) = 0$$

$$P_-(X) = P_+(X) - k\, A''(X) / 30$$

where $P_-(X)$ and $P_+(X)$ are respectively the pressure in the lower and upper sublayer.

  - In an axisymmetric pipe :

$$A(X) = 0$$

- In non - confined flows :

Subsonic flow : $A(X) = 0$; $\qquad P(X) = -\dfrac{1}{\pi} \displaystyle\int \dfrac{A'(\mu)}{X - \mu}\, d\mu$

In natural convection :

$$A(X) = 0 \, ; \qquad P(X) = -\gamma A''(X) \qquad (\gamma \approx 0.115)$$

In mixed convection :

$$P(X) = A(X) \qquad P(X) - \underline{J}\, A(X) = -\frac{1}{\pi} \int \frac{A'(\mu)}{X - \mu} d\mu$$

where J is the Richardson number ($\underline{J} = J / Re^{-1/8}$) and Re the Reynolds number.

Supersonic flows : $P(X) = -A'(X)$

Hypersonic flows : $P(X) = -A(X)$

Whatever the coupling relationship, this problem depends only on A + F ; few mathematical works have been devoted to the general problem : e. g. Audounet and al. [9] proved that, if for $X \leq X_o$, $A + F = 0$, then, in the same domain, the solution of the problem is the upstream solution.

In our case (A(X) = 0), this result allows the beginning of the numerical resolution at the beginning of the bump.

The mathematical problem presents some major difficulties :

- the partial differential equations are non linear
- the flow domain is unbounded
- the pressure gradient is not *a priori* known
- the longitudinal moving equation which is parabolic becomes a backward equation when the velocity sign changes.

For this last reason, the numerical treatment must take into account the downstream condition.

Moreover, if the fow presents a separated region, resolution following the flow direction is ignored when the flow reverses (FLARE approximation [10]), provided that the reverse flow region is limited. But, this approach only yields rather rough results.

For these reasons, we have developped a numerical elliptic model based on the physics of separation [11]. This method is derived from a local relaxation technique developped by Botta and Veldman [12]. We apply this method to this non - linear problem with discretization adapted to the flow direction. We must then give conditions in each boundary of the computational domain : we made so the fundamental assumption of the reattachment of the boundary layer, i.e. far beyound the bump, the condition means that the flow returns to its initial profile :

$$\lim_{X \to \infty} U = \lambda\ Y \quad \text{as} \quad \lim_{X \to \infty} F(X) = 0$$

For convenience (because we choose to keep the pressure in the numerical treatment of the equation of motion, we add a following matching condition on the longitudinal pressure (which is an unknow of the problem, like the velocity, contrary to the classical boundary layer problem):

$$\lim_{Y \to \infty} \left(V + (F'(X) + A'(X))\, U \right) = - \frac{2}{k} \frac{dP}{dX} \qquad (4)$$

This condition, in fact, expresses the compatibility of the equation of motion, when $Y \to \infty$, with the other equations and conditions.

The matching condition (3), which represents the behaviour of U, means that we also have :

$$\lim_{Y \to \infty} \frac{\partial U}{\partial X} = \frac{F'(X) + A'(X)}{2} \qquad (3.a)$$

This condition allows the convergence of the numerical method contrary to the Dirichlet condition (3) in some case.

2. NUMERICAL METHOD

We solve the previous fundamental problem with $A(X) = 0$.

This restriction does not depend on the present problem, because the solution only depends on $A + F$; it simply allows a more complex coupling condition with the outer flow to be avoided, which is not mentioned in this work.

Let hx, hy and i, j be the mesh sizes and index towards X and Y respectively ; we use the centered differences at the point (i,j) for $\frac{\partial U}{\partial Y}$ and $\frac{\partial^2 U}{\partial Y^2}$ in the discretization of the equation of motion.

Numerical tests showed that the centered discretization of the term $\frac{\partial U}{\partial X}$, although it is more precise when there is no boundary layer separation, does not give the convergence of the computing algorithm when separation occurs, by loss of diagonal dominance. We consider the flow direction by using a non - centered approximation of $\frac{\partial U}{\partial X}$ which improves the matrix properties.

The discretization is forward if the direction flow reverses, and vice versa:

$$\left(\frac{\partial U}{\partial X}\right)_{i,j} = \frac{U_{i+1,j} - U_{i,j}}{hx} + o(hx) \qquad \text{if the flow reverses}$$

$$\left(\frac{\partial U}{\partial X}\right)_{i,j} = \frac{U_{i,j} - U_{i-1,j}}{hx} + o(hx) \qquad \text{otherwise}$$

The above approximations applied to (1) yield a non - linear system in the flow domain.This system is iteratively solved, after linearization by taking the non - linear coefficients $U_{i,j}$ and $V_{i,j}$ at the previous iteration.

In the same way, the matching condition (4), written at the previous iteration, allows a first approximation of the pressure gradient for the current iteration.

Consequently, at the current iteration, the discretized equation of motion may be written :

$$k\left[(U_{i,j})^n \left(\left(\frac{\partial U}{\partial X}\right)_{i,j}\right)^{n+1} + (V_{i,j})^n \left(\left(\frac{\partial U}{\partial Y}\right)_{i,j}\right)^{n+1}\right] = - \left(\left(\frac{dP}{dX}\right)_i\right)^n + \left(\left(\frac{\partial^2 U}{\partial Y^2}\right)_{i,j}\right)^{n+1} \quad (5)$$

We deduce the following system of discrete equations :

$$C_W (U_{i-1,j})^{n+1} + C_S (U_{i,j-1})^{n+1} - (U_{i,j})^{n+1} + C_E (U_{i+1,j})^{n+1} + C_N (U_{i,j+1})^{n+1} =$$

$$\frac{hy^2}{(A_{i,j})^n} \left(\frac{dP}{dX}\right)_{i,n} \quad (6)$$

where the coefficients C_W, C_S, C_N, C_E, $(A_{i,j})^n$ are known at the current iteration (n+1).

$$C_W = \frac{khy^2}{2hx(A_{i,j})^n} \left((U_{i,j})^n + |(U_{i,j})^n| \right)$$

$$C_S = \frac{1}{(A_{i,j})^n} \left(1 + \frac{khy}{2} (V_{i,j})^n \right)$$

$$C_E = - \frac{khy^2}{2hx(A_{i,j})^n} \left((U_{i,j})^n - |(U_{i,j})^n| \right) \quad (7)$$

$$C_N = \frac{1}{(A_{i,j})^n} \left(1 - \frac{khy}{2} (V_{i,j})^n \right)$$

$$(A_{i,j})^n = 2 + \frac{khy^2}{hx} |(U_{i,j})^n|$$

In [12], the authors deduce the general form of the eigenvalues of the Jacobi matrix from the discrete homogeneous equation.

These eigenvalues are basic parameters for local relaxation methods,which may be complexe following the signs of C_E, C_W, C_N, C_S :

$$\mu (p1,q1) = 2 (C_E C_W)^{1/2} \cos (p1\pi / N) + 2 (C_N C_S)^{1/2} \cos (q1\pi / M)$$

with $p1 = 1,2, ..., N-1$; $q1= 1,2,..., M-1$

if the domain is described by : $i = 0, \ldots N$ and $j = 0, \ldots M$.

A necessary and sufficient condition of convergence is : $\sup_{(p1,q1)} |\mu_r| < 1$.

In the non - centered case, the coefficients C_E, C_W, C_N, C_S are given by (7) and $C_E C_W = 0$, thus the eigenvalues will be real or purely imaginary :

$$\mu\,(p1,q1) = 2\,(\,C_N C_S)^{1/2} \cos\,(q1\pi / M)$$

To write the eigenvalues of (6), with the coefficients given by (7) , we freeze the system : we consider the associated constant coefficient system by taking the eigenvalue with $q1 = 1$, and we consider the real eigenvalues :

$$\mu_r = \frac{2}{|(A_{i,j})^n|} \left|(\,1 - \frac{k^2 hy^2}{4}\,((V_{i,j})^n)^2\,)\right|^{1/2} \cos\,(\pi / N)$$

We choose as optimum relaxation factor (over - relaxation case) :

$$\omega_{opt} = \frac{2}{(1 + (1-\mu_r)^2)^{1/2}} \geq 1$$

Thus, the longitudinal velocity at the iteration (n+1) is given by :

$$(U_{i,j})^{n+1} = (\,(\underline{U}_{i,j})^{n+1} - (U_{i,j})^n\,)\,\omega_{opt} + (U_{i,j})^n$$

where $(\underline{U}_{i,j})^{n+1}$ is given by (6).

The convergence test of the iterative process is made on the upper bound, in the whole domain Ω, of the absolute value of the difference between two successive iterates of the longitudinal velocity ; the value of the stopping test is 10^{-4} :

$$\sup_{\Omega} |(U_{i,j})^{n+1} - (U_{i,j})^n| < 10^{-4}$$

The iterative process starts with the basic solution not disturbed at the interior of the flow domain Ω :

$$(U_{i,j})^n = \frac{1}{2}(\,(j-1)hy + F((i-1)hx)\,) \qquad (V_{i,j})^n = 0 \qquad (\frac{dP}{dX})_{i,n} = 0$$

We must state the conditions on the boundaries of the domain $\partial\Omega$:
At the wall (lower edge), and at the upstream boundary, we previously wrote the Dirichlet condition for U. However, on the upper boundary, the Dirichlet condition does not allow convergence of the iterative process, so we prescribe the Neumann condition (3.a). Furthermore, a Dirichlet or Neumann's condition on U at the downstream boundary, leads to a numerical instability near the boundary, due to the computation of the continuity equation. This instability disappears when the downstream boundary is left free, i.e. :
- the downstream boundary longitudinal velocity U is computed by linear interpolation from the two previous steps,
- the continuity equation is not computed on the last sections of the flow, the only regularity of the transverse velocity is needed (a non - disturbed return to the initial state).
After convergence of the algorithme, the transverse velocity, at each point of the flow domain, is computed by central discretization of the continuity equation in (i,j-1).
Then, the new pressure gradient is computed by the matching condition on the transverse velocity.

3. NUMERICAL RESULTS

The numerical results, obtained with the present method, are validated in the case of the channel $\lambda = 0.5$ when $A(X) = 0$ with other numerical methods :
- a method developped by F.T Smith which is an extension of Keller and Cebeci [13] method ,
- a finite element method by discretization adapted to the flow direction [14],
- an integral method where a polynomial approximation [15] is used for the longitudinal velocity .

The normalization factor k was choosen equal to unity.
We considered the wall bumps similar to those taken by F.T. Smith in [8] .

$$F_S(X) = h\ X \exp\left(-\frac{X^2}{32}\right) \text{ for } X \geq 0$$

for h = -1 and h = -1.5, which corresponds to expansion.
Flow in a Channel ($\lambda = 0.5$) :

A(X) = 0

We present results on the wall shear stress for the expansion corresponding to h = -1 and h = -1.5 respectively on figures 1 and 2.
We note that for h=-1, the local relaxation method gives good results without boundary layer separation eventhough it was devised to calculate separations.
For h=-1.5, the separation point agrees with F.T. Smith method. However, we note that the amplitude of the peak of maximum negative wall shear stress is more little with parabolic methods than LR method.
The difference with F.T Smith may be due to the numerical viscosity introduced by the FLARE approximation in the parabolic scheme in order to cross the separation point.
The difference with the integral method is certainly due to this of the Reynolds number imposed.

A(X) = (G(X) - F(X))/2 non symmetrical channel

We consider the shapes of the lower and upper walls respectively on the form: F(X)=0 and G(X)= $2F_S(X)$.
The study of the flow near the wall F(X)=0 is identical to the previously problem. The wall shear stress near the lower wall is given on the figures 1 and 2.
Then, for this same bump on the lower wall, the case k = 1 and h = -0.5 does not give separation near the upper plane wall (figure 3).
But we can obtain separation increasing the factor k (k = 14 on figure 4) near the upper uniform wall.

The following results cannot be compare because the structure with A(X) = 0 has not been found until now in the litterature. Indeed the particular case A(X) = 0 corresponds to a particular physical accident.

Natural convection :

A(X) = 0 and λ = 0.6

The last justication of this case is of asymptotic order, given with the mixed convection when J tends to the infinity : we tends to the multistructure in natural convection with A(X) = 0.

The wall shear stress and pressure in the viscous sublayer are presented in figures 5 and 6 respectively.

4. CONCLUSIONS

The results obtained are in good agreement with others.

The elliptic approach developped in this paper, allows significant boundary layer separations to be obtained, without enforcing the crossing of the separation point, contrary to methods which use the FLARE approximation. The method is robust insofar as a large domain can be solved (50 000 points), and fast (70 seconds of computing time for 5000 points) when we search for a first approximation of the separated region. The extension of the validity of the method requires a detailed analysis of several numerical problems, which is out of the scope of the present work. In particular, a coupled utilization of a parabolic approach and of this LR method should save computing time, and consequently enable larger domains to be processed and greatest separations to be obtained. Perspectives of this work is to adapt the LR method with others coupling relationships : in particular in the case of the natural convection where the coupling relation ship P(X) = - γ A"(X) and in the case of an accidented plate with A(X) given by the Hilbert integral.

Wall shear stress

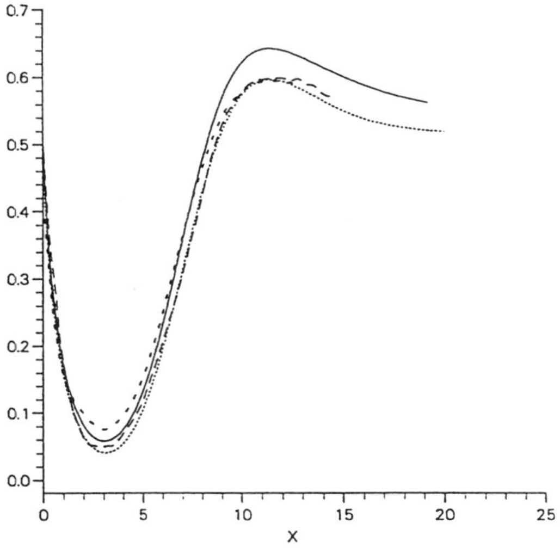

Figure 1 : Wall shear stress for the dilated channel of F.T. Smith with h = -1 and k = 1.

Wall shear stress

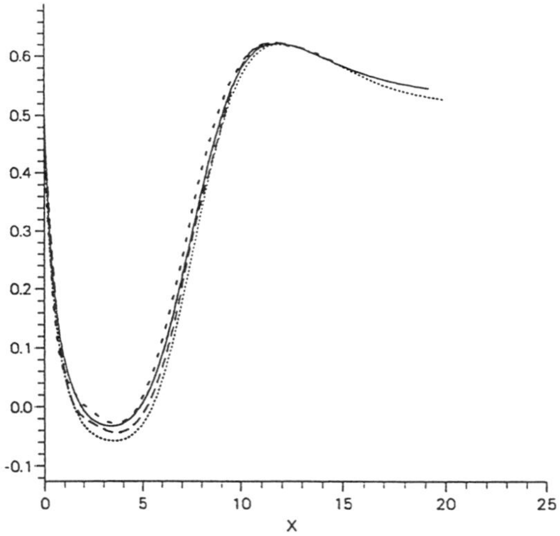

Figure 2 : Wall shear stress for the dilated channel of F.T. Smith with h = -1.5 and k =1.

Legend : - - - - - - - Global Method

_________ Finite Element Method

.... Presented Method

__ __ __ F.T. Smith Method

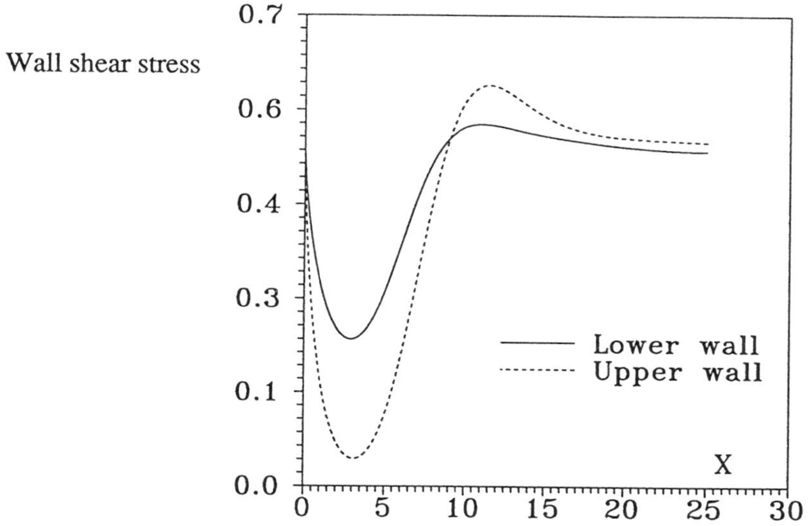

Figure 3 : Wall shear stress for the non symmetric channel with the presented method for h = -0.5 and k = 1.

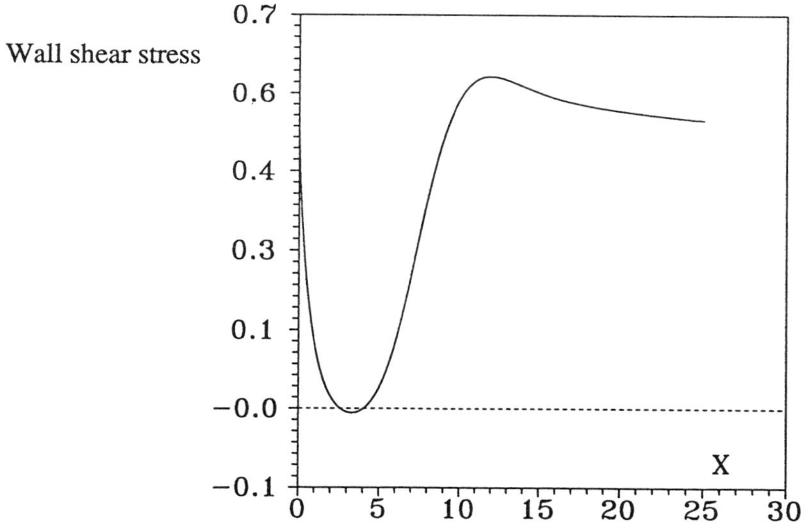

Figure 4 : Wall shear stress for the non symmetric channel on the lower uniform wall with the presented method for k = 14.

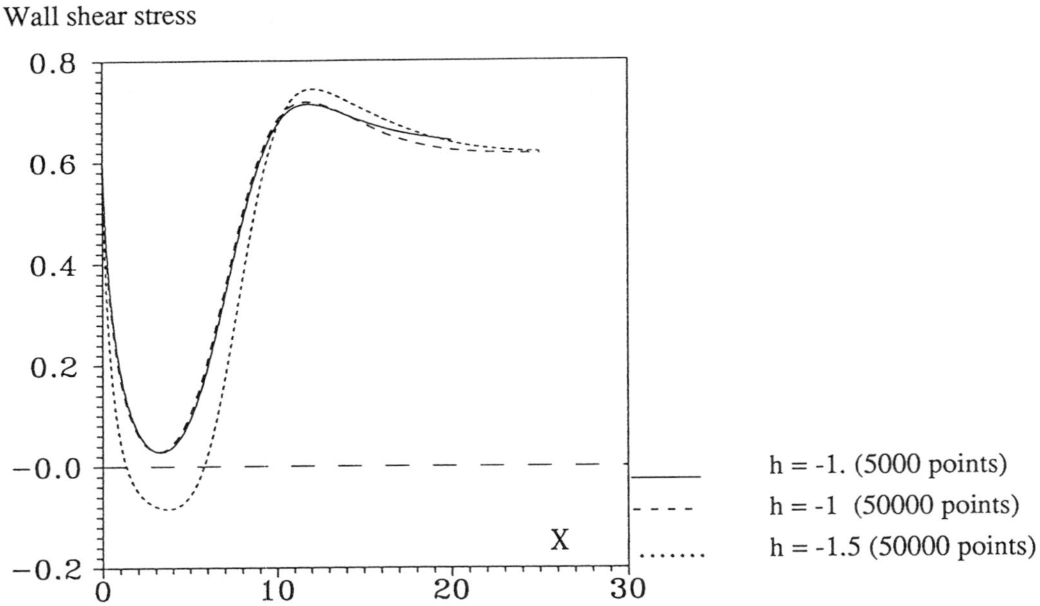

Figure 5 : Natural convection - Wall shear stress for the Smith's bump for k = 1.

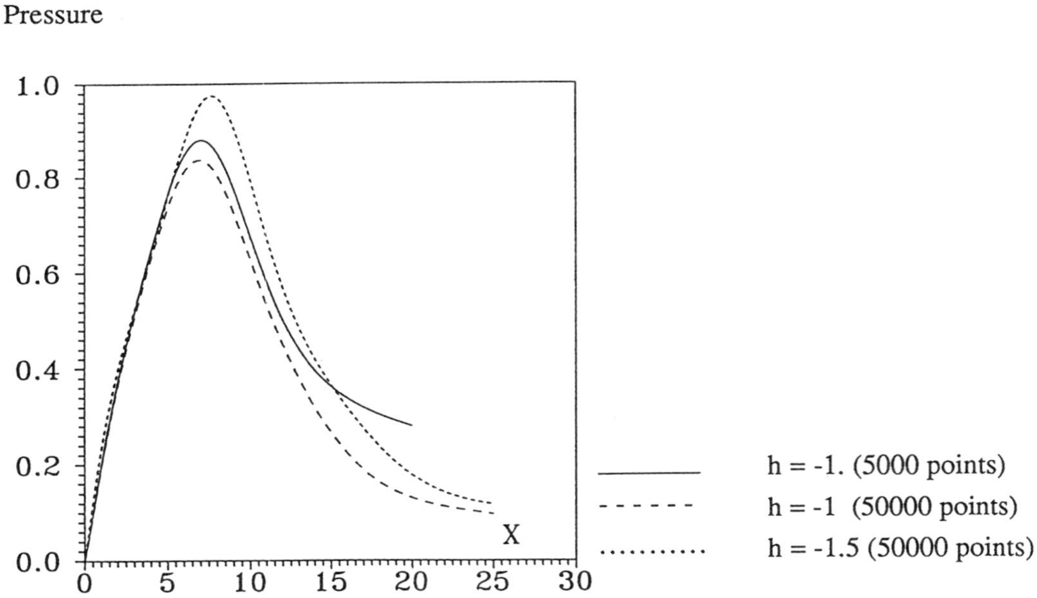

Figure 6 : Natural Convection - Pressure for the Smith's bump for k = 1

REFERENCES

[1] K. STEWARTSON, P.G. WILLIAMS, (1969) Self induced separation, Proc. Roy. Soc. London, A312, p 181 - 206.

[2] P.Y. LAGREE, (1994) Convection mixte à faible nombre de Richardson dans le cadre de la triple couche, C.R. Acad. Sci. Paris, t. 318, Série II, to appear.

[3] F.T. SMITH, (1982) On the high Reynolds number theory of laminar flows. IMA J. Appl. Math. vol 28, p 207 - 281.

[4] J. MAUSS, A. ACHIQ, S. SAINTLOS, (1992) Sur l'analyse conduisant à la théorie de la triple couche, C. R. Acad. Sci. Paris, t. 315, Série II, p 1611 - 1614.

[5] J. MAUSS, (1994) Asymptotic modelling for separating boundary layers, to appear in Springer Verlag - Colloque de Modélisation Asymptotique en Mécanique des Fluides.

[6] EL HAFI & AL., Double deck structure analysis on natural convection, submitted to Mechanics Research Communications ;

[7] S. GHOSH MOULIC, L.S. YAO, (1992) Natural convection near a small protusion on a vertical plate, Int. J. Heat Transfer, vol. 35, No. 11, pp. 2931 - 2940.

[8] F.T. SMITH, Flow through constricted tubes or dilated pipes and channels, Quaterly J. Mech. Appl. Math., vol 29, Pt 3, p 343 - 376 (1976).

[9] J. AUDOUNET, J. MAUSS, S. SAINTLOS, (1991) Sur la théorie des écoulements laminaires à grands nombres de Reynolds, Matapli. SMAI. Bulletin de liaison n° 28. Octobre 1991, p 17 - 25.

[10] T.A. REYHNER, I. FLÜGGE - LOTZ, (1968) The interaction of a shock wave with a laminar boundary layer, Int. J. Non - linear Mechanics, vol 13, p 173 - 199.

[11] S. SAINTLOS, J. MAUSS, A. RIGAL, (1994) Local Relaxation method for boundary layer equations with separation, Int. J. Engng. Sci. Vol. 32, No. 3, pp 409 - 416.

[12] E.F.F. BOTTA, A.E.P. VELDMAN, (1981) On the local relaxation methods and their applications to Convection - Diffusion Equations, J. of Computational Physics, vol 48, p 127 - 149.

[13] H.B. KELLER, T. CEBECI, (1971) Accurate Numerical Methods for boundary layer flows, I. Two dimensionnal laminar flows, Lecture notes in Physics, vol 8, p 92 - 100.

[14] L. PLANTIE, J. AUDOUNET, (1993) Congrès d'Analyse Numérique.

[15] A. NAJÈME, M. ZAGZOULE, J. MAUSS, (1992) Numerical analysis of flow in arterial stenoses, Mechanics Research communications , vol 19, n° 5, p 379 - 384.

<u>Adresses :</u>

S. SAINTLOS : Université Louis Pasteur - Institut de Mécanique des Fluides - URA CNRS 854 - 2, rue Boussingault - F - 67000 Strasbourg - France.

M. EL HAFI : Université Paul Sabatier - L.E.S.E.T.H. - 118, route de Narbonne - 31062 Toulouse cedex.

A. RIGAL : Université Paul Sabatier - Laboratoire d'Analyse Numérique - 118, route de Narbonne - 31062 Toulouse cedex.

J. MAUSS : Université Paul Sabatier - Laboratoire de Modélisation en Mécanique des Fluides - 118, route de Narbonne - 31062 Toulouse cedex.

C SBORDONE

Maximal inequalities and applications to regularity problems in the calculus of variations

1. Introduction

Let Ω be a bounded open domain of $\mathbb{R}^n$, $a(x)$ a bounded measurable function such that $0 < \lambda_0 \leq a(x) \leq \Lambda_0$ and $F = F(t)$ a convex increasing function

$$F : [0, \infty) \to [0, \infty)$$

satisfying the condition

$$pF(t) \leq tF'(t) \leq qF(t) \qquad \forall t \geq 0 \tag{1.1}$$

where $1 < p \leq q < \infty$.

In the following we are interested in the study of *very weak solutions* of the variational equation

$$\operatorname{div}\left(a(x)F'(|\nabla u|)\frac{\nabla u}{|\nabla u|}\right) = 0 \quad . \tag{1.2}$$

Equation (1.2) is the Euler equation of the functional

$$I(v) = \int_\Omega a(x)F(|\nabla v|)dx$$

defined (and eventually infinite) for functions v belonging to the Sobolev space $W^{1,1}(\Omega)$.

The definition of *very weak solution* is best visualized in the particular case $p = q$ (see [15]) where the equation (1.2) reduces to

$$\operatorname{div}\left(a(x)|\nabla u|^{p-2}\nabla u\right) = 0 \tag{1.3}$$

i.e. to the identity

$$\int_\Omega a(x)|\nabla u|^{p-2}\nabla u \nabla \phi = 0 \tag{1.4}$$

for any $\phi \in W_0^{1,\infty}(\Omega)$. In order to give meaning to the integral in (1.4), the assumption $u \in W_{\text{loc}}^{1,p}$ is not necessary. Actually, it will be sufficient to assume

$$u \in W_{\text{loc}}^{1,r}, \qquad \max\{1, p-1\} \leq r < p. \tag{1.5}$$

2nd European Conference on Elliptic and Parabolic Problems, Pont á Mousson, June 13-17, 1994

See [15] for a discussion of this new definition in the case of Euler equation of functionals

$$I(v) = \int_\Omega f(x, Dv)dx, \qquad v \in W^{1,p}(\Omega, \mathbb{R}^N)$$

with

$$\lambda_0|\xi|^p \leq f(x,\xi) \leq \Lambda_0(1+|\xi|^p)$$

Definition 1.1 *A function u verifying (1.5) is called a very weak solution of equation (1.3) if (1.4) holds for any $\phi \in W^{1,\infty}(\Omega)$ with compact support.*

It is immediately seen that, if $u \in W^{1,r}$ then $a(x)|\nabla u|^{p-2}\nabla u$ belongs to $L^{\frac{r}{p-1}}$; hence identity (1.4) remains valid only for test functions ϕ such that $\nabla\phi$ belongs to the dual space $L^{\frac{r}{r-p+1}}$.

Notice that $\dfrac{r}{r-p+1} \geq p$ and

$$W^{1,\frac{r}{r-p+1}} \subset W^{1,p},$$

hence it will not be possible to test identity (1.4) with functions $\phi = \eta u$, $\eta \in C_0^1(\Omega)$ as in the classical case $r = p$.

In [15] (see also [8], [20], [6], [29]) various existence and regularity results for very weak solutions of equation (1.3) are proved. In particular, the following theorem holds

Theorem 1.1 *There exist exponents*

$$1 < r_1 < p < r_2 < \infty$$

$r_i = r_i(n, p, \lambda_0, \Lambda_0)$ such that every very weak solution $u \in W^{1,r_1}_{\mathrm{loc}}(\Omega)$ of equation (1.3) actually belongs to $W^{1,r_2}_{\mathrm{loc}}(\Omega)$.

In the following we want to generalize the notion of a very weak solution to the case of equation (1.2), under the assumption (1.1).

Let us mention some papers which studied higher integrability results for weak solutions of equation (1.2) under non standard growth conditions of the type (1.1) [22], [7], [28].

Namely consider the identity

$$\int_\Omega a(x)F'(|\nabla u|)\frac{\nabla u}{|\nabla u|}\nabla\phi = 0 \tag{1.6}$$

for $\phi \in W^{1,\infty}(\Omega)$ with compact support. The integral in (1.6) has a meaning if $F'(|\nabla u|) \in L^1_{\mathrm{loc}}(\Omega)$, or equivalently, if

$$\frac{F(|\nabla u|)}{|\nabla u|} \in L^1_{\mathrm{loc}}(\Omega)$$

thanks to the assumption (1.1). Following the analogy with the previous case $F(t) = t^p$, we will introduce for $\max\{1, p-1\} \leq r < p$ the functions

$$F_r(t) = F(t)t^{r-p}$$

that verify, for $t \geq 0$,

$$rF_r(t) \leq tF_r'(t) \leq (q-p+r)F_r(t) \quad .$$

We can give now the following

Definiton 1.2 *A function $u \in W^{1,1}_{\mathrm{loc}}(\Omega)$ verifying*

$$F_r(|\nabla u|) \in L^1_{\mathrm{loc}}(\Omega)$$

is called a very weak solution of equation (1.2) if identity (1.6) holds for any $\phi \in W^{1,\infty}(\Omega)$ with compact support.

To simplify the presentation, we will introduce the Orlicz spaces $L_{F_r}(\Omega)$ generated by the convex function $F_r(t)$ (see section 2).

It is immediately seen that if $u \in W^{1,1}_{\mathrm{loc}}(\Omega)$ is a very weak solution such that $|\nabla u| \in L_{F_r}(\Omega)$, then $F'(|\nabla u|)$ belongs to the Orlicz space $L_{G_r}(\Omega)$ where $G_r(t)$ is defined by

$$G_r\left(\frac{F(t)}{t}\right) = F_r(t) \qquad \forall t \geq 0.$$

As a consequence, the gradient $\nabla\phi$ of test functions will vary in the dual space $L_{\tilde{G}_r}(\Omega)$. In section 3 we will identify $\tilde{G}_r$ in some special cases.

In section 5 we will prove the following *uniqueness result* for very weak solutions with compact support

Theorem 1.2 *There exists $1 < r < p$, $r = r(p,n,q)$, such that if $u \in W^{1,1}_{\mathrm{loc}}(\mathbb{R}^n)$ is a very weak solution of*

$$\int_{\mathbb{R}^n} a(x)F'(|\nabla u|)\frac{\nabla u}{|\nabla u|}\nabla\phi = 0 \qquad \forall\phi \in W^{1,\infty}$$

with supp $u \subset\subset \mathbb{R}^n$, $|\nabla u| \in L_{F_r}(\mathbb{R}^n)$ *then* $u \equiv 0$.

While the proof of theorem 1.1 relies on the stability of Hodge decomposition due to T.Iwaniec ([12], [14]), the method we follow here is the same introduced by J.Lewis in [20].

This method relies on some weighted norm inequalities for the maximal function in Orlicz spaces (see section 4).

The weights we deal with are of the form $[M|\nabla u|]^{-\epsilon}$, and ,as was observed also in [6], they belong to certain Muckenhoupt classes. Nevertherless, we prefer to avoid general theorems on weighted maximal inequalities [26], [17] and give direct proofs.

For related results see Dolcini [6], Giachetti-Leonetti-Schianchi [8], Stroffolini [30], Li-McIntosh-Zhang [21], Iwaniec-Scott-Stroffolini [16], Sbordone [29]. In the forthcoming paper by Greco-Iwaniec-Milman [10] a different approach is considered which relies again on the Hodge decomposition of a very general type.

The Hodge decomposition together with maximal inequality revealed a powerful tool in some recent papers on the regularity of Jacobian [14], [2], [13], [9], [11], [23].

2. Notations and preliminary results

In the following we shall denote by Ω an open set of $\mathbb{R}^n$, by Q an open cube in $\mathbb{R}^n$ with sides parallel to the axes and by $Mf(x)$ the cubic maximal function of $f \in L^1_{\text{loc}}(\mathbb{R}^n)$

$$Mf(x) = \sup_{Q \ni x} \fint_Q |f| \tag{2.1}$$

where the supremum is taken over all cubes Q in $\mathbb{R}^n$ containing x and $\fint_Q$ stands for $\frac{1}{|Q|}\int_Q$.

It is well known that M is bounded in $L^p(\mathbb{R}^n)$, when $p > 1$. In 1972 B.Muckenhoupt characterized the weights $w : \mathbb{R}^n \to [0,\infty)$ such that M is bounded in $L^p(\mathbb{R}^n, wdx)$, $p > 1$ as those verifying the A_p – *condition*, that is

$$A_p(w) = \sup_Q \fint_Q w \left(\fint_Q w^{\frac{-1}{p-1}}\right)^{p-1} < \infty \tag{2.2}$$

A crucial property of a weight w in the A_p class is that there exists $\epsilon = \epsilon(n, p, A_p(w))$ such that $w \in A_{p-\epsilon}$ ([26]).

Recent papers deal with the exact expression of ϵ in the case $n = 1$ ([19], [18], [5]).

In the case $p = 1$, (2.2) is replaced by

$$A_1(w) = \sup_Q \fint_Q w \left(\operatorname{ess\,inf}_Q w \right)^{-1} < \infty. \tag{2.3}$$

For $p \geq 1$, the maximal operator is of weak type in $L^p(\mathbb{R}^n, wdx)$ iff $w \in A_p$. Namely, setting $w(E) = \int_E wdx$, the following inequality holds for $\alpha > 0$:

$$w(\{Mf > \alpha\}) \leq c(n)\frac{A_p(w)}{\alpha^p} \int_{\{|f| \geq \frac{\alpha}{2}\}} |f|^p wdx \tag{2.4}$$

(see [3], [26]).

In the following we consider a convex function

$$F \quad : \quad [0, \infty) \to [0, \infty)$$

verifying

$$pF(t) \leq tF'(t) \leq qF(t) \qquad \forall t \geq 0 \tag{2.5}$$

for $1 < p \leq q < \infty$. It is easy to check that the second inequality in (2.5) is equivalent to say that there exists $K > 1$ such that

$$F(2t) \leq KF(t) \qquad \forall t \geq 0 \tag{2.6}$$

the so called Δ_2- condition on F. Moreover, the first inequality in (2.5) is equivalent to the condition

$$t \to \frac{F(t)}{t^p} \qquad \text{increasing.} \tag{2.7}$$

Let us note (see [7]) that (2.5) implies for some constants c_1, c_2, c_3

$$c_1 t^p - c_2 \leq F(t) \leq c_3(t^q + 1) \tag{2.8}$$

but it may be that the exponents p, q appearing in (2.5) are not necessarily the best ones in order (2.8) to hold. For example the convex function

$$F(t) = \begin{cases} et^3 & \text{if } 0 \leq t \leq e \\ t^{4+\sin\log\log t} & \text{if } t \geq e \end{cases}$$

verifies (2.5) with $p = 4 - \sqrt{2}$, $q = 4 + \sqrt{2}$ and (2.8) with $p = 3$, $q = 5$. Moreover if $4 - \sqrt{2} < r < 4 + \sqrt{2}$, then $\frac{F(t)}{t^r}$ does not increase nor does it decrease.

Let us recall that condition (2.5) means that the convex function F and its conjugate $\tilde{F}$

$$\tilde{F}(s) = \sup_{t \geq 0}\{st - F(t)\}$$

verify the Δ_2 condition or that both the Orlicz space L_F and its dual $L_{\tilde{F}}$ are reflexive Banach spaces.

We note as in [11] that on L_F and $L_{\tilde{F}}$ one can introduce a very natural norm, equivalent to the classical Orlicz and Luxemburg norms which behaves very well with respect to Hölder inequality. Namely, if we set

$$F(t) = \int_0^t a^{-1}(\tau)\frac{d\tau}{\tau}$$

$$G(s) = \int_0^s b^{-1}(\sigma)\frac{d\sigma}{\sigma}$$

with $a(x)b(x) = x$ for $x \geq 0$, then $G = \tilde{F}$, and , if we define

$$|||u|||_{L_F} = \inf\left\{\lambda > 0 : \quad \fint_\Omega F\left(\frac{|u|}{\lambda}\right) \leq F(a(1))\right\}$$

$$|||v|||_{L_{\tilde{F}}} = \inf\left\{\mu > 0 : \quad \fint_\Omega \tilde{F}\left(\frac{|v|}{\mu}\right) \leq \tilde{F}(b(1))\right\}$$

then

$$\left|\fint_\Omega uv\right| \leq |||u|||_{L_F} \cdot |||v|||_{L_{\tilde{F}}}$$

Moreover if $F(t) = \dfrac{t^p}{p}$, then we have the precise equality

$$|||u|||_{L_F} = \left(\fint_\Omega |u|^p\right)^{\frac{1}{p}} \quad .$$

In the following we will study the Zygmund spaces $L^p \log^\alpha L$, $p > 1$, $\alpha \geq 0$. These are special Orlicz spaces generated by the functions

$$F(t) = t^p \log^\alpha(e + t) \quad .$$

It is easy to check that the dual space of $L^p \log^\alpha L$ is the space $L^{p'} \log^{\frac{-\alpha p'}{p}} L$ where $p'. = \frac{p}{p-1}$. Finally we recall the useful inequality

$$\tilde{F}\left(\frac{F(t)}{t}\right) \leq F(t).$$

3. Motivation of the definition of very weak solution

For any convex function $F : [0, \infty) \to [0, \infty)$ satisfying the condition

$$pF(t) \leq tF'(t) \leq qF(t) \tag{3.1}$$

with $1 < p \leq q < \infty$ let us consider the family of convex functions

$$F_r(t) = F(t)t^{r-p} \tag{3.2}$$

as $\max\{1, p-1\} < r < p$. It is easy to verify that

$$rF_r(t) \leq tF_r'(t) \leq (q - p + r)F_r(t) \tag{3.3}$$

In view of the applications to the calculus of variations, we are interested in the following:

Problem *Given the form*

$$a(g, h) = \int_\Omega \frac{F(|g|)}{|g|^2} ghdx,$$

assuming that g varies in the space L_{F_r}, determine the set

$$\mathcal{H}_r = \{h \quad : \quad a(g, h) < \infty \quad \text{for} \quad \text{any} \quad g \in L_{F_r}\}$$

Example 3.1 If $r = p$ then $F_r(t) = F(t)$, and denoting with $\tilde{F}$ the conjugate of F:

$$\tilde{F}(s) = \sup_t \{st - F(t)\},$$

we have by Young inequality and (2.9)

$$\frac{F(|g|)}{|g|}|h| \leq \tilde{F}\left(\frac{F(|g|)}{|g|}\right) + F(|h|)$$

$$\leq F(|g|) + F(|h|).$$

In this special case $a(g, h) < \infty$ for $g \in L_{F_r} = L_F$ and $h \in L_F$. So we conclude

$$\mathcal{H}_r = \mathcal{H}_p = L_F$$

Example 3.2 Let $F(t) = t^p$, $\max\{1, p-1\} < r$, then $F_r(t) = t^r$. If h, $g \geq 0$, $g \in L^r$, by Young's inequality

$$a(g,h) = \int_\Omega g^{p-1} h \leq \int_\Omega F_r(g) + \int_\Omega H_r(h)$$

$$\leq \int_\Omega (g^{p-1})^{\frac{r}{p-1}} + \int_\Omega h^{\frac{r}{r-p+1}}$$

In the present case we have

$$\mathcal{H}_r = L^{\frac{r}{r-p+1}}(\Omega)$$

After previous examples it is clear that we can define the function G_r by

$$G_r\left(\frac{F(t)}{t}\right) = F_r(t) \tag{3.4}$$

this will imply, for $g \geq 0$

$$\int_\Omega G_r\left(\frac{F(g)}{g}\right) < \infty$$

if and only if

$$\int_\Omega F_r(g) < \infty$$

and then

$$\mathcal{H}_r = L_{\tilde{G}_r}$$

where $\tilde{G}_r$ is the conjugate of G_r. Let us now compute $F_r(t)$, $G_r(s)$, $\tilde{G}_r(t)$ in a particular case

Example 3.3 Let $p > 1$, $\alpha \geq 0$ and

$$F(t) = t^p \log^\alpha(e+t)$$

It is easy to check that

$$pF(t) \leq tF'(t) \leq (p+\alpha)F(t). \tag{3.8}$$

We wish to prove that if $g \geq 0$

$$\left(g \in L_{F_r} = L^r \log^\alpha L\right) \Rightarrow \left(F'(g) \in L_{G_r} = L^{\frac{r}{p-1}} \log^{-\alpha\frac{r-p+1}{p-1}} L\right) \tag{3.9}$$

Let us first show that there exist constants $0 < c_1 < c_2$ such that for any $t \geq 0$

$$c_1 \log(e+t) \leq \log(e + pt^{p-1} \log^\alpha(e+t)) \leq c_2 \log(e+t) \tag{3.10}$$

Inequalities (3.10) follow elementarly by setting

$$\Phi(t) = \frac{\log(e + pt^{p-1}\log^\alpha(e+t))}{\log(e+t)}$$

and checking that

$$\lim_{t \to 0} \Phi(t) = 1$$

$$\lim_{t \to \infty} \Phi(t) = p - 1$$

Now, let us assume $g \in L_{F_r} = L^r \log^\alpha L$, $g \geq 0$ and let us estimate using (3.8)

$$\begin{aligned} &\int_\Omega F'(g)^{\frac{r}{p-1}} \log^{-\alpha\frac{r-p+1}{p-1}}(e + F'(g)) \leq \\ &\leq (p+\alpha)^{\frac{r}{p-1}} \int_\Omega g^r \log^{\alpha\frac{r}{p-1}}(e+g) \log^{-\alpha\frac{r-p+1}{p-1}}(e+F'(g)) \end{aligned} \tag{3.11}$$

By (3.8), (3.10) we deduce

$$\begin{aligned} \log(e + F'(g)) &\geq \log\left(e + p\frac{F(g)}{g}\right) \\ &= \log(e + pg^{p-1}\log^\alpha(e+g)) \geq \\ &\geq c_1 \log(e+g), \end{aligned}$$

hence

$$\log^{-\alpha\frac{r-p+1}{p-1}}(e + F'(g)) \leq c_1^{-\alpha\frac{r-p+1}{p-1}} \log^{-\alpha\frac{r-p+1}{p-1}}(e+g) \tag{3.12}$$

By (3.11) and (3.12) we arrive at the inequality

$$\int_\Omega F'(g)^{\frac{r}{p-1}} \log^{-\alpha\frac{r-p+1}{p-1}}(e+F'(g)) \leq c(r,p,\alpha) \int_\Omega g^r \log^\alpha(e+g)$$

In order to determine $\tilde{G}_r(t)$, remember that the dual space of

$$L^p \log^\beta L$$

is the space

$$L^{q'} \log^{-\beta \frac{q'}{q}} L \qquad \left(q' = \frac{q}{q-1}\right)$$

so we have

$$\mathcal{H}_r = L_{\tilde{G}_r} = L^{\frac{r}{r-p+1}} \log^{\alpha} L.$$

Remark that if $\alpha = 0$ we recover the case of example 2.

4. Some maximal inequalities

Let Mf denote the Hardy Littlewood maximal function of f, for $f \in L^1_{\mathrm{loc}}(\mathbb{R}^n)$:

$$Mf(x) = \sup_{Q \ni x} \fint_Q |f| \tag{4.1}$$

In this section we will study particular A_p weights of the type

$$w = [Mg]^{-\epsilon} \qquad g \in L^1(\mathbb{R}^n), \quad \epsilon > 0$$

and it will be very useful to know that for ϵ small enough the A_p constants of these weights are independent on ϵ, [6]. For the sake of completeness we prove the following

Lemma 4.1 *Let $p > 1$. Then there exists $c = c(n,p)$ such that*

$$A_{\frac{p+1}{2}}\left([Mg]^{-\epsilon}\right) \leq c(n,p) \tag{4.2}$$

for $0 < \epsilon < \frac{p-1}{4}$, $g \in L^1(\mathbb{R}^n)$.

PROOF. As in [6], we use Coifman-Rochberg theorem [4] which implies that there exists a constant $c = c(n)$ such that for $0 < \sigma < \frac{1}{2}$

$$A_1\left([Mg]^{\sigma}\right) \leq c(n) \qquad \forall g \in L^1(\mathbb{R}^n)$$

Let $q = \frac{p+1}{2}$ and fix $0 < \epsilon < \frac{p-1}{4}$, i.e. $0 < \epsilon < \frac{q-1}{2}$. If $q' = \frac{q}{q-1}$, we have

$$A_q\left([Mg]^{-\epsilon}\right) = \left\{A_{q'}\left([Mg]^{\frac{\epsilon}{q-1}}\right)\right\}^{q-1} \leq \left\{A_1\left([Mg]^{\frac{\epsilon}{q-1}}\right)\right\}^{q-1} \leq c(n)^{q-1}$$

since $\dfrac{\epsilon}{q-1} < \dfrac{1}{2}$.

We are now ready to prove the following:

Theorem 4.1 *Let F verify (1.1). Then there exists $m = m(n,p,q)$ such that*

$$\int_{\mathbb{R}^n} F(Mg)[Mg]^{-\epsilon}dx \le m \int_{\mathbb{R}^n} F(|g|)[Mg]^{-\epsilon}dx \tag{4.3}$$

for $0 < \epsilon < \frac{p-1}{4}$ *and* $g \in L^1(\mathbb{R}^n)$.

PROOF. Set $w = [Mg]^{-\epsilon}$ and use Lemma 4.1 with $p_0 = \frac{p+1}{2}$ to deduce

$$A_{p_0}(w) \le c(n,p) \tag{4.4}$$

By a standard argument, using (2.4) with $p = p_0$, we have

$$\int_{\mathbb{R}^n} F(Mg)wdx = \int_0^\infty F'(s)w(\{Mg > s\})ds$$

$$\le c(n)A_{p_0}(w)\int_0^\infty \frac{F'(s)}{s^{p_0}}\int_{|g|>\frac{s}{2}} |g|^{p_0}wdxds \tag{4.5}$$

$$= c(n)A_{p_0}(w)\int_{\mathbb{R}^n} |g|^{p_0}w\int_0^{2|g|}\frac{F'(s)}{s^{p_0}}dsdx$$

Let us now estimate the inner integral as follows

$$\int_0^{2|g(x)|}\frac{F'(s)}{s^{p_0}}ds \le q\int_0^{2|g(x)|}\frac{F(s)}{s^{p_0+1}}ds = q\int_0^2 \frac{F(t|g(x)|)}{t^{p_0+1}|g(x)|^{p_0}}dt$$

$$= \frac{q}{|g(x)|^{p_0}}\left\{\int_0^1 \frac{F(t|g(x)|)}{t^{p_0+1}}dt + \int_1^2 \frac{F(t|g(x)|)}{t^{p_0+1}}dt\right\}$$

where we used (2.5). Now using that $\frac{F(t)}{t^p}$ is an increasing function, we get

$$\int_0^{2|g(x)|}\frac{F'(s)}{s^{p_0}}ds \le \frac{q}{|g(x)|^{p_0}}\left\{F(|g(x)|)\int_0^1 t^{p-p_0-1}dt + F(2|g(x)|)\int_1^2 t^{-p_0-1}dt\right\} \tag{4.6}$$

From (4.4), (4.5), (4.6), using the Δ_2-condition $F(2t) \le KF(t)$, we get

$$\int_{\mathbb{R}^n} F(Mg)w \le c(n,p,q)\int_{\mathbb{R}^n} F(|g(x)|)w.$$

5. Proof of the main theorem

Let us begin with an extension lemma (see [1], [20], [6])

Lemma 5.1 *Let $u \in W^{1,q}(\mathbb{R}^n)$, $1 < q < \infty$. For any $\lambda > 0$ let*

$$E_\lambda = \{x : \quad M(|\nabla u|)(x) \leq \lambda\} \tag{5.1}$$

then there exists a Lipschitz function $v_\lambda \in Lip(\mathbb{R}^n)$ such that

$$\text{(i)} \qquad v_\lambda = u \qquad \text{on} \quad E_\lambda$$

$$\text{(ii)} \qquad ||\nabla v_\lambda||_\infty \leq c(n)\lambda.$$

We can now pass to the

PROOF (of theo.1.2) Since supp$u \subset\subset \mathbb{R}^n$, for $\lambda > 0$, one can check that the function v_λ defined in Lemma 5.1 has compact support and hence it can be chosen as a test ϕ in the identity

$$\int_{\mathbb{R}^n} F'(|\nabla u|)\frac{\nabla u}{|\nabla u|}\nabla\phi = 0. \tag{5.2}$$

Splitting the integral and using (i), (ii), we obtain

$$\int_{\{M|\nabla u| \leq \lambda\}} F'(|\nabla u|)|\nabla u| \leq c(n)\lambda \int_{\{M|\nabla u| > \lambda\}} F'(|\nabla u|) \tag{5.3}$$

As in [20], fix $\epsilon > 0$, multiply both sides of (5.3) by $\lambda^{-(1+\epsilon)}$, integrate between 0 and $+\infty$, and recall that $F(t) \leq tF'(t)$. So, we can write

$$\int_0^\infty \lambda^{-(1+\epsilon)} \int_{\{M|\nabla u| \leq \lambda\}} F(|\nabla u|)dx d\lambda \leq$$

$$\leq c(n) \int_0^\infty \lambda^{-\epsilon} \int_{\{M|\nabla u| > \lambda\}} F'(|\nabla u|)dx d\lambda \tag{5.4}$$

By Fubini (5.4) becomes

$$\int_{\mathbb{R}^n} F(|\nabla u|) \int_{M|\nabla u|}^\infty \lambda^{-1-\epsilon} d\lambda dx \leq c(n) \int_{\mathbb{R}^n} F'(|\nabla u|) \int_0^{M|\nabla u|} \lambda^{-\epsilon} d\lambda dx \tag{5.5}$$

and by assumption (4.4)

$$\frac{1}{\epsilon}\int_{\mathbb{R}^n} F(|\nabla u|)[M|\nabla u|]^{-\epsilon} \leq \frac{c(n)}{1-\epsilon}\int_{\mathbb{R}^n} F'(|\nabla u|)[M|\nabla u|]^{1-\epsilon}$$

$$\leq \frac{c(n)}{1-\epsilon} q \int_{\mathbb{R}^n} \frac{F(|\nabla u|)}{|\nabla u|}[M|\nabla u|]^{1-\epsilon} \tag{5.6}$$

$$\leq \frac{c(n)}{1-\epsilon} q \int_{\mathbb{R}^n} F(M|\nabla u|)[M|\nabla u|]^{-\epsilon}$$

By Theorem 4.1, the preceding inequality implies for $0 < \epsilon < \frac{p-1}{4}$

$$\int_{\mathbb{R}^n} F(M|\nabla u|)[M|\nabla u|]^{-\epsilon} \leq c(n,p,q)\epsilon \int_{\mathbb{R}^n} F(M|\nabla u|)[M|\nabla u|]^{-\epsilon} \tag{5.7}$$

Choose $0 < \epsilon_0 < \frac{p-1}{4}$ such that $c(n,p,q)\epsilon_0 = \frac{1}{2}$ and set $r = p - \epsilon_0$. Then if we suppose $|\nabla u| \in L_{F_r}(\mathbb{R}^n)$ we have

$$\int_{\mathbb{R}^n} \frac{F(M|\nabla u|)}{[M|\nabla u|]^{\epsilon_0}} \leq c \int_{\mathbb{R}^n} \frac{F(|\nabla u|)}{|\nabla u|^{p-r}}$$

by the maximal theorem in the Orlicz class $L_{F_r}(\mathbb{R}^n)$, and so the right hand side of (5.7) with $\epsilon = \epsilon_0$ is finite and then necessarily

$$\int_{\mathbb{R}^n} \frac{F(M|\nabla u|)}{[M|\nabla u|]^{\epsilon_0}} = 0$$

or $|\nabla u| = 0$ a.e.. Since $u \in W^{1,1}(\mathbb{R}^n)$ was a compact support function, we have

$$u \equiv 0 \qquad a.e. \cdot$$

Remark 5.1 If the support of u is not compact, the function v_λ given by Lemma 5.1 is not admissible since it is not zero at infinity. To get a correct proof one should first multiply u by a cut-off function between Q_R and Q_{2R}. This will give an equation with some extra terms, but the Lipschitz approssimation will have compact support contained in Q_{4R}. Passing to the limit as $R \to \infty$ we get the result. See [6] who treats p-harmonic type equations with degenerate coefficients.

ACKNOWLEDGEMENTS I would like to thank C.Bandle, J.Bemelmans, M.Chipot, J.Saint Jean Paulin and I.Shafrir for organizing this nice conference in Metz.

References

[1] -E.Acerbi, N.Fusco-*Semicontinuity problems in the calculus of variations*- Arch. Rational Mech. Anal., **86**(1984),125-145

[2] -H.Brezis, N.Fusco, C.Sbordone-*Integrability for the Jacobian of orientation preserving mappings*- J. Funct. Anal. **115, 2** (1993), 425-431

[3] - S.M.Buckley-*Estimate for operator norms on weighted spaces and reverse Jensen inequalities*- Trans.Amer.Math.Soc. **340, 1** (1993), 253-272

[4] -R.R.Coifman, R.Rochberg-*Another characterisation of BMO*- Proc. Amer.Math.Soc. **79, 2** (1980), 249-254

[5] -L.D'Apuzzo, C.Sbordone-*Reverse Hölder inequality. A sharp result*- Rend. Mat. **10** (1990), 357-366

[6] - A.Dolcini-*A uniqueness result for very weak solutions of p-harmonic type equations*- to appear

[7] -N.Fusco, C.Sbordone-*Higher integrability of the gradient of minimizers of functionals with nonstandard growth conditions*- Comm. Pure Applied Math. **XLIII** (1990), 673-683

[8] -D.Giachetti, F.Leonetti, R.Schianchi-*On very weak solutions of nonlinear equations*- Proc. Royal Soc. Edinburgh (to appear)

[9] -L. Greco-*A remark on the equality* det Df= Det Df- Differential and integral equat. **6, 5** (1993) ,1089-1100

[10] -L. Greco, T.Iwaniec, M.Milman-in preparation

[11] -L. Greco, T.Iwaniec, G. Moscariello-*Limits of the improved integrability of the volume forms*- to appear

[12] -T.Iwaniec-*p-Harmonic tensors and quasiregular mappings*- Annals of Math., **136** (1992), 589-624

[13] -T.Iwaniec, A.Lutoborski-*Integral estimates for null Lagrangians*- Arch. Rat. Mech. Anal. **125** (1993), 25-79

[14] -T.Iwaniec,C.Sbordone-*On the integrability of the Jacobian under minimal hypothesis*-Arch. Rat. Mech. Anal. **119** (1992), 129-143

[15] -T.Iwaniec,C.Sbordone- *Weak minima of variational integrals*-J. Reine Angew Math. **454** (1994), 143-161.

[16] -T.Iwaniec,C.Scott, B.Stroffolini-to appear

[17] -R.A.Kerman, A.Torchinsky-*Integral inequalities with weights for the Hardy maximal function*-Studia Math. **71**(1981), 277-284

[18] -J.Kinnunen-*Sharp results on reverse Hölder inequalities*- Annal.Acad.Sci. Fennicae, ser A, Math. Dissertation, **95** (1994), 1-34

[19] -A.A.Korenovskii-*The exact continuation of a reverse Hölder inequality and Muckenhoupt's condition*- Math.Notes **52, 5-6** (1993), 1192-1201

[20] -J. Lewis-*On very weak solutions of certain elliptic systems*- Comm.P.D.E **18, 9-10** (1993), 1515-1537

[21] -C.Li, A.McIntosh, K.Zhang-*Higher integrability and reverse Hölder inequality-* to appear

[22] -P.Marcellini-*Regularity of minimizers of integrals of the calculus of variations with. nonstandard growth conditions-*Arch. Rational Mech. Anal., **105**(1989),267-284

[23] -M.Milman-*Integrability of Jacobians of orientation preserving maps: interpolation methods-* Comptes Rendus Acad. Sciences **317**, Serie I,n.6, (1993), 539-543

[24] -G.Modica-*Quasiminimi di classi di funzionali del calcolo delle variazioni-* Annali Mat. Pura e Appl. **142** (1985), 121-143

[25] -G.Moscariello-*On the integrability of the Jacobian in Orlicz spaces-*Math. Japonica, **40, 1** (1994), 1-7

[26] -B.Muckenhoupt-*Weighted norm inequalities for the Hardy maximal function-* Trans. Amer. Math. Soc. **165** (1972), 207-226

[27] -C.Sbordone-*On some integral inequalities and their applications to the calculus of variations-* Boll. Un. Mat. It. ser VI, vol.V (1986), 73-94

[28] -C.Sbordone-*Quasiminima of degenerate functionals with non polynomial growth-* Rend. Sem. Mat. Fis. Milano **LIX** (1989), 173-184

[29] -C.Sbordone-*Very weak minimizers of integrals of the Calculus of Variations-* Rend. Sem. Mat. Univ. Pol. Torino **52, 1**, (1994), 65-69

[30] -B.Stroffolini-*On weakly A-harmonic tensors-* Studia Math. (to appear)

Author's address:
Carlo Sbordone
Dip. Matematica e Appl. "R.Caccioppoli"
Universitá Monte S.Angelo- Via Cintia
80126 Napoli
e-mail: sbordone@na.infn.it

M SCHECHTER AND K TINTAREV

Critical points for pairs of functionals and semilinear elliptic equations

The paper considers a version of the calculus of variations dealing with mappings from a Banach space to $\mathbf{R}^n$. As an application, we study a semilinear elliptic boundary value problem without any symmetry assumptions on the nonlinearity or any geometric assumtions on the domain, and show that it has infinitely many positive solutions.

We establish existence of infinitely many positive solutions of the following Dirichlet problem. Let $\Omega \in \mathbf{R}^n, n > 2$, be a bounded domain with smooth boundary and let $q \in C(\mathbf{R})$ be nonnegative and satisfy an inequality

$$|q(x,s)| \leq C(1+|s|), x \in \Omega$$

with some $C > 0$. We are looking for positive solutions of the Dirichlet boundary value problem of the form

$$-\Delta u = q(x,u), u|_{\partial\Omega} = 0. \tag{0.1}$$

We impose sufficient conditions on q such that (0.1) has infinitely many solutions (which will be positive by the maximum principle). These conditions, roughly speaking, require that $q(x,s)$ will be a sum of two terms, $\lambda_0 s$ and $p(x,s)$, where λ_0 is the smallest eigenvalue of the linear problem

$$-\Delta\phi = \lambda\phi, \phi|_{\partial\Omega} = 0. \tag{0.2}$$

and $p(x,s)$ is a fast oscillating function of an arbitrarily small amplitude. The exact conditions on p will be formulated in Section 2. The main result is Theorem 2.4.

Our proof consists of the interpretation of (0.1) as a critical point equation for a mapping from a Hilbert space $H_0^1(\Omega)$ to $\mathbf{R}^2$ and the study of Lagrange multipliers involved in such critical point equations. In Section 1 we present an elementary exposition of the calculus of variations for n-tuples of functionals. In Section 2 we consider the application to (0.1).

1.Critical points for n-tuples of functionals

Let E be a real infinite-dimensional Banach space and assume that $f \in C^1(E; \mathbf{R}^n)$, i.e. f has a continuous Frechet derivative on E. One says that $x \in E$ is a critical point for f with a critical value $\omega \in \mathbf{R}^n$ if there exists a $\nu \in \mathbf{R}^n \setminus \{0\}$ such that

$$\nu \cdot f'(x) = 0 \text{ and } f(x) = \omega. \tag{1.1}$$

Applications to physical problems often deal with functionals which are linear combinations of several terms. For example, an energy functional may be a sum of kinetic energy and several potential energies reflecting different physical mechanisms, whose values, contributing to the total energy, are important by themselves. In this sense (1.1) is an elaboration of the critical point equation for the function

$$G_\nu(x) = \nu \cdot f(x) \tag{1.2}$$

with the critical value $c = \nu \cdot \omega$. When one replaces (1.2) by (1.1) one is often able to track the critical points by looking, instead of the level or sublevel sets of G_ν, at preimages of sets in $\mathbf{R}^n$ under f. The benefits of such view can be illustrated by the case which is the prime concern of this paper, namely, when G_ν is a functional unbounded from above and from below, that is $G_\nu(E) = \mathbf{R}$, this does not prevent the functional f from being semibounded in the following sense:

$$V := \overline{f(E)} \neq \mathbf{R}^n. \tag{1.3}$$

In such case some critical points for G_ν with some ν will be extreme points of f, that is, critical points with critical values on the boundary of the image V. Precisely,

PROPOSITION 1.1 *Let $x \in E$ be such that $f(x) \in \partial V$. Then x is a critical point for f.*

This statement can be seen, for example, as a corollary of the implicit function theorem: if x were not a critical point, the function f would map a neighborhood of x onto a neighborhood of $f(x)$, which means $f(x) \notin \partial V$.

The assumption that a boundary point of the image is attained is not any easier to verify in applications than the existence of a minimum for a single functional. Generally, semibounded maps provide only critical sequences:

PROPOSITION 1.2 *For every $\omega \in \partial V$ there exist a sequence $x_j \in E$ and a bounded sequence $\nu_j \in \mathbf{R}^n \setminus \{0\}$ which does not converge to zero, such that*

$$\nu_j \cdot f(x_j) \to 0 \text{ in } E^*, f(x_j) \to \omega. \tag{1.4}$$

Propositions 1.1 and 1.2, unfortunately, cannot be applied to the study of critical points for a single functional (1.2) with the given Lagrange multipliers ν since they do not say anything about the values of ν that one may obtain. However, Lagrange multipliers happen to be equal to generalized normals to the boundary of image V.

We start with the definition of porximal normals used in non-smooth analysis (cf.[1]).

DEFINITION 1.3 *Let $V \subset \mathbf{R}^n$ be a closed set. One says that $(\omega, \nu) \in N^0(V)$ (or that $\nu \in N^0_\omega(V)$) if $(\omega, \nu) \in T^*V$ and there exists a function $\psi \in C^2(\mathbf{R}^n; \mathbf{R})$ such that*

$$\psi(\omega) = \min_V \psi \text{ and } \nabla\psi(\omega) = \nu. \tag{1.5}$$

It is well-known that $N^0_\omega(V)$ is nonempty on a dense subset of V and if the boundary of V is smooth, the set $N^0(V)$ is a usual conormal bundle.

LEMMA 1.4 *Let $(\omega, \nu) \in N^0(V)$. Then there exists a sequence $x_j \in E$ and a sequence $\nu_j \in \mathbf{R}^n$ such that*

$$\nu_j \cdot f'(x_j) \to 0 \text{ in } E^*, f(x_j) \to \omega, \nu_j \to \nu. \tag{1.6}$$

Proof. Consider the semibounded functional $\psi \circ f$. Without loss of generality one can assume that ψ attains its minimum on V only at ω. Then $\psi \circ f$ will have a minimising critical sequence x_j with

$$f(x_j) \to \omega \tag{1.7}$$

As a critical sequence it will satisfy the equation $(\psi \circ f)'(x_j) \to 0$ in E^* which by the chain rule means $\nabla\psi(f(x_j)) \cdot f'(x_j) \to 0$. Set

$$\nu_j := \nabla\psi(f(x_j)$$

and note that by (1.6) $\nu_j \to \nabla\psi(\omega)$. □

Requirement of convergence of the critical sequence given by Lemma 1.4 constitutes the n-dimensional version of the Palais-Smale condition:

$$(PS_{\nu,\omega}) \qquad \nu_j \cdot f'(x_j) \to 0 \text{ in } E^*, f(x_j) \to \omega, \nu_j \to \nu \neq 0 \Rightarrow \{x_j\} \text{ has a limit point.}$$

Obviously, if $(PS_{\nu,\omega})$ holds for (ω, ν) in some closed set A, then the set $\overline{N(V)} \cap A$ consists of pairs of critical values for f with corresponding Lagrange multipliers. In application $(PS_{\nu,\omega})$ is considerably weaker than the conventional (PS) condition for the functional (1.2). Indeed, it deals with sequences on which every component of f is bounded, which in applications often corresponds to bounded sequences.

Association of the set of Lagrange multipliers with $N(V)$ has its inconveniencies because of discontinuity, but a further study of geometry of $N(V)$ yields more critical points that fill the gap. For this we can refer the reader to [4,6,7]. To simplify the situation we will now consider the case associated with semilinear elliptic equations, with $n = 2$, E being a Hilbert space H and $f(u) = (\frac{1}{2}\|u\|^2, g(u))$. We will also assume that the map $(g'(u), u)$ is continuous with respect to weak convergence. This immediately implies that g itself is weakly continuous. Let

$$S_t := \{u \in H : \|u\|^2 = 2t\}.$$

THEOREM 1.5 *Let f be as above. Let*

$$\gamma(t) := \sup_{u \in S_t} g(u). \tag{1.8}$$

Then $\gamma(t)$ is a continuous non-decreasing function in $[0,\infty)$. For every $t > 0$ the function (1.8) has left and right hand derivatives $\gamma'_{\pm}$ satisfying

$$0 < \gamma'_{-}(t) \leq \gamma'_{+}(t),\ t > 0. \tag{1.9}$$

If $\gamma_{+}(t) \neq 0$, then there is a $u \in \Sigma_t := \{u \in S_t : g(u) = \gamma(t)\}$, such that

$$g'(u) = \gamma'_{+}(t)u. \tag{1.10}$$

If $\gamma_{-}(t) \neq 0$, then there is a $u \in \Sigma_t := \{u \in S_t : g(u) = \gamma(t),\}$ such that

$$g'(u) = \gamma'_{-}(t)u. \tag{1.11}$$

This statement is but a slight modification of the preceding arguments and its proof is found in [4,5].

2. Infinitely many solutions

From the continuity of γ and (1.9) we have immediately

LEMMA 2.1 *If $0 < a < c < b$ and $\gamma(a) \leq a, \gamma(b) \leq b, \gamma(c) \geq c$, then there exists a point $d \in [a,b]$ such that $\gamma'(d)$ exists and equals 1.*

COROLLARY 2.2 *If there are sequences θ_j, τ_k, such that $\theta_j \to \infty, \tau_k \to \infty, \gamma(\theta_j) \leq \theta_j$ and $\gamma(\tau_k) \geq \tau_k$, then there are infinitely many solutions of*

$$u = g'(u). \tag{2.1}$$

We apply this method to the problem (0.1) with $q(x,s) = \lambda_0 s + p(x,s)$. Let $H = H_0^1(\Omega)$ and

$$g(u) = \int_\Omega (\frac{1}{2}\lambda_0 u^2(x) + P(x,u(x)))dx,$$

where

$$P(x,s) := \int_0^s p(x,\sigma)d\sigma.$$

Let ϕ_0 be the positive solution of (0.2) with $\|\phi_0\|_H = 1$. Our basic assumption are

(A) There are sequences $\{\sigma_j\}, \{\tau_k\}$ such that $\sigma_j \to \infty, \tau_j \to \infty$,

$$\int_\Omega P(x, \sqrt{2\sigma_j}\phi_0(x))dx \to -\infty, \int_\Omega P(x, -\sqrt{2\sigma_j}\phi_0)dx \to -\infty, \tag{2.2}$$

$$\text{and either } \int_\Omega P(x, \sqrt{2\tau_k}\phi_0)dx \geq 0 \text{ or } \int_\Omega P(x, -\sqrt{2\tau_k}\phi_0)dx \geq 0; \tag{2.3}$$

and

(B) The function p is uniformly bounded on $\Omega \times \mathbf{R}$.

Note that the assumption (A) remains true if one changes the values of p on any set of the form $\Omega \times [s_1, s_2]$ and if the new p satisfies (B). This means that these conditions do not exclude the case when $q(x,s) = \lambda_0 s + p(x,s) > 0, s \neq 0$. In that case the following theorem states existence of infinitely many *positive* solutions. Note also the set of functions p or P satisfying conjectures (A),(B), forms a positive cone.

THEOREM 2.3. *Under the above hypotheses the problem (0.1) has an infinite number of solutions.*

Proof. In order to apply Corollary (2.2) we note that the functional

$$(g'(u), u)_H = \int_\Omega [\lambda_0 u^2(x) + p(x, u(x))u(x)]dx$$

is weakly continuous due to the compactness of the embedding of H in L^2. We shall show that under hypotheses (A), (B), there are sequences $\{\sigma_j\}, \{\tau_k\}$ satisfying the hypotheses of Corollary 2.2. This will produce an infinite number of solutions of

$$(u, v)_H = (g'(u), v)_H, \tag{2.4}$$

which, given smooth Ω, translates into classical solutions of (0.1). Suppose that the first inequality of (2.3) holds. Let

$$\phi_t := \sqrt{2t}\phi_0. \tag{2.5}$$

Then $\|\phi_t\|_H^2 = 2t = \lambda_0 \|\phi_t\|_{L^2(\Omega)}^2$. Thus

$$\gamma(t) \geq g(\phi_t) = t + \int_\Omega P(x, \phi_t(x))dx. \tag{2.6}$$

Thus by the first inequality in (2.3),

$$\gamma(\tau_k) \geq \tau_k, \ k = 1, 2, \ldots \tag{2.7}$$

If the second inequality in (2.3) holds the argument above can be repeated with

$$\phi_t := -\sqrt{2t}\phi_0.$$

Let $\epsilon > 0$ be given. For arbitrary $t > 0$ there is a $u_t \in S_t$ such that $\gamma(u_t) \leq g(u_t) + \epsilon$. One can represent this function as

$$u_t = \sqrt{2t}[\pm \cos\theta_t \phi_0 + \sin\theta_t w_t], \tag{2.8}$$

where $w_t \perp \phi_0, \|w_t\|_H = 1$. Representation (2.8) holds with $\cos\theta_t \geq 0$ and an appropriate choice of the sign $\pm$. Thus

$$\gamma(t) \leq t(\cos^2\theta_t + \frac{\lambda_0}{\lambda_1}\sin^2\theta_t) + \int_\Omega P(x, u_t(x))dx + \epsilon, \tag{2.9}$$

with $\lambda_1 > \lambda_0$. Let now

$$\phi_t := \pm\sqrt{2t}\phi_0$$

with the same choice of sign as in (2.8). From (2.9), by (2.6) and with v_t being some convex combination of u_t and ϕ_t, one has that

$$t(1 - \frac{\lambda_0}{\lambda_1})\sin^2\theta_t - \epsilon \leq \int_\Omega [P(x, u_t) - P(x, \phi_t)]dx = \int_\Omega p(x, v_t)(u_t - \phi_t)dx \leq$$

$$\leq C\sqrt{t}\int_\Omega [(1 - \cos\theta_t)|\phi_0| + |\sin\theta_t||w_t|]dx \leq C\sqrt{t}|\sin\theta_t|. \tag{2.10}$$

Hence

$$\sin^2\theta_t \leq C/t. \tag{2.11}$$

In particular, we have $\theta_t \to 0$ as $t \to \infty$. We see from (2.9),(2.11) that

$$\int_\Omega [P(x, u_t) - P(x, \phi_t)]dx \leq C. \tag{2.12}$$

Consequently, by (2.9),

$$\gamma(t) - t \leq \int_\Omega P(x, u_t)dx - (1 - \frac{\lambda_0}{\lambda_1})\sin^2(\theta_t) + \epsilon \leq \int_\Omega [P(x, u_t) - P(x, \phi_t)]dx + \int_\Omega P(x, \phi_t)dx + \epsilon.$$

Therefore, by (2.12) and since ϵ is arbitrary,

$$\gamma(t) - t \leq C + \int_\Omega P(x, \phi_t(x))dx.$$

We now see that (2.2), even though the sign of ϕ_t is not determined, implies $\gamma(\sigma_k) - \sigma_k \to -\infty$ as $k \to \infty$. The theorem now follows from Corollary 2.2. □

EXAMPLE 2.4 We would like to construct a $p(x,s) = p(s)$ satisfying hypotheses (A) and (B). Let $M = \sup_\Omega \phi_0$. Let $r_k^\pm$ be any two sequences with terms satisfying the following relations: $r_1 = 1$,

$$|\{x \in \Omega : \phi_0(x) \leq 5Mr_{m-1}/(4r_m)\}| \leq |\{x \in \Omega : \frac{1}{2}M \leq \phi_0(x) \leq M\}| \tag{2.13}$$

and

$$\frac{r_m}{r_{m-1}} > 5. \tag{2.14}$$

Let $\chi \in C_0^1(0,1)$ be a nonnegative function equal to 1 on $(\frac{1}{4}, \frac{3}{4})$. Finally, let

$$P(s) = \sum_{k=1}^{\infty}(-1)^k r_k^+ \chi(\frac{s}{r_k^+ M} - \frac{1}{4}) + \sum_{k=1}^{\infty}(-1)^k r_k^- \chi(\frac{-s}{r_k^- M} - \frac{1}{4}). \tag{2.15}$$

Note that this is the sum of functions with disjoint supports and that (B) is satisfied. Then for even m,

$$\int_\Omega P(r_m^+\phi_0)dx \geq r_m^+|\{x \in \Omega : \frac{1}{2}M \leq \phi_0(x) \leq M\}| - r_{m-1}^+|\{x \in \Omega : \phi_0(x) \leq 5Mr_{m-1}^+/r_m^+\}|$$

$$\geq \frac{4}{5}r_m^+|\{x \in \Omega : \frac{1}{2}M \leq \phi_0(x) \leq M\}| \geq C5^m. \tag{2.16}$$

A similar estimate will hold if one replaces the argument of P in (2.16) by $-r_m^-\phi_0$. For odd m one similarly gets

$$\int_\Omega P(\pm r_m^\pm \phi_0)dx \leq -C5^m.$$

This verifies (A).

One of the authors (M.S.) would like to thank the Mathematics Department of Uppsala University for their hospitality during the writing of this paper.

REFERENCES

1. F.H.Clarke, Methods of Dynamics and Non-Smooth Optimization, CBMS-NSF Regional Conference Series 57, SIAM, Philadelphia 1989.

2. D.G.De Figuiredo, P.-L.Lions, R.D.Nussbaum, A priori estimates and existence of positive solutions of semilinear elliptic equations, J.Math.Pures et Appl.,61 (1982) 41-63.

3. R.S.Palais, Morse theory on Hilbert manifolds, Topology 2 (1963), 299-340

4. M.Schechter, K.Tintarev, Spherical maxima in Hilbert space and semilinear elliptic eigenvalue problems, Diff.Int.Equ. 3 (1990) 889-899.

5. M. Schechter, K.Tintarev, Eigenvalues for semilinear boundary value problems, Arch. Rat. Mech. Anal.113 (1991),197-208.

6. M.Schechter, K.Tintarev, Functions with polynomial growth and critical points, preprint

7. M. Struwe, The existence of surfaces of constant mean curvature with free boundaries, Acta Math. 160 (1988), 19-64

8. K.Tintarev, Level set maxima and quasilinear elliptic problems, Pacif J. Math. 153 (1) (1992), 185-200.

9. K. Tintarev, Mountain impasse theorem and spectrum of semilinear elliptic problems, Trans. Amer. Math. Soc. 336 (1993) 621-629.

10. K.Tintarev, Mountain pass and impasse for non-smooth functionals woth constraints, J.Nonlin.Anal.- Theory, Methods Appl.(to appear)

Martin Schechter,

University of California, Irvine CA92717, USA

Kyril Tintarev,

Uppsala University, Box 480, Uppsala 751 06, Sweden

J VON BELOW

Front propagation in diffusion problems on trees

This contribution is concerned with the occurence of travelling front solutions in reaction - diffusion problems on tree like ramified media. In the model problem to be considered here, the interaction on the branches of the tree is governed by the autonomous equations $u_{jt} = a_j u_{jx_jx_j} + f(u_j)$ allowing two equilibria, while at the vertices we require continuity and certain consistent Kirchhoff conditions for the incident flows. It turns out that, under the continuity condition at ramification nodes only, there is a front continuum parametrized by a single branch speed. Especially, up to inverting the orientation and up to translation, there is only one front corresponding to the respective minimal branch speeds. Hence, given Kirchhoff conditions can be realizable, redundant or impossible of fulfillment. Conceivably, we impose only Kirchhoff conditions with constant coefficients, and we deduce the necessary and sufficient conditions for their validity. In the remaining parts we investigate the attractivity and stability properties of the equilibria.

For more general problems, including inconsistent Kirchhoff conditions and more general equations, as well as for a general parabolic theory on networks we refer to the forthcoming monography [9]. Other types of transition, also in connection with hyperbolic equations and related higher dimensional problems can be found e.g. in [1], [15], [17].

1. Trees, Networks and Vertex Transition Conditions

Let us recall, e.g. from [4], [8], [9] the notion of a c^ν-network $\mathcal{T}$. Let Γ denote a simple and connected topological graph with finite sets of vertices $E = \{e_i \,|\, 1 \le i \le n\}$ and branches $K = \{k_j \,|\, 1 \le j \le N\}$ Thus, by definition, $\Gamma = (E, K)$ consists in a collection of N Jordan curves k_j with the following properties: Each k_j has its endpoints in the set E, any two vertices in E can be connected by a path with arcs in K, and any two branches $k_j \neq k_h$ satisfy $k_j \cap k_h \subset E$ and $|k_j \cap k_h| \le 1$. Then the union $\mathcal{T} = \bigcup_{j=1}^N k_j \subset \mathbb{R}^m$ is called the *c^ν-network* belonging to Γ, if all arc length parametrizations π_j belong to $C^\nu([0, l_j], \mathbb{R}^m)$. The arc length parameter of an branch k_j is denoted by x_j. Endowed with the induced topology $\mathcal{T}$ is a connected and compact space in $\mathbb{R}^m$. A *tree* $\mathcal{T}$ is a c^ν-network whose underlying topological graph Γ contains no circuits. We will always assume that $\mathcal{T}$ is a tree with $\nu \ge 3$. The valency of a vertex e_i is denoted by γ_i. We distinguish the *ramification nodes* $E_r = \{e_i \in E \,|\, \gamma_i > 1\}$ from the *boundary vertices* $E_b = \{e_i \in E \,|\, \gamma_i = 1\}$. The orientation of $\mathcal{T}$ is given by the incidence matrix $\mathcal{D} = (d_{ij})_{n \times N}$ with

$$d_{ij} = \begin{cases} 1 & \text{if } \pi_j(l_j) = e_i, \\ -1 & \text{if } \pi_j(0) = e_i, \\ 0 & \text{otherwise.} \end{cases}$$

With respect to a given orientation $\mathcal{D}$, the *indegree* γ_i^+ and the *outdegree* γ_i^- of the vertex $e_i \in E$ are defined as $\gamma_i^+ = |\{k_j \in K \,|\, d_{ij} = 1\}|$ and $\gamma_i^- = |\{k_j \in K \,|\, d_{ij} = -1\}|$. A *sink* is a vertex e_i with vanishing outdegree, while a *source* is a vertex e_i with vanishing indegree. Set $E_b^+ = \{e_i \in E_b \,|\, \gamma_i^+ = 1\}$ and $E_b^- = \{e_i \in E_b \,|\, \gamma_i^- = 1\}$.

For functions $u : \mathcal{T} \times J \to \mathbb{R}$, where $J = [0, T]$ is some time interval, we define $u_j = u \circ (\pi_j, id) : [0, l_j] \times J \to \mathbb{R}$ and use the abbreviations $u_j(e_i, t) := u_j(\pi_j^{-1}(e_i), t)$, $u_{x_j}(e_i, t) := \frac{\partial}{\partial x_j} u_j(\pi_j^{-1}(e_i), t)$ etc. As special subspaces of $C(\mathcal{T} \times J)$ we introduce $C^\mu(\mathcal{T} \times J) = \{u \in C(\mathcal{T} \times J) \,|\, \forall j \in \{1, ..., N\} : u_j \in C^\mu([0, l_j] \times J)\}$ for $\mu \leq \nu$.

The basic transition condition at the ramification nodes is given by the *continuity condition*

$$k_j \cap k_s = \{e_i\} \implies u_j(e_i, t) = u_s(e_i, t). \tag{1}$$

With respect to different transition conditions we decompose the vertex set E into two disjoint parts

$$E = E_1 \uplus E_K \quad \text{with} \quad E_r \subseteq E_K \quad \text{and} \quad E_1 \subseteq E_b.$$

At each vertex $e_i \in E_1$ either an inhomogeneous Dirichlet condition is imposed or, in view travelling waves, no boundary condition is prescribed, and e_i is considered as a free end. At the vertices $e_i \in E_K$ we impose a dynamical Kirchhoff condition

$$\sum_{j=1}^{N} d_{ij} a_j u_{jx_j}(e_i, t) + \sigma_i u_t(e_i, t) = 0, \tag{CDK}$$

which includes the classical Kirchhoff law

$$\sum_{j=1}^{N} d_{ij} a_j u_{jx_j}(e_i, t) = 0. \tag{CK}$$

Here the coefficients have been chosen consistent with the diffusion rates a_j. Note that condition (CK) corresponds to the mass conservation at the node and ensures dissipativity when imposed in E_K, cf. [16]. Under (CDK) the condition

$$\sigma_i \geq 0 \qquad \text{in } E_K \tag{2}$$

leads to dissipativity, and a linear and semilinear global parabolic existence and regularity theory can be established, cf. [7],[8],[9]. Under nondissipative Kirchhoff conditions blow up, excitation and nonuniqueness phenomena can occur. Anyhow, continuous travelling waves in networks obey canonically certain dynamical Kirchhoff conditions that are in general inconsistent and not dissipative, see Remark 2 below. With respect to the decomposition $E = E_1 \uplus E_K$ we define the *parabolic network interior* Ω_p and the *parabolic network boundary* ω_p as $\Omega_p = (\mathcal{T} \backslash E_1) \times (0, T]$ and $\omega_p = (\mathcal{T} \times \{0\}) \cup (E_1 \times (0, T])$ respectively, cf. [7].

2. Existence of Travelling Front Solutions

For given positive diffusion constants a_j we consider reaction - diffusion equations on the branches of a c^3-tree $\mathcal{T}$ of the form

$$u_{jt} = a_j u_{jx_jx_j} + f(u_j) \qquad \text{on } k_j \quad \text{for } 1 \leq j \leq N \tag{3}$$

subject to the basic transition condition (1). For these equations we are looking for *travelling front solutions* $u \in C^{2,2}(\mathcal{T} \times \mathbb{R})$, that satisfy, by definition, on each branch k_j the travelling wave conditions

$$u_j(x_j, t) = \varphi_j(x_j - \tau_j t), \ \tau_j > 0, \ \varphi_j \in C^2(\mathbb{R}) \tag{4}$$

and the front conditions

$$\lim_{z \to \infty} \varphi_j(z) = A, \quad \lim_{z \to -\infty} \varphi_j(z) = B, \ \ A \leq \varphi_j \leq B, \ \ A < B. \tag{5}$$

Troughout, the reaction term is subject to the hypotheses

$$f(A) = f(B) = 0, \ f > 0 \text{ in } (A, B), \ f \in C^1([A, B]), \ f'(A) > 0, \ f'(B) < 0. \tag{6}$$

Mostly, we impose the additional hypothesis

$$f(u) \leq f'(A)(u - A) \quad \text{in } [A, B]. \tag{7}$$

In particular, any front is a continuous nonstationary travelling wave on the tree $\mathcal{T}$ with positive branch speeds. Note that the orientation $\mathcal{D}$ can always be chosen such that all branch speeds are positive. Thus, instead of considering arbitrary branch speeds for a given orientation, we can throughout assume that $\tau_j > 0$ for $1 \leq j \leq N$ under the requirement that we have to find an appropriate incidence matrix $\mathcal{D}$ of Γ. From (1) and (4) we immediately deduce the incidence relation

$$k_j \cap k_s = \{e_i\} \implies \varphi_j(\varepsilon_{ij} - \tau_j t) = \varphi_s(\varepsilon_{is} - \tau_s t), \ \varepsilon_{ij} = l_j \frac{1 + d_{ij}}{2}. \tag{8}$$

Thus, the continuity in E_r implies that for any fixed branch k_h each φ_j is of the form $\varphi_h(z) = \varphi_j\left(C(h,j) + \tau_j \tau_h^{-1} z\right)$, where $C(h,j)$ is some constant depending on the branch speeds and lengths along a suitable path containing k_h and k_j. This shows that a nonstationary travelling wave is completely determined by its profile on one single branch k_j and by the speeds $\tau_1, \ldots, \tau_N$.

First, consider a single branch k_j corresponding to the subset $[0, l_j]$ on the real line. It is well known that there exists a continuum $[\tau_j^*, \infty)$ of speeds belonging to front solutions of (3) on $\mathbb{R}$, cf. [2], [11], [12]. An easy modification of the argument in [12] yields that the minimal speed τ_j^* satisfies

$$2\sqrt{a_j f'(A)} \leq \tau_j^*$$

with equality for instance under condition (7). The travelling fronts with (3) - (5) on k_j correspond to heteroclinic orbits (φ_j, ψ_j) joining the stationary solutions $(B, 0)$ and $(A, 0)$ of the first order system

$$\begin{cases} \varphi_j' = \psi_j, \\ \psi_j' = -\frac{\tau_j}{a_j}\psi_j - \frac{1}{a_j}f(\varphi_j). \end{cases} \tag{9}$$

The front on the branch k_j is obtained by restricting $\varphi_j(z)$ to $z + \tau_j t \in [0, l_j]$. Clearly, the wave property ensures that φ_j on $\mathbb{R}$ and its restriction to k_j determine each other uniquely.

Now assume that u is a travelling front solution of the equations (3). Then the incidence relation (8), the equations (9) and the connectedness of $\mathcal{T}$ yield the following necessary condition on the branch speeds.

$$\frac{\tau_h}{\tau_j} = \sqrt{\frac{a_h}{a_j}} \qquad \text{for } 1 \leq h, j \leq N \tag{10}$$

Under (7), condition (10) is especially valid for all minimal speeds τ_j^*. Now we can state the following existence criterion.

Theorem 1: *Suppose the conditions (6) and (7) are fullfilled. Set* $\tau_j^* = 2\sqrt{a_j f'(A)}$. *Then for any branch* k_j *and any* $\tau_j \geq \tau_j^*$ *there exists a travelling front solution* u *of the equations (3), where the remaining branch speeds are given by* $\tau_h = \tau_j\sqrt{a_h a_j^{-1}}$. *Especially, the minimal branch speed on the whole tree* $\mathcal{T}$ *is given by*

$$\tau^* = \tau^*(\mathcal{T}, f, a_1, \ldots, a_N) := 2\sqrt{f'(A)} \min_{1 \leq j \leq N} \sqrt{a_j}.$$

Proof: Choose $\tau_j \geq \tau_j^*$. By the results in [12], there exists a heteroclinic orbit (φ_j, ψ_j) of the system (9) joining the stationary points $(B, 0)$ and $(A, 0)$. Let ε_{ij} and τ_h be as defined in (8) and the assertion and set

$$\varphi_h(z) = \varphi_j\left(\varepsilon_{ij} - \frac{\tau_j}{\tau_h}\varepsilon_{ih} + \frac{\tau_j}{\tau_h}z\right) \qquad \text{for } k_j \cap k_h = \{e_i\}.$$

Then (φ_h, ψ_h) is a heteroclinic orbit belonging to $-\tau_h\varphi' = a_h\varphi'' + f(\varphi)$. The connectedness of Γ guarantees that all $u_j(x_j, t) := \varphi_j(x_j - \tau_j t)$ constitute a well defined function u satisfying (3) and (5) as required. ◇

Without (7) one gets the existence of fronts with a minimal branch speed satisfying $\tau^* \geq 2\sqrt{f'(A)}\min_j \sqrt{a_j}$. Note further that for fixed diffusion rates and the reaction term f, the travelling front is completely determined up to translation by its profile and its speed on one branch already by imposing only the continuity condition in E_r. By (10), the travelling fronts in Theorem 1 correspond in a one to one way to the half line

$$\tau\left(1, \sqrt{\frac{a_2}{a_1}}, \ldots, \sqrt{\frac{a_N}{a_1}}\right), \tau \geq \tau_1^*$$

in the speed parameter cone. There is of course the other half line of negative speeds parametrized by $-\tau_1 \leq -\tau_1^*$ in the same way. Note further that the branch speeds of a fixed front depend monotonically on the diffusion rates, especially

$$\min_j \tau_j = \frac{\tau_1}{\sqrt{a_1}} \min_j \sqrt{a_j}, \qquad \max_j \tau_j = \frac{\tau_1}{\sqrt{a_1}} \max_j \sqrt{a_j}.$$

Remark 1. If one considers different reaction terms f_j in (3), then the formula (10) becomes

$$\frac{\tau_h}{\tau_j} = \sqrt{\frac{a_h}{a_j} \frac{\tau_j \varphi_j' + f_j(\varphi_j)}{\tau_j \varphi_j' + f_h(\varphi_j)}} \qquad \text{for } 1 \leq h, j \leq N.$$

This implicit a priori constraint emphasizes that one should vary the diffusion coefficients rather than to allow different reaction terms f_j. Moreover, all the systems corresponding to $\varphi'' = -\tau_h^2(\tau_j a_h)^{-1}\varphi' - \tau_h^2(\tau_j^2 a_h)^{-1} f_h(\varphi)$ with $1 \leq h, j \leq N$ must have a common heteroclinic orbit joining $(B, 0)$ and $(A, 0)$. This condition seems hardly applicable, unless the reaction terms are all equal.

3. Travelling Fronts under Kirchhoff Conditions

So far, the fronts have not yet been required to fulfill any Kirchhoff condition. In view of Theorem 1, they constitute additional conditions either to the diffusion rates or to the unknown speeds. These conditions can possibly be redundant, but also be impossible to satisfy. Though the transition conditions do not depend on the orientation of Γ, the existence of a travelling front under dissipative Kirchhoff conditions imposes certain restrictions to the orientation.

Obviously, the Neumann condition at a boundary vertex is only realizable by stationary waves, therefore the consideration of (CK) is only of interest in E_r. Introduce the decomposition $E_3 = \{e_i \in E_K \,|\, \sigma_i \neq 0\}$, $E_2 = E_r \setminus E_3$. It makes no sense to impose Dirichlet conditions at boundary vertices to a travelling front, unless one prescribes the condition $u(e_{i_0}, t) = \varphi_j(\varepsilon_{i_0 j} - \tau_j t)$ at one vertex $e_{i_0} \in E_b$ and thereby determines the front completely. But dynamical Kirchhoff conditions at boundary vertices are reasonable under the same restriction as at ramification nodes, see (13) below. So, in general, we will not impose any boundary conditions at $E_1 \cup (E_b \cap E_2)$.

We first observe by (8):

$$k_j \cap k_s = \{e_i\} \implies u_{jx_j}(e_i, t) = \frac{\tau_s}{\tau_j} \varphi_s'(\varepsilon_{is} - \tau_s t).$$

Combining this with (10) yields:

$$(CK) \quad \text{in} \quad E_r \iff \sum_{j=1}^{N} d_{ij}\tau_j = 0 \quad \text{in } E_r \iff \sum_{j=1}^{N} d_{ij}\sqrt{a_j} = 0 \quad \text{in } E_r. \qquad (11)$$

If a continuous nonstationary travelling wave exists in $\mathcal{T}$ and satisfies (CK), then neither a sink nor a source can occur in E_r.

In the dynamical case we find that

$$(CDK) \quad \text{in} \quad E_K \iff \sigma_i = \frac{a_1}{\tau_1^2}\sum_{j=1}^{N} d_{ij}\tau_j \quad \text{in} \quad E_K, \qquad (12)$$

since by (8) and (10) at each $e_i \in E_K$, (CDK) is equivalent with

$$\sigma_i = \frac{\sqrt{a_1}}{\tau_1}\sum_{j=1}^{N} d_{ij}\sqrt{a_j} = \frac{a_1}{\tau_1^2}\sum_{j=1}^{N} d_{ij}\tau_j. \qquad (13)$$

Here, under (2), no sources are possible if a nonstationary wave exists on the tree. The formula (10) shows also that imposing (CDK) with $\sigma_i \neq 0$ at one vertex only, at most one branch speed vector is admissible. Set

$$\delta_i = \sum_{j=1}^{N} d_{ij}\sqrt{a_j}.$$

If a travelling front fulfills (CDK), then necessarily

$$\sigma_i\delta_h = \sigma_h\delta_i, \quad \delta_i \geq 2\sigma_i\sqrt{f'(A)} \qquad \text{for all } e_i, e_h \in E_3. \qquad (14)$$

Conversely, for given numbers σ_i subject to (14) for $e_i \in E_3$ and $\delta_i = 0$ for $e_i \in E_2$, branch speeds τ_j exist subject to (10) and $\tau_j \geq \tau_j^*$. Thus, we have the following

Theorem 2: *Suppose the conditions (6) and (7) are fulfilled and a dissipative Kirchhoff condition (CDK) is prescribed. Then the branch equations (3) have a travelling front solution $u \in C^{2,2}(\mathcal{T} \times \mathbb{R})$ satisfying (CDK) if and only if*

$$\delta_i \geq 2\sigma_i\sqrt{f'(A)} \quad \text{in } E_K, \qquad \delta_i = 0 \quad \text{in } E_2, \qquad \text{and} \qquad \sigma_i\delta_h = \sigma_h\delta_i \quad \text{in } E_3.$$

In this case the branch speeds are given by (10) and (12). If in addition $E_3 \neq \emptyset$, then the branch speed vector $(\tau_1, \ldots, \tau_N)$ is uniquely determined. On the other hand, in the case $E_r = E_2 = E_K$, there is a continuum of front solutions as in Theorem 1 satisfying the condition (CK) in E_r if and only if $\delta_i = 0$ in E_r.

Introduce the *total network speed change* as $\tau_{\text{eff}} = \tau_{\text{eff}}(\mathcal{T}, a_1, \ldots, a_N, f) := \tau_{\text{eff}}^+ - \tau_{\text{eff}}^-$, where

$$\tau_{\text{eff}}^+ = \sum_{e_i \in E_b^+} \sum_{j=1}^{N} d_{ij}\tau_j, \qquad \tau_{\text{eff}}^- = -\sum_{e_i \in E_b^-} \sum_{j=1}^{N} d_{ij}\tau_j.$$

As the column sums of $\mathcal{D}$ vanish, condition (CK) in E_r implies $\tau_{\text{eff}} = 0$, while for a dissipative condition (CDK) we find $\tau_{\text{eff}} \leq 0$. Locally, at a single vertex e_i, (CDK) is dissipative if and only if $\sum_{j=1}^{N} d_{ij}\tau_j \geq 0$, and it is nondissipative there if and only if $\sum_{j=1}^{N} d_{ij}\tau_j < 0$. This clearly emphasizes the damping effect of the dissipative dynamical Kirchhoff condition and the excitation caused by the nondissipative one, that is illustrated also by the following

Example: Suppose u is a travelling front on a tree $\mathcal{T}$ satisfying a condition (CDK) in E_K. Consider the following special cases.

Case I: $\gamma_i^+ = 1$ for all $e_i \in E \setminus E_b^-$ (orientation with black arrows in Fig. 1). Then $|E_b^-| = 1$. If (CDK) is dissipative in E_K, then the maximal branch speed occurs only on the branch k_1 incident with E_b^-, while the minimal branch speed occurs only on branches incident with E_b^+. Moreover, $\max_j \tau_j \geq \tau_{\text{eff}}^+$. If (CDK) is nondissipative in E_K, then we can only state the speeding up $\tau_{\text{eff}}^+ \geq \tau_{\text{eff}}^- = \tau_1$.

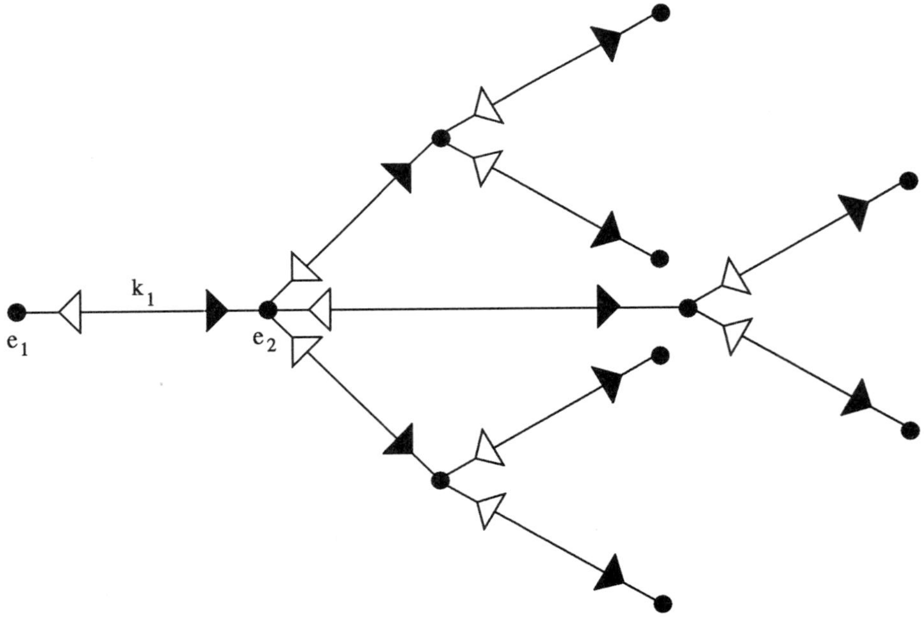

Fig.1 Speed dissipation under dissipative dynamical Kirchhoff condition and speeding up under nondissipative dynamical Kirchhoff condition.

Case II: $\gamma_i^- = 1$ for all $e_i \in E \setminus E_b^+$ (orientation with white arrows in Fig. 1). Then $|E_b^+| = 1$. If (CDK) is nondissipative in E_K, then the maximal branch speed occurs only on the branch k_1 incident with E_b^+, while the minimal branch speed occurs only on branches incident with E_b^-. Moreover, $\max_j \tau_j \geq \tau_{\text{eff}}^-$. If (CDK) is dissipative in E_K, then we can only state the speed damping $\tau_1 = \tau_{\text{eff}}^+ \leq \tau_{\text{eff}}^-$.
Note that any tree bears orientations as in Case I and as in Case II.

A travelling front is called *isotachic*, if its branch speeds of are all equal, what is equivalent with $a := a_1 = \ldots = a_N$. If such a front exists, then the tree has to bear rather strong symmetries. With (11) we conclude that an isotachic travelling front exists and fulfills (CK) if and only if

$$\gamma_i^+ = \gamma_i^- \qquad \text{in } E_r. \tag{15}$$

It follows immediately from (11) that a tree having a ramification node of odd valency cannot admit an isotachic front. But trees with even valencies in E_r can always be orientated in such a way that (15) holds. This follows by induction on $|E_r|$ and by the dual direction principle [3.15]. Hence, (11) and Theorem 2 yield the following

Corollary: *Suppose $\mathcal{T}$ is tree with even valencies in E_r. Then there exists an orientation $\mathcal{D}$ of Γ such that $\mathcal{T}$ admits isotachic front solutions of the branch equations $u_{jt} = a u_{jx_jx_j} + f(u_j)$ under the Kirchhoff Law (CK).*

For an application, let us consider a very simple neuron model allowing front propagation under a dissipative a Kirchhoff condition (CDK). Schematically, a neuron consists of the dendritic tree D, the soma e_2 and the axon k_1, see Fig. 1 with white arrows. It is considered as a c^3-tree $\mathcal{T}$ with underlying graph Γ composed by $\Delta = (\{e_2, \ldots, e_n\}, \{k_2, \ldots, k_N\})$ and $(\{e_1, e_2\}, \{k_1\})$ corresponding to D and to the axon, respectively. Both are connected at the soma which is considered as a distinguished ramification node e_2. The orientation of Γ is such that $\gamma_i^+ = 2$, $\gamma_i^- = 1$ for $i \geq 3$, $d_{21} = -1$, and, necessarily $\gamma_2^+ = \gamma_2 - 1$. The conductivities on the dendritic branches are all assumed to be 1, while $a = a_1$ is a parameter modelling a conductivitiy jump between the axon and the dendritic tree. The boundary vertex on the axon is considered as a free end at which no condition is imposed, thus $E_K = E \backslash \{e_1\}$. Assume $\gamma_2^+ \geq \sqrt{a}$, and choose

$$0 < \sigma = \sigma_i \leq \left(2\sqrt{f'(A)}\right)^{-1} \quad \text{for } i \geq 3, \qquad \sigma_2 = \sigma\left(\gamma_2^+ - \sqrt{a}\right).$$

Then, by Theorem 2, $\mathcal{T}$ admits a travelling front for (3) under the dissipative Kirchhoff condition (CDK), which is isotachic on D with $\tau = \sigma^{-1}$ and has the speed $\tau_1 = \sqrt{a}\sigma^{-1}$ on the axon. Thus, if there at least two dendrites incident with the soma, the tree $\mathcal{T}$ can serve as a model for front propagations under a dissipative dynamical consistent Kirchhoff condition with a conductivity jump on the axon bounded from above by $a \leq (\gamma_2^+)^2$.

More generally, Theorems 1 and 2 can be applied to reaction terms $f(u)$ with a sign change as for fronts on the real line considered in [11.5]. In this way one can handle e.g. branch equations of Fisher type

$$u_{jt} = a_j u_{jx_jx_j} + u_j(1-u_j)(u_j-\mu), \qquad 0<\mu<1,$$

and extend Cohen's model [10] of a local toxin propagation to a tree like structure of the neuron as discussed above.

Remark 2: Any continuous travelling wave u satisfies the special dynamical Kirchhoff condition

$$u_t(e_i,t) + \sum_{j=1}^{N} d_{ij} \frac{d_{ij}\tau_j}{\gamma_i} u_{jx_j}(e_i,t) = 0 \qquad (e_i \in E)$$

that, in general, is inconsistent, and that can only be dissipative locally at a sink, and that is never dissipative in the whole set E_K for $|E_K| \geq 2$. Moreover, if $(\Gamma, \mathcal{D})$ is orientated such that E_r contains neither sinks nor sources, then the wave u satisfies the special dissipative Kirchhoff law

$$\sum_{d_{ij}=1} \frac{\tau_j}{\gamma_i^+} u_{jx_j}(e_i,t) = \sum_{d_{ij}=-1} \frac{\tau_j}{\gamma_i^-} u_{jx_j}(e_i,t) \qquad (e_i \in E_r).$$

4. Attractivity of Equilibria

Let us now investigate the attractivity and stability properties of the equilibrium $u_\infty \equiv B$ with respect to solutions under dissipative Kirchhoff conditions. For domains D, it is well known that B is the global attractor for all solutions $\not\equiv A$ with values in $[A,B]$ and subject to Neumann boundary conditions, cf. [13], [14]. Here, we want to show the corresponding result on trees under (CDK). In general, different conditions (CDK) can lead to different global attractors [9.8], but in our situation we will show that the equilibrium u_∞ is a global attractor under any (CDK). By definition, any front tends to B as $t \to \infty$. Obviously the constant solutions $u \equiv A$ and u_∞ of (3) satisfy any condition (CDK). Thanks to the comparison principle [6],[9.6], any solution $v \in C(\mathcal{T} \times [0,T]) \cap C^{2,1}_{CDK}(\Omega_p)$ of (3) with $A \leq v \leq B$ in ω_p satisfies the same inequality in $\mathcal{T} \times [0,T]$ when (2) holds. Moreover, using the global existence results [8], we can set $T=\infty$. For $E_1 \neq \emptyset$, u_∞ cannot attract all solutions with values in $[A,B]$ unless one requires that $v\big|_{E_1}$ tends to B as $t \to \infty$. Therefore, let us consider only the case $\omega_p = \mathcal{T} \times \{0\}$ i.e.

$$E = E_K.$$

Assume each f_j satisfies (6) and is defined in (B, ∞) such that

$$f_j < 0 \text{ in } (B, \infty), \quad f_j \in C^1([A, \infty)). \tag{16}$$

Introduce the problem

$$\begin{cases} a)\ v \in C(\mathcal{T} \times [0, \infty)) \cap C^{2,1}_{CDK}(\mathcal{T} \times (0, \infty)), \\ b)\ v_{jt} = a_j v_{jx_jx_j} + f_j(v_j) \qquad \text{on } k_j \quad \text{for } 1 \leq j \leq N, \\ c)\ v\big|_{\mathcal{T}\times\{0\}} \geq A, \quad v\big|_{\mathcal{T}\times\{0\}} \not\equiv A. \end{cases} \tag{17}$$

Note that we can apply the spectral analysis [5] and observe that u_∞ is locally asymptotically stable in $C^{2,1}_{CDK}(\mathcal{T} \times [0, \infty))$. But we can deduce the global attracting property of the equilibrium u_∞ more directly by using a Lyapunov type argument. Applying the strong minimum principle [7] locally to $v_{jt} - a_j v_{jx_jx_j} \geq 0$, any solution v of (17) is seen to satisfy

$$v > A \quad \text{in } \mathcal{T} \times (0, \infty). \tag{18}$$

Theorem 3: *Suppose the conditions (2), (6) and (16) hold and (CDK) is imposed in $E_K = E$. Then any solution v of (17) satisfies*

$$\lim_{t\to\infty} \max_{\mathcal{T}} |v(\cdot, t) - B| = 0.$$

Proof: We introduce a Lyapunov functional of energy type as in the domain case [13], [14] in order to show the global asymptotic stability of u_∞, namely

$$\mathcal{E}(v) = \sum_{j=1}^{N} \int_0^{l_j} \left(\frac{1}{2} a_j v_{jx_j}^2 - F_j(v_j) \right) dx_j, \qquad F_j(z) = \int_A^z f_j(s) ds.$$

In the considered function class, $\mathcal{E}$ is bounded from below by (18). Finally, due to (CDK), we obtain along solutions of (17)

$$\frac{d}{dt}\mathcal{E}(v) = -\sum_{j=1}^{N} \int_0^{l_j} v_{jt}^2 dx_j - \sum_{i=1}^{n} \sigma_i v_t^2(e_i). \qquad \diamond$$

The assertion of Theorem 3 is false under nondissipative Kirchhoff conditions. Take for instance $a_j \equiv 1$ and $f(u) = f_j(u) = \sin u$. Let $\{(\varphi(z), \psi(z)) |\ z \in [0, 2l]\}$ be a portion of a periodic orbit near $(0,0)$ in the phase plane of the Hamiltonian system

$$\begin{cases} \varphi' = \psi, \\ \psi' = -f(\varphi), \end{cases}$$

such that

$$0 < \varepsilon \leq \varphi(z) < \pi \quad \text{for } z \in [0, 2l] \text{ and } \varphi(0) = \varphi(2l) = \varepsilon, \ \psi(2l) = -\psi(0).$$

Take Γ to be a star graph with ramification node e_1 and $2m$ branches satisfying $d_{1j} = -1$ for $1 \leq j \leq m$ and $d_{1j} = 1$ for $m+1 \leq j \leq 2m$. Let $\mathcal{T}$ be a c^3-tree belonging to Γ with $l_j \equiv l$. Define $u \in C^{2,1}(\mathcal{T} \times \mathbb{R})$ by

$$u_j(x_j, t) = \begin{cases} \varphi(l - x_j) & \text{for } 1 \leq j \leq m, \\ \varphi(2l - x_j) & \text{for } m+1 \leq j \leq 2m. \end{cases}$$

Then u is a stationary solution of the branch equations (17b) and obviously does not approach the constant B. At the boundary vertices u satisfies a Neumann condition, while at e_1, u satisfies the nondissipative and inconsistent Kirchhoff condition

$$\sum_{j=1}^{N} d_{1j} d_{1j} u_{jx_j}(e_1, t) + \sigma_i u_t(e_1, t) = 0$$

with arbitrary σ_i. Since at the amplitude the distance between $(\varphi(l), 0)$ and $(B, 0)$ remains constant and can be chosen arbitrarily small, $u_\infty \equiv B$ cannot be locally attractive.

References:

[1] F. Ali Mehmeti and S. Nicaise, Nonlinear interaction problems, *Nonlinear Analysis, Theory, Meth. & Appl.* **20** (1993) 27 - 61.

[2] D. G. Aronson and H. F. Weinberger, Nonlinear diffusion in population genetics, combustion, and nerve pulse propagation, in: Partial Differential Equations and related Topics, *Lect. Not. Math.* **446**, J. Springer Verlag Berlin, 1975, pp. 5 - 49.

[3] M. Behzad, G. Chartrand and L. Lesniak - Forster, Graphs and digraphs, Prindle, Weber & Schmidt Boston, 1979.

[4] J. von Below, Classical solvability of linear parabolic equations on networks, *J. Differential Equ.* **72** (1988) 316 - 337.

[5] J. von Below, Sturm - Liouville eigenvalue problems on networks, *Math. Meth. Applied Sciences* **10** (1988) 383 - 395.

[6] J. von Below, Comparison theorems for parabolic network equations with dynamical node conditions, *Sem.-ber. Funktionalanalysis Tübingen* **18**(1990) 17 - 22.

[7] J. von Below, A maximum principle for semilinear parabolic network equations, in: J. A. Goldstein, F. Kappel, and W. Schappacher (eds.), Differential equations with applications in biology, physics, and engineering, *Lect. Not. Pure and Appl. Math.* **133**, M. Dekker Inc. New York, 1991, pp. 37 - 45.

[8] J. VON BELOW, An existence result for semilinear parabolic network equations with dynamical node conditions, in: C. Bandle, J. Bemelmans, M. Chipot, M. Grüter and J. Saint Jean Paulin (eds.), Progress in partial differential equations: elliptic and parabolic problems, *Pitman Research Notes in Math. Ser.* **266**, Longman Harlow Essex, 1992, pp. 274 - 283.

[9] J. VON BELOW, Parabolic network equations, *to appear.*

[10] H. COHEN, Nonlinear diffusion problems, in: A. Taub (ed.), *Studies in Applied Mathematics* **7**, Prentice - Hall Inc. Englewood Cliffs, N. J., 1971, pp. 27 - 63.

[11] K. P. HADELER, Hyperbolic travelling fronts, *Proc. Edinburgh Math. Soc.* **31** (1988) 89 - 97.

[12] K. P. HADELER and F. ROTHE, Travelling fronts in nonlinear diffusion equations, *J. Math. Biology* **2** (1975) 251 - 263.

[13] D. HENRY, Geometric theory of semilinear parabolic equations, *Lect. Not. Math.* **840**, J. Springer Verlag Berlin, 1981.

[14] P. HESS, Periodic - parabolic boundary value problems and positivity, *Pitman Research Notes Math.* **247**, Longman Publ. Essex, 1991.

[15] J. E. LAGNESE, G. LEUGERING and E. J. P. G. SCHMIDT, Modelling of dynamic networks of thin thermoelastic beams, *Math.Meth. Applied Sciences* **16** (1993) 327 - 358.

[16] G. LUMER, Connecting of local operators and evolution equations on networks, in: Potential Theory Copenhagen 1979, (Proceedings), *Lect. Not. Math.* **787**, J. Springer Verlag Berlin, 1980, pp. 219 - 234.

[17] S. NICAISE, Polygonal interface problems, *Meth. Verf. Math. Physik* **39**, Verlag Peter Lang Frankfurt 1993.

Joachim von Below
Lehrstuhl für Biomathematik
Universität Tübingen
Auf der Morgenstelle 10
72076 Tübingen
Germany

G WOLANSKY

A Gamma-limit approach for viscosity stationary solutions of a model convection equation

A variational formulation and Gamma-limit technique is used for investigating stationary shock-wave solutions of a model convection equation with explicit dependence on the position. The jump and entropy conditions across a discontinuity are obtained naturally as necessary conditions for the criticallity of an energy functional at a Borel measure associated with the solution. This energy functional is a Gamma-limit of a sequence of Free-energy functionals corresponding to the viscosity equation. Due to a certain symmetry of the energy functional we obtain an additional, dual entropy condition. It is shown that the jump condition and both entropy conditions are necessary for the existence of a stationary shock train. Somewhat stronger conditions are given, sufficient for the existence of a stable, stationary shock train.

1 Introduction

Consider the equation

$$\frac{\partial W}{\partial t} + a(W,x)\frac{\partial W}{\partial x} = 0 \tag{1}$$

defined over $(x,t) \in I \times R^+$, where $I \equiv [0,1]$ and a is a continuously differentiable function of both variables. Eq (1) is associated with the initial-boundary conditions:

$$W(0,t) = 0 \quad ; \quad W(1,t) = 1 \quad \forall t \geq 0 \tag{2}$$

and

$$W(\cdot,0) \in C^1(I) \text{ is monotone non-decreasing} \tag{3}$$

The solutions of (1) will be considered as viscosity solutions, obtained by an artificial viscosity. That is, W is considered as the limit, as $\varepsilon \to 0$, of W_ε where the last one is a solution of

$$\frac{\partial W_\varepsilon}{\partial t} + a(W_\varepsilon, x)\frac{\partial W_\varepsilon}{\partial x} = \varepsilon\frac{\partial^2}{\partial x^2}W_\varepsilon \tag{4}$$

In this paper, we are interested in stationary viscosity solutions of (1) and the stability thereof. The case $a \equiv a(W) + \lambda$ where λ is a constant parameter is well known (see. e.g. Lax, 73). A monotone non-decreasing, piecewise constant stationary solution W_0 must satisfy the jump condition corresponding to zero velocity at a point of discontinuity $x_0 \in I$, namely

$$f\left(W_0^-(x_0)\right) - f\left(W_0^+(x_0)\right) + \left(W_0^+(x_0) - W_0^-(x_0)\right)\lambda = 0 \tag{5}$$

where f is the primitive of a and $W_0^-(x_0)$ $(W_0^+(x_0))$ are the left (right) values at the discontinuity x_0. In addition, the generalized entropy condition [Lax, 73] should be satisfied as well:

$$f\left(\alpha W_0^+(x_0) + (1-\alpha)W_0^-(x_0)\right) \geq (1-\alpha)f(W_0^-(x_0)) + \alpha f(W_0^+(x_0)) \tag{6}$$

for any $0 \leq \alpha \leq 1$, namely the graph of f have to be above the line of slop λ connecting $\{W^-(x_0), f(W^-(x_0)\}$ and $\{W_0^-(x_0), f\left(W_0^-(x_0)\right)\}$. As a consequence of (5) and (6), it follows that, under the condition $a \equiv a(W) + \lambda$, a piecewise constant, viscosity (entropy) stationary solution of (1) which satisfies both end condition $W_0(0) = 0$, $W_0(1) = 1$ can, in general, support at most one discontinuity. Indeed, suppose there exists a piecewise constant stationary solution W_0 admitting two discontinuity points $0 < x_1 < x_2 < 1$. Then $W_0^-(x_1) = 0$, $0 < W_0^+(x_1) = W_0^-(x_2) = M < 1$ and $W_0^+(x_2) = 1$. By (5) and (6) it follows that the graph of f should be above the line of slop $f(1) - f(0)$ connecting $\{0, f(0)\}$ and $\{1, f(1)\}$, and intersect this curve tangentially at $W = M$. Under the above condition, a stationary solution W_0 may exists only if $\lambda = f(1) - f(0)$. Evidently, these conditions are structurally unstable and can be removed by a small perturbation of the parameters.

The situation turns out to be different in the case (1), where the dependence of a on x is nontrivial.

In Sec. 2 we introduce the free energy functional associated with (4) and use it to introduce a variational formulation for stationary, stable solutions. In Sec. 3 we define the stationary viscosity solutions of (1) and stability thereof. It is proved that stationary, stable viscosity solutions are obtained as strict local minimizers of an energy functional. This energy functional is defined over a set of probability Borel measures on I and is obtained as a Γ-limit of the free-energy functionals as $\varepsilon \to 0$.

In Sec. 4 we study piecewise constant, stationary viscosity solutions, i.e a train of shock-waves

$$W(x) = \sum_{1}^{N} m_i H(x - x_i) \tag{7}$$

where $H(x) = 1$ if $x \geq 0$, $H(x) = 0$ otherwise, $m_i > 0$ and $\sum_1^N m_i = 1$. Using the results of Sec. 3 we obtain necessary conditions for (7) to be a stationary viscosity solutions of (1). These conditions are divided into two classes: The first class, denoted kinematical conditions, are the jump condition (5) and the entropy condition (6), which must hold for any of the shocks positioned at x_i, $1 \leq i \leq N$, separately. The second class, denoted "dual conditions", has no counterpart in the case of x-independent case $a = a(W)$.

In Sec. 5 we discuss sufficient conditions under which a shock-train of the form (7) is a stationary, stable viscosity solution. Using the result of Sec. 3, these sufficient conditions correspond to W being a local minimizer of the energy functional. It turns out that these sufficient conditions correspond to minimizing a function of the $2N-1$ independent variables $x_1, .., x_N$, $m_1, ..., m_{N-1}$ in an appropriately defined simplex. The details of the proof are given in a separate publication (Wolansky, 94).

2 Variational formulation of viscosity solution

Consider a monotone non-decreasing solution of

$$\frac{\partial W}{\partial t} + a(W, x)\frac{\partial W}{\partial x} = \varepsilon \frac{\partial^2 W}{\partial x^2} \tag{8}$$

where a is a C^1 function of both variables. Let

$$\rho(\cdot,t) \equiv \frac{\partial W(\cdot,t)}{\partial x} \tag{9}$$

Since W is a smooth solution for $\varepsilon > 0$, we may differentiate (8), using (9), to obtain

$$\frac{\partial \rho}{\partial t} + \frac{\partial}{\partial x}\left(\rho \frac{\partial}{\partial x} U_\rho(x,t)\right) = \varepsilon \frac{\partial^2}{\partial x^2}\rho \tag{10}$$

where

$$U_\rho(x,t) \equiv \int_0^x a(W_\rho(s,t))ds \tag{11}$$

and

$$W_\rho(x,t) = \int_0^x \rho(s,t)ds \tag{12}$$

by (9) and (2). In order to preserve the second end conditions (2) and (3), we associate the no-flux boundary condition for (10):

$$\rho \frac{\partial}{\partial x} U_\rho(x,t) - \varepsilon \frac{\partial}{\partial x}\rho = 0 \tag{13}$$

at $x = 0$ and $x = 1$, as well as the following restrictions on the initial data

$$\rho(x,0) \geq 0 \quad ; \quad \int_0^1 \rho(x,0)dx = 1$$

It can easily be shown that (10) with (13) are equivalent to (8).

A direct calculation show that

$$\frac{d}{dt}\Psi_\varepsilon\left(\rho_{(\cdot,t)}\right) \leq 0 \tag{14}$$

where the equality holds iff $\rho(\cdot,t)$ satisfies (13) $\forall x \in I$ and

$$\Psi_\varepsilon(\rho) \equiv \varepsilon \int_0^1 \rho \ln \rho dx + \int_0^1 f(W_\rho, x) \ , \tag{15}$$

f is the primitive of a with respect to it's first variable W.

The functional Ψ_ε can be interpreted as the *Free Energy* $E - ST$ associated with the mass distribution ρ of a mechanical system governed by the force law $\partial U_\rho/\partial x$. Here ε is related to the temperature T, $-\int_0^1 \rho \ln \rho$ is related to the entropy S while the energy E is associated with $\int_0^1 f(W_\rho)$. Indeed, in the case of the Burger's equation $f(W) = -1/2W^2$, $V \equiv 0$ we obtain via an integration by parts:

$$\int_0^1 f(W)dW = \frac{1}{2}\int_0^1 \rho U_\rho + C$$

where U_ρ is interpreted as the potential induced by a gravitational type interaction law:

$$\frac{\partial^2 U_\rho}{\partial x^2} = \rho \ , \ U_\rho(0) = U_\rho(1) = 0$$

(c.f Wolansky, 92).

Define the functional $\Psi_\epsilon \ : \ \Gamma \to R^+ \cup \{\infty\}$ over the set

$$\Gamma \equiv \left\{\rho \in L_1([0,1]),\ \rho \geq 0,\ \int_0^1 \rho dx = 1\right\}$$

Lemma 1

Let ρ_0 be a local minimizer of Ψ_ϵ in Γ. Then ρ_0 is a stationary solution of (10) (alt.

$$W_0(x) = \int_0^x \rho_0$$

is a stationary solution of (8)). If ρ_0 is a strict local minimizer, then ρ_0 (alt. W_0) is an asymptotically stable stationary solution of (10) (alt. (8)).

The proof of Lemma 1 follows from the preceding argument and the lower-semi continuity of Ψ_ϵ over Γ, since Ψ_ϵ satisfies all the postulates of a Lyapunov functional (cf., e.g., Henry, 81).

3 Viscosity stationary solutions

The object of the present section is to obtain a characterization of stable, stationary solutions of (1), (2), namely

(16) $$a(W,x)\frac{\partial W}{\partial x} = 0$$

where W is monotone non-decreasing function, continuously differentiable with respect to both variables, satisfying

(17) $$W(0) = 0 \ ; W(1) = 1 \ .$$

Consider the space $\mathcal{A}$

$$W = W(x) \in \mathcal{A} \Longleftrightarrow W \text{ is a non-decreasing, right-continuous over } I$$

$$W(0) \geq 0 \ ; \ W(1) = 1$$

For any $W \in \mathcal{A}$ we associate the unique probability, Borel measure $\mu = \mu_W$ so that

$$\mu_W([0,x]) \equiv W(x) \tag{18}$$

$\forall x \in I$. Moreover, for any probability Borel measure μ there exists a unique $W_\mu \in \mathcal{A}$ so that

$$W_\mu(x) = \mu([0,x]) \tag{19}$$

Thus, there exists a 1-1 correspondence between $\mathcal{A}$ and the set of probability Borel measures on I. We denote the last set by

$$\overline{\Gamma} \equiv \{\mu \in C^*([0,1]),\ \mu \geq 0,\ \mu([0,1]) = 1\} \subset C^*(I) \tag{20}$$

Associating with the weak-* topology, we can use the above 1-1 correspondence to define a topology over $\mathcal{A}$, i.e the topology induced by the weak-* topology over $\overline{\Gamma}$. We will refer to the above topology for both sets as the "weak topology".[1]

Definition

1) *$W \in \mathcal{A}$ is a viscosity solution of (16), (17) iff there exists a stationary solution W_ε of*

$$a(W_\varepsilon, x)\frac{\partial W_\varepsilon}{\partial x} = \varepsilon \frac{\partial^2 W_\varepsilon}{\partial x^2} \tag{21}$$

$$W_\varepsilon(0) = 0 \ ; \ W_\varepsilon(1) = 1 \tag{22}$$

for any $\varepsilon > 0$ sufficiently small, and

$$\lim_{\varepsilon \to 0} W_\varepsilon = W$$

[1] The topology over $\mathcal{A}$ is equivalent to the Skorkhood topology on the space $D(0,1)$ of piecewise continuous functions, (c.f. [Pathasarathy, Ch. VII, Sec. 6, 1967]), restricted to the set of monotone non-decreasing functions

holds in the weak topology.

2) *A viscosity solution W is called regular if $W(0) = 0$ and $\lim_{x \nearrow 1} W(x) = 1$.*

3) *A viscosity solution W is stable iff for any weak neighborhood $U_1 \subset \mathcal{A}$ of W there exists a weak neighborhood U_2, $W \in U_2 \subset U_1$, so that for any $\hat{W} \in U_2 \cap C^1(I)$, the orbit $\hat{W}_\varepsilon(\cdot, t)$ starting at $\hat{W}_\varepsilon(\cdot, 0) = \hat{W}$ will stay in U_1 for any $t \in R^+$ provided $\varepsilon \leq \bar{\varepsilon} = \bar{\varepsilon}(\hat{W}, U_1)$*

Define the *energy functional* Ψ_0 over Γ as

$$\Psi_0(\mu) \equiv \int_0^1 f(W_\mu, x) ds \tag{23}$$

In order to ease the notation, we will assume

$$a(W, x) = f\prime(W) - V\prime(x) \tag{24}$$

here and thereafter, where both f and V are $C^2([0,1])$. Under (24), the energy is reduced, via integration by parts, into

$$\Psi_0(\mu) \equiv \int_0^1 f(W_\mu) ds + \int_0^1 V(x) d\mu \tag{25}$$

THEOREM 1

i) *Ψ_0 is a continuous functional on the set $\overline{\Gamma}$ equipped with the weak topology.*

ii) *Let $\mu_0 \in \overline{\Gamma}$ be a strict local minimizer of Ψ_0 with respect to the weak topology, that is, $\exists\ U \subset C^*$ an open set, $\mu_0 \in U$ and*

$$\Psi_0(\mu) > \Psi_0(\mu_0) \quad \forall \mu \in U \cap \overline{\Gamma} - \{\mu_0\} .$$

Then the associated $W_0 \equiv W_{\mu_0}$ is a stationary, stable viscosity solution of (16, 17).

For the proof of Theorem 1, based on a Γ-limit argument (Aubin, 84), see Wolansky (94).

4 Necessary conditions for viscosity-shock solutions

Consider the simplex in R^{2N-1}

$$\Lambda_N \equiv \left\{ 0 \leq x_1 \leq x_2 \leq \ldots \leq x_N \leq 1;\ \ m_i \geq 0, ; 1 \leq i \leq N;\ \sum_1^N m_i = 1 \right\}$$

for each $z \in \Lambda_N$ we associate $\mu_z \in \overline{\Gamma}$

$$\mu_z = \sum_1^N m_i \delta(x - x_i) \tag{26}$$

and set

$$\Phi_0(z) \equiv \Psi_0(\mu_z) = \sum_0^N f(M_i)(x_{i+1} - x_i) + m_i V(x_i) \tag{27}$$

where $x_0 \equiv 0$, $x_{N+1} \equiv 1$ and

$$M_0 = 0, \; M_i = \sum_1^i m_j \; .$$

This presents an identification between an atomic Borel measure of N points μ and a point $z \in \Lambda_N$. According to definition, the above identification is not a unique one, since one may identify such μ with $z\prime$ in the simplex Λ_{N+1} as follows: $x\prime_i = x_i$ for $i \leq k-1$, $m\prime_i = m_i$ for $i \leq k-2$, $x\prime_i = x_{i-1}$, $m\prime_i = m_{i-1}$ for $N+1 \geq i \geq k+1$ while either:

$$x_{k-1} = x\prime_{k-1} < x\prime_k < x\prime_{k+1} = x_k; \quad m\prime_k = 0 \tag{28}$$

or

$$x\prime_k = x\prime_{k+1} = x_k, \; 0 < m\prime_k < m_k, \; m\prime_{k+1} = m_k - m\prime_k \; . \tag{29}$$

Given $z \in \Lambda_N$, denote

$$W_z(x) \equiv \sum_0^N m_i H(x - x_i) \tag{30}$$

where $H(x) = 1$ for $x > 0$, $H(x) = 0$ otherwise.

Suppose $\mu_0 \in \overline{\Gamma}$ is of the form (26). There are two sets of necessary conditions for μ_0 to be a local minimizer for Ψ_0.

Kinematical conditions

Let $z\prime$ be any representation of N - points Borel measure μ_0 in Λ_{N+1}. Then

$$\delta_{x_i}\Phi_0(z\prime) \geq 0 \tag{31}$$

holds for any $1 \leq i \leq N+1$, where $\delta_{x_i} \to 0$ is any permissible variation in the x_i components of Λ_{N+1}, i.e. $z_i \in \Lambda_{N+1} \to x_i + \delta_{x_i} \in \Lambda_{N+1}$. This condition yields:

i) If $x_{i-1} < x_i < x_{i+1}$ then both δ_{x_i} and $-\delta_{x_i}$ are permissible variation. Hence

$$\frac{\partial \Phi_0}{\partial x_i} = 0 \tag{32}$$

which yields

$$f(M_{k-1}) - f(M_k) + (M_k - M_{k-1})V\prime(x_k) = 0 \ . \tag{33}$$

ii) Consider Φ_0 over Λ_{N+1} and the representation of μ_0 as a point $z\prime$ on the boundary of Λ_{N+1} given by (29). Since $z\prime_k \to z\prime_k + \delta_{x_i}$ is permissible only for $\delta_{x_i} > 0$, then the right derivative of Φ_0 with respect to $x\prime_k$ should be nonnegative

$$\frac{\partial \Phi_0}{\partial x\prime_k} \geq 0 \tag{34}$$

i.e

$$m\prime_k V\prime(x\prime_k) - f(M\prime_k) + f(M\prime_{k-1}) \geq 0 \tag{35}$$

where $M\prime_k = \sum_1^k m\prime_j$ so $M\prime_{k-1} = M_{k-1}$ while $x\prime_k = x_k$ by definition. Since $M_{k-1} < M\prime_k < M_k$ we can set $M\prime_k = \alpha M_k + (1-\alpha)M_{k-1}$ for any $0 < \alpha < 1$. Substituting these in (33) and (35) we obtain

$$f\left(\alpha M_k + (1-\alpha)M_{k-1}\right) \geq (1-\alpha)f(M_{k-1}) + \alpha f(M_k) \ . \tag{36}$$

Eq. (36) is the generalized entropy condition (6) at any of the shocks $W(x_i) = M_i$. It means that the graph of f have to support from below the polygonal contour whose slops balance the drift field $V\prime(x_i)$ corresponding to any of the shocks. This is demonstrated at Fig [A].

iii) If $x_1 = 0$ (res. $x_N = 1$) and $m_1 > 0$ (res. $m_N > 0$) then (30) corresponds to non-regular viscosity solution for which $W_z(0) > 0$ (res. $\lim_{x \nearrow 1} W_z(x) < 1$). In thus case, $z \in \partial\Lambda_N$ and the only permissible variations of x_1 (res. x_N) are $x_1 \to x_1 + \delta$ (res. $x_N \to x_N - \delta$) where $\delta > 0$. This yields

(37)
$$m_1 V\prime(0) - f(m_1) \geq 0$$

(res.

(38)
$$f(M_{N-1}) - f(1) + (1 - M_{N-1}) V\prime(1) \leq 0)$$

Dual conditions

On top of the above kinematical conditions (31), we present an additional set of conditions originated by the requirement

(39)
$$\delta_{m_i} \Phi_0(z\prime) \geq 0$$

where $z\prime$ is a representation of μ_0 in Λ_{N+1}. As before, we split this condition into two cases:

i) $m_k > 0$, hence $m_k \to m_k + \delta_{m_k}$ is permissible for both positive and negative, sufficiently small δ_{m_k}, provided the constraint $\sum m_i = 1$ holds. This yields

(40)
$$\frac{\partial \Phi_0}{\partial m_k} = \lambda$$

where λ is a Lagrange multiplier (independent of k) which follows from the constraint. Eq (40) can be read as

(41)
$$V(x_i) + \sum_{j=i}^{N} f\prime(M_j)(x_{j+1} - x_j) = \lambda$$

for any $1 \leq k \leq N$ for which $m_k > 0$.

ii) Consider Φ_0 over Λ_{N+1} and the representation of μ_0 as a point $z\prime$ on the boundary of Λ_{N+1} given by (28). Since $m\prime_k \to m\prime_k + \delta_{m_k}$ is permissible only for $\delta_{m_k} > 0$, then the right derivative of Λ_{N+1} with respect to $m\prime_k$ should be greater or equal λ. Taking into account the constraint $\sum m_i = 1$ and denoting $x_k < x \equiv x\prime_k < x_{k+1}$ we obtain

(42)
$$V(x) + f\prime(M_k)(x_{k+1} - x) + \sum_{j=k+1}^{N} f\prime(M_j)(x_{j+1} - x_j) \geq \lambda$$

for any $x \in [0,1]$. Denote

$$S(x) = V(x) + \int_x^1 f\prime(W_z(s))\, ds \tag{43}$$

where W_z is given, in terms of μ_z, by (30), then both (41) and (42) can be read

$$S(x) \geq \lambda \; ; \; \forall\, 0 \leq x \leq 1 \, , \tag{44}$$

where the equality in (44) is obtained at $x = x_i$ for $i = 1 \, , \, .., N$.

In Fig [B] we sketch the Graph of V against the graph of $S_0 = -\int_x^1 f\prime(W(s))ds$. The latter is a polygonal curve whose slops are given by $f\prime(M_i)$ at any of the shocks. Notice that conditions (43, 44) state that the graph of V should support the above polygonal curve from below.

The apparent duality between the kinematical condition (36) and (43, 44), manifested in Figs [A] and [B], requires some explanation. Given $\mu \in \overline{\Gamma}$ we may define a dual Borel measure $\mu^{-1} \in \overline{\Gamma}$ as follows: Define

$$W_\mu^{(-1)}(s) \equiv \text{Inf}\{x \geq 0 \; ; W_\mu(x) \geq s\}$$

Then $W_\mu^{(-1)}$ is right continuous monotone non-decreasing function which corresponds to a unique Borel measure $\mu^{(-1)} \in \overline{\Gamma}$ via

$$\mu^{(-1)}([0,s]) = W_\mu^{(-1)}(s)$$

One can easily observe that $\Psi_0(\mu) = \Psi_0^{(-1)}(\mu^{(-1)})$ where $\Psi_0^{(-1)}$ is defined as Ψ_0 (25) with the role of f and V interchanged

$$\Psi_0^{(-1)}(\mu) \equiv \int_0^1 V\left(W_\mu\right)) \, ds + \int_0^1 f(x) d\mu$$

so μ_0 is a critical point of Ψ_0 iff $\mu_0^{(-1)}$ is a critical point of $\Psi_0^{(-1)}$. This observation explains the apparent similarity between Figs [A] and [B].

THEOREM 2 ; necessary conditions

Let W_z (30) be a stationary train of viscosity shock waves. Then conditions (31) and (39) must hold.

Proof of Theorem 2

We assume that W_z is a regular viscosity solution. The proof for non regular shock train follows along similar lines.

The kinematical conditions (33) and (36), equivalent to (31) follows from the local conditions which must hold for any stationary shock. Indeed, rescaling a stationary solution of (8) near a shock positioned at x_k by $x \to x_k + \varepsilon^{-1}(x - x_k)$ and setting $\zeta = \varepsilon^{-1}(x - x_k)$ we obtain

$$f\prime(W)\frac{\partial W}{\partial \zeta} + V\prime(x_k + \varepsilon\zeta)\frac{\partial W}{\partial \zeta} = \frac{\partial^2 W}{\partial \zeta^2} . \tag{45}$$

Letting $\varepsilon \to 0$ we obtain

$$f\prime(W)\frac{\partial W}{\partial \zeta} + V\prime(x_k)\frac{\partial W}{\partial \zeta} = \frac{\partial^2 W}{\partial \zeta^2} \tag{46}$$

subject to

$$\lim_{\zeta \to -\infty} = M_{k-1}$$

$$\lim_{\zeta \to \infty} = M_k .$$

The constant $V\prime(x_k)$ appearing in (46) is used to balance the velocity of the shock, and (33) is nothing but the jump condition, while (36) is the generalized entropy condition for a single shock (Lax, 73). Both conditions must, evidently, be satisfied for each of the isolated shocks separately.

Let now W_ε be a stationary solution of (8) and set

$$S_\varepsilon(x) = \int_x^1 f\prime(W_\varepsilon(s))\, ds + V(x) . \tag{47}$$

Then

$$W_\varepsilon(x) = \gamma_\varepsilon \int_0^x e^{-\frac{S_\varepsilon(s)}{\varepsilon}} ds \tag{48}$$

where $\gamma_\varepsilon > 0$ is a constant defined in order to meet the boundary conditions $W_\varepsilon(1) = 1$. If W_z is a viscosity solution, given by the weak limit of W_ε for $\varepsilon \to 0$, then S is the a.e limit of S_ε. Moreover, by (47) and the uniform bound over $f\prime(W_\varepsilon)$ we also have

$$\lim_{\varepsilon \to 0} S_\varepsilon(x) = S(x) \tag{49}$$

uniformly on $[0,1]$. Set $\lambda_\varepsilon = \varepsilon \ln \gamma_\varepsilon$ and consider $S_\varepsilon - \lambda_\varepsilon$. If $\exists\, x$ such that $S_\varepsilon(x) - \lambda_\varepsilon < -\delta < 0$ for some $\delta > 0$ and arbitrary sufficiently small $\varepsilon > 0$, then by the equi-continuity of S_ε (c.f (49) and (47)) there exists a neighborhood η of x for which

$$S_\varepsilon(y) - \lambda_\varepsilon < -\frac{\delta}{2} \; ; \; \forall y \in \eta \; . \tag{50}$$

Hence $W_\varepsilon(\overline{x}) \geq \mathcal{O}(|\eta|)e^{\delta/2\varepsilon}$ for the right-hand point $\overline{x}$ of η, where $|\eta|$ stands for the length of the interval η. This contradicts $W \leq 1$ on $[0,1]$. Thus

$$\liminf_{\varepsilon \to 0} S_\varepsilon(x) - \lambda_\varepsilon \geq 0 \tag{51}$$

$\forall x \in [0,1]$.

Consider now S_ε for sufficiently small $\varepsilon > 0$ in a neighborhood of $x = x_k$. By the already proven conditions (33) and (36) we observe that x_k is a strict local minimizer of S. Assume

$$S_\varepsilon(x_k) - \lambda_\varepsilon > \delta \; . \tag{52}$$

By (48) and the equi-continuity of S_ε it follows that

$$W_\varepsilon(\underline{x}) - W_\varepsilon(\overline{x}) \leq e^{-\delta/2\varepsilon}|\eta| \tag{53}$$

where $\underline{x}$ (res. $\overline{x}$) is the left (res. right) limit of an interval η containing x_k. This contradicts the definition of W_z as a viscosity solution since $\lim_{\varepsilon\to 0} W_\varepsilon(x) = W_z(x)$ a.e. while $W_z(\underline{x}) - W_z(\overline{x}) = m_k > 0$ □

5 Sufficient conditions for stable viscosity solutions

We now turn to sufficient conditions for stable viscosity solutions of the form (30). By Theorem 2 we obtain

Corollary

W_z given by (30) is a stable viscosity solution of (16) provided the corresponding μ_z (26) is a local strict minimizer of Ψ_0 on $\overline{\Gamma}$. If $x_1 > 0$ and $x_N < 1$, then W_z is a regular solution satisfying (17) as well.

The object of this section is to find more explicit conditions for μ_z to satisfy the assumption of the Corollary. In terms of the function Φ_0 defined over Λ_{N+1}, (27), it seems that the following is a reasonable sufficient condition:

Conjecture

μ_z is a strict local minimizer of Ψ_0 on $\overline{\Gamma}$ if Φ_0 attains a local minimum at the set of representations $z' \subset \Lambda_{N+1}$ of $z \in \Lambda_N$.

Below we prove some weaker version of the Conjecture. Following the arguments of Sec. 4, we obtain sufficient conditions for the assumption of the conjecture:

(54) $$(M'_k - M_{k-1})V'(x_k) - f(M'_k) + f(M_{k-1}) > 0$$

for any $M_k > M'_k > M_{k-1}$ (c.f. (35)),

(55) $$S(x) > \lambda \; ; \; \forall x \neq x_k \quad , 1 \leq k \leq N$$

while

(56) $$S(x_k) = \lambda \; ; \; 1 \leq k \leq N$$

(compare (44)). In addition, the jump condition (33) is satisfied (namely, equality in (54) holds if $M'_k = M_k$) and the quadratic form associated with the second derivative of Φ_0 is positive definite on the hyperplane in Λ_N admitting the constraint $\sum_1^N m_i = 1$, i.e, for any vector $\zeta = (\zeta_1, \ldots \zeta_{2N})$

(57) $$\sum_{i=1}^{2N} \sum_{j=1}^{2N} \frac{\partial^2}{\partial z_i \partial z_j} \Phi_0(z)\zeta_i\zeta_j > c\left(\sum_{i=1}^{N} \zeta_i^2 + |\zeta \cdot l|^2\right)$$

where $c > 0$ and l stands for the vector

$$l_i = 0 \quad 1 \leq i \leq N \; ; \; l_i = 1 \quad N+1 \leq i \leq 2N$$

In the case of non-regular shock train we require the strong inequality in (37, 38), namely $m_1 V'(0) - f(m_1) > 0$, (res. $f(M_{N-1}) - f(1) + (1 - M_{N-1}) V'(1) < 0$)), provided $x_1 = 0$, $m_1 > 0$ (res. $x_N = 1$, $m_N > 0$).

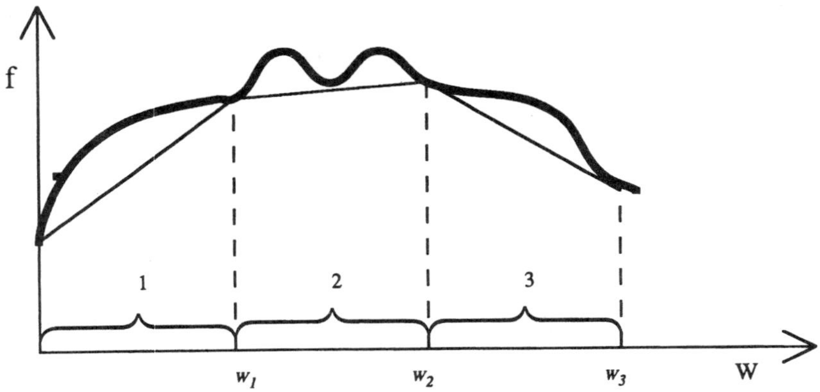

Fig A

A: The graph of $f(W)$ (shifted by an appropriate constant) is presented by the heavy curve. The slops of the polygonal (light) curves are determined by the local jump conditions across the shocks:

(1) $\frac{f(W_1)-f(0)}{W_1} = V'(x_1)$ for $0 < W < W_1$.

(2) $\frac{f(W_2)-f(W_1)}{W_2-W_1} = V'(x_2)$ for $W_1 < W < W_2$.

(3) $\frac{f(1)-f(W_2)}{1-W_2} = V'(x_3)$ for $W_2 < W < 1$.

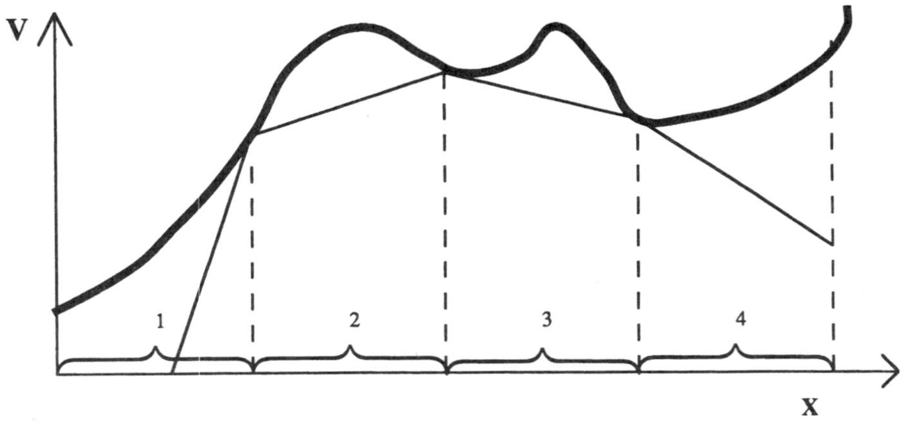

Fig B

B: The graph of $V(x)$ (shifted by an appropriate constant) is presented by the heavy curve. The polygonal (light) curve is the graph of $S_0(x) = -\int_x^1 f'(W(s))ds$ and the slops are:

(1) $S_0'(x) = f'(0)$ for $0 < x < x_1$.

(2) $S_0'(x) = f'(W_1)$ for $x_1 < x < x_2$.

(3) $S_0'(x) = f'(W_2)$ for $x_2 < x < x_3$.

(4) $S_0'(x) = f'(1)$ for $x_3 < x < 1$.

Notice that the jump condition (33) together with the strict inequality (54) yields the strict entropy condition

$$f(\alpha M_k + (1-\alpha)M_{k-1}) > (1-\alpha)f(M_{k-1}) + \alpha f(M_k) \tag{58}$$

provided $0 < \alpha < 1$ (compare (36)). Condition (57) yields, in particular $V''(x_k) > 0$, $1 \leq k \leq N$ if $N \geq 1$, while $f''(M_k) > 0$, $1 \leq k \leq N$ is necessarily valid for $N \geq 2$ (c.f. the constraint $\sum_1^N m_i = 1$).

THEOREM 3 ; sufficient conditions

Conditions (54 - 57) and the jump conditions (33) are sufficient for W_z given by (30) to be a stable, stationary viscosity solution.

The proof of Theorem 3 is obtained by Theorem 1 together with the proof that the conditions of Theorem 3 are sufficient for μ_0 given by (26) to be a strict local minimizer of Ψ_0 in $\overline{\Gamma}$. The (quite technical) details of the proof are given in (Wolansky, 94).

References

Aubin, J.P & Ekeland, I.: Applied Nonlinear Analysis, Wiley (1984)

Billingsley, P.: Convergence of Probability Measures, Wiley, N.Y, 1968

Henry, D.: Geometric Theory of Semilinear Parabolic Equations, Springer Lecture Notes in Math., 840 (1981)

Lax, P.D., Hyperbolic systems of conservation laws and the mathematical theory of shock waves, Regional conf. series in Appl. Math, No.11, SIAM PUB., 1973

Pathasarathy, K.R.: Probability Measures on Metric Space, Acad. Press, 1967

Wolansky, G.: On steady distribution of self-attracting clusters under friction and fluctuations, Arch. Rational Mech. Anal. 119, 355-391, (1992)

Wolansky, G.: Dual entropy conditions for a train of stationary shock waves, Preprint, (1994)

Department of Mathematics, Technion 32000, Haifa, Israel